RISK COMMUNICATION

RISK COMMUNICATION
A Handbook for Communicating Environmental, Safety, and Health Risks

Fifth Edition

Regina E. Lundgren
Andrea H. McMakin

IEEE PRESS

Library of Congress Cataloging-in-Publication Data:

Lundgren, Regina E., 1959–
  Risk communication : a handbook for communicating environmental, safety, and health risks / Regina E. Lundgren, Andrea H. McMakin. – Fifth edition.
      pages cm
  Includes bibliographical references.
  ISBN 978-1-118-45693-4 (paper)
  1. Risk communication–Handbooks, manuals, etc. I. McMakin, Andrea H., 1957– II. Title.
  T10.68.L86 2013
  658.4'08–dc23
                                        2012051127

Printed in the United States of America

ISBN: 9781118456934

11

CONTENTS

PART III   PUTTING RISK COMMUNICATION INTO ACTION

PART IV   EVALUATING RISK COMMUNICATION EFFORTS

20 EVALUATION OF RISK COMMUNICATION EFFORTS   301

PART V   SPECIAL CASES IN RISK COMMUNICATION

21 EMERGENCY RISK COMMUNICATION   313

22 INTERNATIONAL RISK COMMUNICATION   349

LIST OF FIGURES

LIST OF TABLES

PREFACE

The first edition of this book came about because Regina Lundgren had always been fascinated with communication. She started writing novels in the third grade. When she was asked on her first day at the University of Washington what she hoped to do with her degree in scientific and technical communication, she replied, "I want to write environmental impact statements." When Patricia Clark hired her to work at the Pacific Northwest National Laboratory to do just that, she was overjoyed.

Her fascination with communication led her to pursue an interest in risk communication. That in turn took her from leading the public relations function for an 800-person environmental research and development organization to developing her own consulting and training firm. Since then, she has been on a panel for the first workshop on risk communication for weapons of mass destruction events; developed the risk communication plan for the most sophisticated cancer cluster investigation in the nation's history; crafted one of the first state-level risk communication plans for public health preparedness; and taught countless scientists, engineers, and communicators to share complex scientific and technical information among other projects for clients in government, industry, and academia.

Her earlier work at the Pacific Northwest National Laboratory put her in contact with Andrea McMakin, an accomplished risk communicator who had led environmental risk communication efforts that touched several states. Andrea's master's degree in communication, experience in training scientists and engineers to communicate, and first-hand knowledge of working with the news media as both writer and facilitator made her the perfect coauthor from the second edition of this book to this fifth edition.

The previous editions of this book have been used by practitioners, students, and teachers of risk communication across the United States and in at least 20 other countries. Readers' suggestions and new experiences have helped us make the new edition even more useful in terms of content. We added new information on research and lessons learned from some of the major disasters in the last decade. We updated and expanded information on social media, technology-based applications, and public health campaigns.

This book was not written in a vacuum. We owe much of our own knowledge to our forebears in risk communication, including Vince Covello, Peter Sandman, Billie Jo Hance, Caron Chess, Baruch Fischhoff, Paul Slovic, Roger Kasperson, and Jim Creighton. Several other experts in science, management, and communication have inspired us by personal example: Pete Mellinger, Emmett Moore, Jack Robinson, Lori Ramonas, Bob Gray, Judith Bradbury, Kristi Branch, Geoff Harvey, Bill Hanf, Marilyn Quadrel, Dan Strom, Darby Stapp, Barb Wise, and Randal Todd.

Regina would like to thank Laurel Grove, who brilliantly edited the first edition, and Kristin Manke, who provided a professional index for the book. She would also like to thank Ann Lesperance for her invaluable insights into the use of social media in emergency contexts. Most of all, Regina would like to thank her husband Larry and sons Ted and William, who always support her in all she does.

Andrea wishes to acknowledge the advice and review of several experts, including *L.A. Times* reporter David Shaw; science journalists Bill Cannon, Karen Adams, and Mary Beckman; radio reporter Charles Compton; media specialists Geoff Harvey, Greg Koller, and Staci West; Portland State University professor Char Word; statistician Greg Piepel; graphic artist Mike Perkins; and information technology specialist Don Clark. She also thanks the many communication and public health researchers and information specialists who answered questions and corrected errors.

Regina and Andrea gratefully acknowledge peer review of the second edition by two luminaries in the risk communication field: Caron Chess, Director of Rutgers University's Center for Environmental Communication, and Susan Santos, founder and principal of the health and environmental management and risk communication consultancy FOCUS GROUP. Their insights and suggestions helped us think through several issues while staying true to the experiences of our readers.

We also thank Steve Welch for having the vision to continue publishing the book, Mary Hatcher for requesting the fifth edition, the other staff at Wiley-IEEE Press for their help and encouragement, and the reviewers from the IEEE who provided suggestions to improve this edition.

We welcome comments and suggestions from readers; please send them to us in care of our publisher, Wiley-IEEE Press, at pressbooks@ieee.org.

REGINA E. LUNDGREN
ANDREA H. MCMAKIN

ABOUT THE AUTHORS

Regina E. Lundgren is an independent consultant in risk communication, public involvement, and science and strategic communication. For over 25 years, she has specialized in communicating environmental, safety, and health risks to lay audiences. Her communication materials have won national and international awards. She serves her clients in government and industry in a variety of roles, from consultant to trainer to project manager. She developed the risk communication plan for the most sophisticated cancer cluster investigation in the nation's history and consulted on the first state-level system to communicate disaster risks to the public. She conducted research on the mental models approach to risk communication. She is a frequent speaker and trainer to industry groups, professional societies, and government organizations. She has a degree in scientific and technical communication from the University of Washington and a certificate in regulatory analysis from the Harvard School of Public Health. You can learn more at her website at http://www.rlriskcom.com.

Andrea H. McMakin is a marketing communication specialist at the Pacific Northwest National Laboratory, a U.S. Department of Energy national laboratory in Richland, Washington. For more than 20 years, she has been involved with risk communication programs in national and global security, climate change, health and environmental impacts, worker chemical exposure, and risk perception research. Her work has been published and cited in technical journals, scientific and trade publications, and major regional newspapers. She holds a Master of Arts degree in Communication from Washington State University.

1

INTRODUCTION

Risk communication encompasses many types of messages and processes. It is the poster warning food workers to handle food safely to prevent the spread of *Escherichia coli* bacteria. It is the emergency response worker rallying a community to evacuate in the middle of the rising flood. It is the community representatives sitting down with industry to discuss the siting and operation of a hazardous waste incinerator. Risk communication involves people in all walks of life—parents, children, legislative representatives, regulators, scientists, farmers, industrialists, factory workers, and writers. It is part of the science of risk assessment and the process of risk management.

> Risk communication involves people in all walks of life—parents, children, legislative representatives, regulators, scientists, farmers, industrialists, factory workers, and writers. It is part of the science of risk assessment and the process of risk management.

This book was written for those who communicate health, safety, and environmental risks in the United States, primarily:

- The writers, editors, and communication specialists who prepare the messages, coach the speakers, and facilitate public involvement
- The scientists, engineers, and health care professionals who must communicate the results of risk assessments
- The organization representatives who must present a risk management decision
- Those new to the field of risk communication and anyone being asked to communicate risk for the first time

Risk Communication: A Handbook for Communicating Environmental, Safety, and Health Risks,
Fifth Edition. Regina E. Lundgren and Andrea H. McMakin.
© 2013 The Institute of Electrical and Electronics Engineers, Inc., and Regina E. Lundgren and
Andrea H. McMakin. Published 2013 by John Wiley & Sons, Inc.

Because each of these readers may have different needs and questions concerning risk communication, we have divided the book into five parts. Each part or chapter within a part is relatively self-contained; a reader can choose to read some chapters and to skip others of less interest. Part I gives background information necessary to understand the basic theories and practices of risk communication and provides a basis for understanding information in the other parts. Part II tells how to plan a communication effort. Part III gives more in-depth information on different methods of communicating risk and describes how each differs from its counterparts in other areas of communication. Part IV discusses how to evaluate risk communication efforts, including how to measure success. Part V offers advice on special cases in risk communication: emergencies, public health campaigns, and international communication. A list of additional resources, a glossary, and an index are also provided. To emphasize key points, each chapter concludes with a summary section. Chapters that discuss how to apply risk communication (as opposed to those that deal with more theoretical aspects like principles and ethics) end with a checklist, which can be used to help plan and develop your risk communication efforts.

TO BEGIN

Many of the terms used in this book are defined in ways that differ slightly from usage in other branches of science or communication. A glossary is provided, but as a beginning, we want to explain exactly what we mean by risk communication and how it differs from other forms of technical communication.

Technical communication is the communication of scientific or technical information. Audiences can range from children in a sixth-grade science class, to workers learning a new procedure on a piece of equipment, to scientists reviewing the work of peers. The purpose of technical communication can be to inform, educate, or even occasionally persuade.

> Risk communication comes in many forms. In this book, we generally divide risk communication along functional lines, distinguishing between care communication, crisis communication, and consensus communication.

Risk communication is a subset of technical communication. As such, it has its own characteristics. At its most basic, it is the communication of some risk. (In this book, it is used to mean the communication of health, safety, or environmental risks.) The audience can be similar to those described for technical communication, but it can also be a wide cross section of the United States. For example, information to present the risk of not wearing seatbelts could have as an audience anyone who will ever ride in a car.

Sometimes, the risk being communicated is frightening to a particular segment of the audience. Other times, the audience is unaware of or even apathetic to the risk. In still other cases, the organization communicating the risk is not credible to a portion of the audience or the audience finds the way the risk is being managed to be unacceptable. The strong emotions, or the lack thereof, audiences associate with a risk can make it difficult to communicate.

The purpose of risk communication can also differ from that of technical communication. In dangerous situations, such as floods and tornadoes, risk communication may have to motivate its audience to action. In other situations, the purpose is more appropriately to inform or to encourage the building of consensus (more on this in Chapter 5). Another difference between risk communication and technical communication is that risk communication more often involves two-way communication, that is, the organization managing the risk and the audience carry on a dialogue. In technical communication, most efforts are designed to disseminate information, not to receive information back from the audience or to include the audience in the decision-making process. An example of two-way technical communication is scientists reviewing the work of peers.

Risk communication comes in many forms (see Figure 1-1). In this book, we generally divide risk communication along functional lines, distinguishing between care communication, consensus communication, and crisis communication, which are described in more detail later in this chapter. While these three forms have elements in common with other

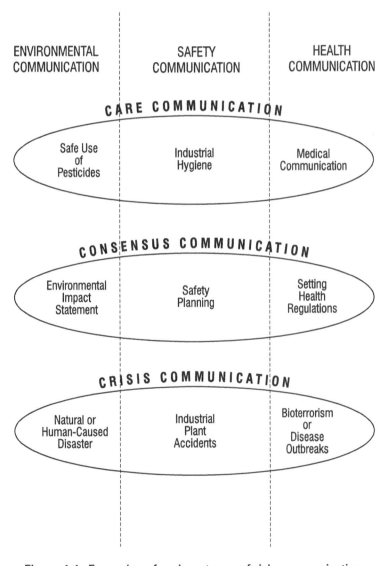

Figure 1-1. Examples of various types of risk communication.

forms of technical communication, they always have circumstances that require different tactics, or ways of communicating, to effectively deliver their messages to and involved their respective audiences. For example, consensus communication involves much more audience interaction than do care or crisis communication. Risk communication can also be divided topically: for example, into environmental, safety, and health risk communication.

Care communication is communication about risks for which the danger and the way to manage it have already been well determined through scientific research that is accepted by most of the audience. Another distinguisher is that, generally, those charged with communicating have little return on investment other than the betterment of human lives. Think of the American Heart Association and local public health agencies.

> Care communication is communication about risks for which the danger and the way to manage it have already been well determined through scientific research that is accepted by most of the audience.

Two subsets of care communication are health care communication (sometimes called health education or health marketing), which seeks to inform and advise the audience about health risks such as smoking or AIDS, and industrial risk communication, which involves informing workers about potential safety and health risks in the workplace. Industrial risk communication can be further divided into ongoing communication about industrial hygiene and individual worker notification, which informs workers of the findings of retrospective mortality studies, in which the mortality rates of a set of workers have been evaluated against standards. Examples of these are the longitudinal studies to determine whether painting radium watch dials was hazardous to the workers (that is, whether they had a higher rate of mortality compared with standards).

Consensus communication is risk communication to inform and encourage groups to work together to reach a decision about how the risk will be managed (prevented or mitigated). An example would be a citizen advisory panel and the owner/operator of the local landfill working together to determine how best to dispose of hazardous chemicals found at the landfill. Consensus communication of risk is also a subset of stakeholder participation, which encourages all those with an interest (stake) in how the risk is managed to be involved in consensus building. Often, the agency or organization with the greatest financial stake funds this process. (Stakeholder participation is also generally called public engagement, public involvement, public participation, stakeholder involvement, public consultation, and audience interaction.) Stakeholder involvement, however, can go far beyond risk communication, into the realms of conflict resolution and negotiation. These realms encompass entire disciplines in themselves and, hence, are beyond the scope of this book.

> Consensus communication is risk communication to inform and encourage groups to work together to reach a decision about how the risk will be managed (prevented or mitigated).

Crisis communication is risk communication in the face of extreme, sudden danger—an accident at an industrial plant, the impending break in an earthen dam, or

> Crisis communication is risk communication in the face of extreme, sudden danger—an accident at an industrial plant, the impending break in an earthen dam, or the outbreak of a deadly disease.

the outbreak of a deadly disease. This type can include communication both during and after the emergency. (Communication during planning on how to deal with potential emergencies would be either care or consensus communication, depending on how much the audience is involved in the planning.)

THE RISK COMMUNICATION PROCESS

An overview of the risk communication process will also help explain the concepts presented elsewhere in this book. The process begins with a hazard, a potential or actual danger to the environment or human health or safety. Examples include an oil spill (environment), cigarette smoking (health), and a loose stair tread in an office building (safety). Usually by law but sometimes by commitment, some organization is responsible for managing the risks posed by this hazard, that is, preventing or mitigating any damage (decreasing the probability or lessening the consequences). In the case of a land-based oil spill, the U.S. Environmental Protection Agency, among other organizations, must develop regulations to prevent occurrence and oversees cleanup if preventive measures fail. The American Lung Association has a commitment to eradicate cigarette smoking. The Occupational Safety and Health Administration requires that organizations maintain safe work environments.

Risk management usually begins with a risk assessment. Just how dangerous is the risk? How much of a hazardous chemical has to spill into a river before the water's natural self-cleansing ability is overwhelmed? Can AIDS be spread by contact with infected health care practitioners? How does the way workers use a forklift affect their risks of being injured or of injuring another? Risk assessment is a scientific process that characterizes risk and assesses the probability of occurrence and outcomes. Based on probabilities, it usually tries to answer questions such as the following:

- Who, or what ecosystems, will be harmed?
- How many of them will be harmed?
- How will they be harmed and by how much?
- How long will the harm continue?

Sometimes the risk assessment has a benefit component attached (risk/benefit analysis). This kind of analysis seeks to determine whether any benefits attached to the risk would balance against the harm caused. For example, does the benefit of the potential advancement of science balance against the potential harm of experimenting with radioactive materials? This kind of analysis may or may not include factors other than the strictly scientific evaluation of the risk and benefit.

Information from the risk assessment is used by risk managers to decide what to do about the risk. Their decisions, and often the process by which they decide, are usually communicated to the people who would be or are affected by the risk or to those interested in the risk for other reasons (ethical issues, for example). Sometimes the risk managers try to encourage this audience to take action (care or crisis communication), sometimes they need to educate the audience about the risk so that the audience has the information needed to make a decision (care communication), and sometimes they need to discuss the risk with the audience so that a consensus on a course of action can be reached with all parties speaking the same language (consensus communication).

In the case of consensus communication, the decision about how risks are to be managed is made through stakeholder involvement. This type of management requires risk communication that seeks to:

- Determine stakeholder perceptions of a variety of factors including the risk, the organization in charge of managing the risk, and the process being used to reach the decision
- Inform, not persuade (except in the context of an agreed-upon negotiation)
- Balance the needs of competing stakeholders
- Assist in reaching a resolution that all parties can live with

For example, the process of using an environmental impact statement to evaluate a set of alternative actions often begins with a series of stakeholder meetings to encourage individuals and groups to help define what should be evaluated (this part of the process is called scoping). Care communication and crisis communication also need to identify stakeholder perceptions and concerns; however, in these cases, the information is used to develop messages that will inform the audience and will encourage them to take some course of action. An example of this is the U.S. Environmental Protection Agency's program to communicate the dangers of radon in the home (for example, Weinstein and Sandman 1993).

At any point during the process, the organization that has been communicating may evaluate its risk communication effort to determine successes and failures. What should be changed next time? What was most effective for this audience, in this situation? Is there anything that can be generalized to apply to other situations and audiences?

> Where potential personal harm is concerned, the believability of information provided depends greatly on the degree of trust and confidence in the risk communicator. If the communicator is viewed as having a compromised mandate or a lack of competence, credence in information provided tends to be weakened accordingly. Or if the particular risk has been mismanaged or neglected in the past, skepticism and distrust may greet attempts to communicate risks.
> —Roger E. Kasperson (1986, p. 277).

AUDIENCES, SITUATIONS, AND PURPOSES

The ideas and techniques given in the rest of the book are tools. They are what we and other risk communicators have found to work for a given audience in a given situation with a given purpose. While a growing body of research lays out guidelines for effective risk communication, the differing dynamics among audiences, situations, and purposes makes finding the one "right solution" impossible, even if there is one right solution to find. Wherever possible, we have cited the work of others as confirmation of our own findings and those of other practitioners in the field. Citations for the

> While a growing body of research lays out guidelines for effective risk communication, the differing dynamics among audiences, situations, and purposes makes finding the one "right solution" impossible, even if there is one right solution to find.

research discussed in the text can be found at the end of each chapter. Other sources of information in the area of risk communication can be found in the Resources section at the back of the book.

Many of the resources listed discuss such issues as credibility of the organization communicating or managing the risk, fairness of the risk in the audience's eyes, and trust among parties. These issues will be dealt with only as they relate to specific points in the rest of this book; however, they are important issues that heavily affect the ability to communicate risk effectively. Unfortunately, they are often outside the control of most of us who actually communicate risk. When we step in front of an audience, policies made by those far above us and sometimes years in the past have already either forged a bond of trust with the audience or broken it. Likewise, our credibility as risk communicators will depend on the credibility of other risk communicators who previously faced the same audience.

Although we cannot change the past, we can be aware of past mistakes or successes and make sure that our own efforts are trustworthy, credible, and fair, insofar as we have the authority to make them so. And we must champion the cause of trustworthy, credible, and fair risk management decisions in our own organizations, both because it is ethical and because it is the only way to ensure successful communication.

REFERENCES

Kasperson, R. E. 1986. "Six Propositions on Public Participation and Their Relevance for Risk Communication." *Risk Analysis*, 6(3):275–281.

Weinstein, N. D. and P. M. Sandman. 1993. "Some Criteria for Evaluating Risk Messages." *Risk Analysis*, 13(1):103–114.

I

UNDERSTANDING RISK COMMUNICATION

To understand risk communication, you will need to understand the approaches to communicating risk, the laws that shape the way we communicate risk today, the constraints to effective risk communication, the ethical issues, and the basic principles of risk communication, which have evolved out of the approaches, laws, constraints, and ethics. Additional sources of information are listed in Resources at the back of the book.

> Learning about risk occurs not in isolated individuals but in a social dynamic, with multiple sources of information, channels of information flow, confirmatory and challenging mechanisms, and linkage with other social issues.
> —Roger E. Kasperson (1986, p. 131).

2

APPROACHES TO COMMUNICATING RISK

There are a number of approaches to the process of risk communication and its components, including how messages are sent and received, how conflicts are managed, and how decisions are made. Some of these approaches are communication research methods in themselves, some grew out of research in fields other than communication, and still others are based on traditions across disciplines.

Why should those who are communicating risks learn about the various approaches? Each approach views risk communication from a slightly different perspective, just as different audiences view a risk from different perspectives. The more risk communication perspectives the communicators understand, the more likely they will be able to choose approaches that will meet the needs of their particular situation and audience, and the more likely that their risk communication efforts will succeed.

Are all approaches equally valid? Each approach was developed to illuminate a particular perspective on risk communication. Depending on how broad that perspective is, the approach may be applicable to a variety of situations and audiences. Some approaches, although still widely used in communicating risk, may be outdated given the situations and audiences that face communicators today. For example, the traditional communication method developed by Claude Shannon in 1948 is still used occasionally today to structure risk communication efforts despite the fact that more sophisticated

> Are all approaches equally valid? Each approach was developed to illuminate a particular perspective on risk communication. Depending on how broad that perspective is, the approach may be applicable to a variety of situations and audiences.

Risk Communication: A Handbook for Communicating Environmental, Safety, and Health Risks,
Fifth Edition. Regina E. Lundgren and Andrea H. McMakin.

models have been developed, which include the two-way communication that is important to risk communication.

The following discussion of approaches to risk communication presents an overview of 14 of the most common approaches as well as implications for those who are communicating risk and how the approach might be used in various situations.

COMMUNICATION PROCESS APPROACH

> Risk communication is a form of communication that, like other forms, is represented by the traditional model of communication. That is, there is a source of communication that generates a message that goes through a channel to a receiver.

Risk communication is a form of communication that, like other forms, is represented by the traditional model of communication (Shannon 1948). That is, there is a source of communication that generates a message that goes through a channel to a receiver. For example, a regulatory agency (the source) may decide that a chemical poses an unacceptable risk to the public (the message) and issue a press release (the channel) published as a story by the news media (another channel) that is read by members of the local community (the receivers). Various studies in risk communication have looked at individual components of this model (sources, messages, etc.) to see how changes in any component affect the others. For example, researchers at the Center for Mass Media Research at Marquette University found that the receiver relied more heavily on different channels for information based on personal emotions such as worry in the wake of a parasite outbreak in drinking water in Milwaukee (Griffin et al. 1994).

The implications for risk communicators are that each of the model components needs to be considered when developing risk communication efforts. Will the source of the message be credible with the intended audience? Have the messages been developed in such a way as to be easily understood by the receivers? What channels (methods) are available that reach the intended audiences? What attitudes from the receiver will affect how the message is perceived? Can we plan for effective feedback to evaluate not only the risk communication process but the decision process as well? Additional information in Parts II–V will help risk communicators answer these types of questions for care, consensus, and crisis communications (see Chapter 1 for a description of these types).

NATIONAL RESEARCH COUNCIL'S APPROACH

In the 1980s, the U.S. National Research Council funded an extensive study in the effective communication of risk (NRC 1989). The multiagency panel of experts came to several conclusions. One was that risk communication can be defined as the "interactive process of exchange of information and opinions among individuals, groups, and institutions concerning a risk or potential risk to human health or the

> Risk communication can be defined as the "interactive process of exchange of information and opinions among individuals, groups, and institutions concerning a risk or potential risk to human health or the environment."
> —NRC (1989).

environment." The panel saw risk communication as a process by which scientific organizations both disseminate technical information and gather information about the opinions and concerns of nonscientific groups.

More recently, the National Research Council sponsored a second group of experts to look at how risk assessment (which they called characterization), management, and communication could be improved (NRC 1996). This group found that risk assessment should be directed toward informing decisions and solving problems, and that this consideration of the social context of the risk should start from the very beginning of the risk assessment and continue through management and communication. This group called for early and interactive involvement with those at risk.

The implication for those who communicate risk is that any form of successful risk communication must incorporate the "exchange of information and opinions" and the participation of the stakeholder groups from the beginning. How this exchange is accomplished will vary for each type of risk communication (care, consensus, or crisis). The audience is necessarily involved in exchanging information with those who are communicating and managing the risk in consensus communication. The exchange can be incorporated into care communication by, at the very least, soliciting audience feedback before and after risk information is distributed. The exchange may be the most difficult in crisis communication. In a crisis, by definition, there is almost never a time to bring together representatives of the audience to determine their needs and concerns. One way to solve this problem is to exchange information with the potential audience (those who may be affected by the crisis, for example, the community surrounding a chemical plant) as part of emergency planning efforts.

MENTAL MODELS APPROACH

The concept of how people understand and view various phenomena, or their mental models of the situation, is grounded in cognitive psychology and artificial intelligence research (Geuter and Stevens 1983). The mental models approach as applied to risk communication comes largely from researchers at Carnegie-Mellon University

> In the absence of evidence, no one can predict confidently how to communicate about a risk. Effective and reliable risk communication requires empirical study.
> —Granger Morgan et al. (2002, p. 182).

(Morgan et al. 2002). The approach was used successfully in other forms of technical communication, such as computer documentation, before being used to focus risk communication efforts.

Using this approach, communicators begin by determining to which audience the risk communication efforts will be directed. They then interview members of that audience to determine how the audience views the risk. For example, in one of the more publicized efforts, for the U.S. Environmental Protection Agency's radon information program, researchers interviewed members of the audience using open-ended questions that gradually became more focused as the interview progressed (from "tell me everything you know about radon" to "tell me more about how it affects you"). Answers from all the participants were used to compile a "mental model," a view of how the audience saw radon, its exposure routes, and dangers. This mental model was compared with the expert model, the model that scientists use to evaluate radon. Researchers followed up with a more focused questionnaire to verify differences between the two models. Risk communication messages were then designed to address the gaps or inconsistencies in the audience's knowledge (Morgan et al. 1992). The intent was not to convince the public to think like scientists but to identify the information the public would need to make an informed decision.

The implication for those who communicate risks is that to really communicate with your audience, you must understand what your audience already believes about the risk. Risk communication messages that do not address key audience concerns or account for existing beliefs will fail. At the very least, all three forms of risk communication (care, consensus, and crisis) must incorporate some audience analysis.

For care communication, which often has as its audience a wide cross section of the population, the communicators will need to understand lifestyle characteristics of each portion of the audience to tailor risk messages. For example, teenagers have been found to be more likely to engage in drunk driving than other age groups. How do their beliefs differ from those of other age groups to make them more likely to drive while drunk?

For consensus communication, the communicators will need to understand the concerns and beliefs of the audience before they can hope to agree on a solution. For example, how do different Native American tribes view the environment and how will these views affect their stance on hazardous waste cleanup?

For crisis communication, communicators need to understand the cultures of the audience to be able to discuss ways to mitigate a crisis. A specific example is the illness that struck the Navajo in northern New Mexico and Arizona in 1993 (dubbed the "mystery illness" by the news media because the cause eluded health care professionals for some time). The illness was found to be spread by breathing airborne particles of infected mouse dung, which would seem to be a relatively easy source to avoid. However, because sweeping was part of the religious and cultural activities of the Navajo, and sweeping raised the dust in which the dung was often found, the source proved to be a more insidious one to combat. See Chapter 8 for more information on understanding your audience.

CRISIS COMMUNICATION APPROACH

The crisis communication approach holds that those who are communicating the risk should use every device to move the audience to appropriate action. For example, in a flood, they should construct messages that cause the audience to evacuate to higher ground while refraining from hindering the work of rescue groups. As one risk communication professional explained, "You only give the audience the information they need to get them to leave. Anything else is extraneous." Given this goal, passing on such information as probability of risk and alternatives seems pointless. This approach holds that the organization knows what is best for the audience and should act as a firm parent in enforcing its opinion.

In an era when those who are affected by a risk are demanding more and more involvement, this approach seems outdated in the extreme. As its name suggests, the only type of risk communication for which it might be valid is crisis communication. However, even in extreme public health emergencies like a bioterrorist attack, the public's demand for information is likely to be higher than this approach would satisfy. Research funded by the U.S. Centers for Disease Control and Prevention found that, in the case of terrorist attacks using a variety of weapons of mass destruction ranging from a radiation dispersal device to plague, audiences wanted to know, at a minimum about the incident, the threat agent, and the health issues (Becker 2004). The implication for those who are communicating in a crisis is that persuasion, which is discussed in Chapter 5, is justified in this case and that information given to those at risk should be limited. Our own experience and a growing body of evidence suggest that people are more likely to change behavior when they know the "why," not just the "what" or "how."

This model has a second aspect that arises from public relation practices. In some cases, the purpose of crisis communication is seen as protecting the image of a particular organization, often the one that might bear some responsibility for the crisis (for example, a fruit juice manufacturer that learns a batch of its product has been contaminated with *Escherichia coli* bacteria after several children have been hospitalized). This book looks at crisis communications solely as a way to reach those at risk and to help them avoid or mitigate the crisis. However, effectively communicating during a crisis has been shown to improve an organization's image over the long term, even when that organization is in some way responsible for the crisis (see, for example, Peters et al. 1997).

CONVERGENCE COMMUNICATION APPROACH

Everett Rogers (Rogers and Kincaid 1981) developed a theory that communication (including risk communication) is an iterative long-term process in which the values (culture, experiences, and background) of the risk communication organization and the audience affect the process of communication. The organization issues information, and the audience processes it to the extent possible and issues its own information ("we don't trust you," "what is this stuff?" and "do you want me to do something?"). The organization then processes that information and responds by issuing additional or modified information. The two groups continue to cycle information back and forth, slowly converging onto common ground.

The implication for those communicating risk is that the audience must be involved in the risk communication process and that the process must be a dialogue, not a monologue on the part of the organization. Continuous feedback and interpretation are necessary for the communication to be effective. This is true for care and crisis communication, but particularly for consensus communication. For care communication, the dialogue may be with a sample of the intended audience. For crisis communication, the dialogue may be with community members involved in emergency planning efforts. For consensus communication, of course, the dialogue is with the group with which you are trying to reach an agreement.

> The audience must be involved in the risk communication process, and the process must be a dialogue, not a monologue on the part of the organization.

THREE-CHALLENGE APPROACH

This approach gets its name from Rowan (1991), who views risk communication as three challenges:

1. **Knowledge challenge.** The audience needs to understand the technical information surrounding the risk assessment.
2. **Process challenge.** The audience needs to feel involved in the risk management process.

3. **Communications skills challenge.** The audience and those who are communicating the risk need to communicate effectively.

Those who are communicating the risk must meet each of these challenges for risk communication efforts to succeed.

The implication for those communicating the risk (whether through care, consensus, or crisis communication) is that both those who are communicating risk and the audience must have excellent communication skills. If the audience's skills are lacking, those who are communicating the risk will have to compensate with techniques designed to increase comprehensibility.

To meet the knowledge challenge, the technical information will have to be presented in a variety of ways: in information materials (pamphlets, fact sheets, and technical reports), in visual representations of risk (graphics, such as simple diagrams, pie charts, and conceptual drawings), through face-to-face communication (presentations with vivid projected graphics and handouts), through stakeholder participation (small group discussions with facilitators who are knowledgeable about the risk), and in technology-assisted communication (websites and interactive models of risk).

To meet the process challenge, the audience will have to be included in how the risk is being managed. For care communication, the audience may be involved by choosing among a variety of preventive or mitigating measures (the risk of dying from heart disease can be reduced by changing to a high-fiber, low-fat diet; exercising; and/or stopping smoking). For consensus communication, the audience may be involved by helping to develop the way the risk management decision will be made (do all stakeholders vote or will representatives develop a decision?), making the decision (working within a group to develop a consensus), and implementing the decision (developing the policies and procedures or actually doing what was decided). For crisis communication, the audience may be involved by helping to develop emergency preparedness plans or by assisting in evacuations or other management strategies.

To meet the communications skills challenge, those who are communicating the risk may need to interview the audience to help audience members focus their thoughts or meet with audience members in smaller groups so that members can help each other communicate.

SOCIAL CONSTRUCTIONIST APPROACH

> The social constructionist model suggests that we do not, or should not, develop policy in private by some rhetorical means and then, through rhetoric, attempt to impose that policy on our fellows.
> —Craig Waddell (1995, p. 201).

Similar to the National Research Council's approach, this approach focuses on the flow of technical information and values, beliefs, and emotions (Waddell 1995). Many approaches consider that, during a risk assessment, the scientific community provides technical knowledge while the audience or stakeholders provide values, beliefs, and emotions through feedback on the risk communication effort or in the risk management process. The social constructionist approach holds that in fact both inputs come from both sides. In other words, the scientific community has values, beliefs, and emotions that subtly affect how risks are assessed and communicated, and the stakeholders often have technical knowledge that could affect the risk assessment and communication process.

The implication for risk communicators is that social context and culture can influence beliefs and actions of all parties. Understanding this context and facilitating an exchange of information, attitudes, values, and perceptions in both directions ("expert" to "stakeholder" and vice versa) can help build better risk decisions, whether in care, consensus, or crisis communication.

HAZARD PLUS OUTRAGE APPROACH

Noted risk communication expert Peter Sandman, in a number of articles, books, and videos, popularized the approach originally developed through research by Baruch Fischhoff and Paul Slovic that risk should be viewed as hazard plus outrage (Sandman 1987). That is, the audience's view of risk (as opposed to that of the experts assessing the risk) reflects not just the action (hazard) but, even more important, what emotions they feel about the action (their outrage). Think of the hazard part of the risk as the expert's assessment of the risk—the factors considered in the risk assessment, which may lack emotional aspects of the risk. Think of the outrage part of the risk as the nonexpert's assessment of the risk—the average person's assessment of the risk, which may lack some or all of the factors considered in the risk assessment. If both assessments are in agreement, that is, the expert and nonexpert agree that the risk is substantial (high hazard and high outrage) or insubstantial (low hazard and low outrage), then there is a lesser chance of controversy. If the two assessments do not agree (high hazard and low outrage or low hazard and high outrage), there is a greater chance of controversy.

> [Hazard and outrage] are connected by the fact that outrage is the principal determinant of perceived hazard. When people are upset, they tend to think they are endangered; when they're not upset, they tend to think they're not endangered.
> —Peter Sandman (2003, p. 26).

The implication for those who are communicating risk (whether in care, consensus, or crisis communication) is that a presentation of technical facts will not necessarily give most audiences the information they want. Indeed, the audience will probably not even listen to those facts until their concerns and feelings have been addressed. You cannot ignore the outrage part of the risk and focus solely on the hazard. This does not mean that you should pretend to agree with the audience's concerns, which would be disastrous to building the levels of credibility needed to communicate risk. Instead it means that you must understand the audience's feelings and make sure that the information addressing those feelings is included in the risk message.

MENTAL NOISE APPROACH

Vincent Covello, another luminary in the field of risk communication, cautions his students in course work and workshops to beware of mental noise when communicating risks. The approach is also used by some portions of the U.S. Department of Defense and in public health circles.

This approach holds that when people perceive themselves at risk, their ability to hear and process information decreases dramatically. Under such circumstances, the ability to attend to and retain information is estimated to be 80% less than normal. This limitation is particularly true in sudden, unexpected crises. For example, those responding to the bombing of the Oklahoma City federal building in April 1995 found that both verbal and written information had to be provided, sometimes multiple times, for the victims' families to understand what had happened and what they should do next (Blakeney 2002).

The implication for those communicating risk is that risk information must be carefully packaged and presented, particularly in crisis communication. Proponents of the approach advise that no more than three key messages, repeated frequently, should be used, along with reinforcement of verbal and written communications with visuals, and ruthless removal of jargon, technical terms, and acronyms. For additional information on designing risk information materials, see Chapter 13.

SOCIAL NETWORK CONTAGION APPROACH

Organizational studies have looked at the way social networks influence behavior and attitudes in the workplace. The results of these studies suggest that who we spend time with affects how we view the world. More recently, this approach has been applied to risk communication efforts (for example, see Scherer and Cho 2003).

> We must look beyond individuals to their communities when communicating risk.

The approach holds that, when faced with a risk, people adopt the behaviors and attitudes of others in their social network as a response to the risk. Their network does not need to influence them intentionally; the change may come about gradually as a result of shared time and similar perceptions in other areas. The stronger the social tie and the more frequent the interaction, the more likely will be the cohesiveness of reaction to the risk. These social ties may also be built on social media websites such as Facebook.

The implication for those communicating risk is that we must look beyond individuals to their communities when communicating risk. In trying to encourage people to modify risky behaviors (for example, in care communication), getting their social network to endorse or reinforce less risky actions might be more effective than targeting individuals. Involving all members of a social network might be the only way to gain participation in some areas for consensus communication. In crisis communication, partnering with key social leaders may spread the news the fastest. For more information on understanding the needs of the audience, see Chapter 8.

SOCIAL AMPLIFICATION OF RISK APPROACH

This approach grew from a social science perspective and was furthered by internationally known researchers Roger and Jean Kasperson (for example, Kasperson et al. 1988), as well as Paul Slovic and James Flynn. Slovic, Flynn, and others helped add the concept of stigma to the approach (for example, in Flynn et al. 2011).

The basic theory is that social activities will magnify the consequences of a risk event, often in unexpected ways. Think of a pebble thrown into a lake; the ripples spread far beyond the pebble's initial impact on the water. So too, a risk can spread to impact business sales, regulatory agendas, community opposition, and legal action. These consequences can in turn result in an industry or a community being stigmatized: seen as negative or bad by others who associate with it.

The news media in particular has been credited with amplifying risk consequences. For example, work at several universities in the United Kingdom (Birmingham, East Anglia, Surrey, and Queen's University in Belfast) found that whereas media coverage often brought in secondary issues associated with risk, government agencies charged with communicating risk usually failed to recognize or address such issues (Department of Health 2003). This dichotomy resulted in stories about the risk growing in number and outrage, without providing any support to its resolution.

Veteran researchers William Leiss at the University of Ottawa and Douglas Powell at Kansas State University took the approach one step further after studying high-profile cases of risk communication such as mad cow disease and silicone breast implants (Leiss and Powell 2005). They theorized that a risk information "vacuum" is most likely to blame for the social amplification of risks. That is, when experts refuse to provide information, a hungry public will fill the void, often with rumor, supposition, and less-than-scientific theories. Silence from experts and decision makers, particularly regulatory agencies, breeds fear and suspicion among those at risk and makes later risk communication much more difficult.

The approach has several implications for risk communicators. One is to actively plan for and respond to the social ripples that arise from a risk being identified, particularly in consensus and crisis communication. Such planning should incorporate a thorough understanding of audience needs. Another implication is to follow one of the primary principles of risk communication: ensure that information is communicated early, often, and fully. This is particularly important for crisis communication. Be the expert and share the information, before other organizations and individuals rise up to fill that gap with information that, at a minimum, may make risk communication efforts more difficult and, at the worst, could well end up endangering public health or safety. For more principles of risk communication, see Chapter 6.

SOCIAL TRUST APPROACH

The social trust approach also grew from social science research, particularly that of George Cvetkovich and Tim Earle. This approach holds that a person's trust in an institution (for example, a government agency) is built on an understanding of the institution's goals, motives, and actions in relationship to the person's values. In other words, if I perceive, based on observed behaviors, that the organization managing the risk has the same values I have, I will place my trust in them to appropriately manage the risk. Research has found that the higher the trust, the lower the estimate of risk and the higher the estimate of benefits (Cvetkovich and Winter 2001).

Another aspect of this approach is called the asymmetry principle. Researchers found that it is more difficult to create or earn trust than to destroy it. Studies show that if people do not trust an organization, negative information associated with that organization reinforces their distrust, whereas positive information is discounted. Fortunately, the converse is also true: If people

> Trust and confidence are being shown as increasingly important to how people perceive risks and how they respond to risk management strategies.

trust an organization, positive information will reinforce that trust, and negative information will be discounted (Cvetkovich et al. 2002).

When the control of risk is not at the personal level, trust becomes a major and perhaps the most important variable in public acceptance of the risk management approach (Cvetkovich and Winter 2001). This premise is also supported by work by Vincent Covello, who has been credited with the idea that when people perceive themselves to be at risk, they understand and put into practice only those messages that come from sources they perceive as trustworthy and credible. Some researchers believe that this perception of trust and credibility comes from the audience's perceptions of the organization's ability to care, commitment to resolving the risk, competence, and honesty. The single biggest contributor to increasing trust and credibility is the organization's ability to care or show empathy.

Implications for those communicating risk is that information alone, no matter how carefully packaged and presented, will not communicate risk effectively if trust and credibility are not established first. Trust and confidence are being shown as increasingly important to how people perceive risks and how they respond to risk management strategies. Unfortunately, how an organization is perceived is often beyond the control of risk communicators. Some techniques for building trust and credibility can be found in Chapter 4 and Chapter 5.

EVOLUTIONARY THEORY APPROACH

Some of the more recent research in risk communication comes via the theory of evolution (for example, Tucker et al. 2008). The idea is that the way we evolved from hunter-gatherers shapes how we process, perceive, and can understand risks. Natural selection resulted in people who value fairness, equity, justice, prudence, and generosity, and who fear breaking social contracts. In general, humans are hardwired to cooperate in sharing resources. Biomathematics experts W. Troy Tucker and Scott Ferson propose six categories of risk that humans have routinely faced: disease, paternity, accidents, intergroup competition (war), subsistence failure, and cooperation failure. People trade risks among these categories. For example, stakeholders may insist on more stringent regulations for a new landfill, resulting in more time on the job for workers (and more potential for accidents), if it decreases the chance of groundwater contamination, which could prevent farming in some areas (subsistence failure).

Risk communicators, then, could expect that risks that may be perceived as unfair, unjust, costly, wasteful, or harmful to a single group of people may be viewed as more dangerous and less acceptable by stakeholders. Other research bears out these characteristics. For example, see Chapter 4 and the subsection on Hostility and Outrage.

EXTENDED PARALLEL PROCESS MODEL APPROACH

This model, which grew out of health risk communication research, looks at how and why people respond to fear-based messages. Such messages require the audience to evaluate the threat and the ability of the recommended action to resolve the threat. To evaluate the threat, the audience looks at relevancy and severity. If they deem the hazard as irrelevant or insignificant to them, they will likely take no action to mitigate it. If they decide that it is relevant to them or severe enough not to want to take chances, they consider their own abilities and the efficacy of the approach. If they believe that they are capable of taking the action and that the action will work, they generally take the recommended action. However, if they believe themselves incapable of taking the action or see the action as useless, they will act to control fear instead of controlling the hazard. Audiences generally control fear through denial or hostility (Witte et al. 2001).

The implication for risk communicators, particularly in care communication, is to make sure that messages are relevant to the intended audience and clearly show the severity of the hazard. The audience must be capable of taking the recommended actions (physically, emotionally, socially, and financially), and they must believe that the action will work. Without such underpinnings, messages may well evoke more hostility than action. For more information on crafting messages, see Chapter 9.

SUMMARY

No one approach to risk communication can be applied equally well to all the purposes, audiences, and situations for which risk is being communicated. Instead approaches to risk communication come from a variety of disciplines, each of which can provide insight to those who are communicating the risk. Understanding the various approaches and their implications can provide us with a repertoire of ways to develop our risk communication efforts, giving us a greater chance of success than if we were communicating without this knowledge.

REFERENCES

Becker, S. M. 2004. "Emergency Communication and Information Issues in Terrorist Events Involving Radioactive Materials." *Biosecurity and Bioterrorism: Biodefense Strategy, Practice, and Science,* 2(3):195–207.

Blakeney, R. L. 2002. "Providing Relief to Families after a Mass Fatality: Roles of the Medical Examiner's Office and the Family Assistance Center." *OVC Bulletin*, November 2002. U.S. Department of Justice, Office of Justice Programs, Office for Victims of Crime, Washington, DC.

Cvetkovich, G. and P. L. Winter. 2001. "Social Trust and the Management of Risks to Threatened and Endangered Species." Presented at the Annual Meeting of the Society for Risk Analysis, December 2–5, 2001, Seattle, Washington.

Cvetkovich, G., M. Siegrist, R. Murray, and S. Tragesser. 2002. "New Information and Social Trust: Asymmetry and Perseverance of Attributions about Hazard Managers." *Risk Analysis*, 22(2):359–367.

Department of Health (of the United Kingdom). 2003. "The Social Amplification of Risk: The Media and the Public." http://www.hse.gov.uk/research/crr_pdf/2001/crr01329.pdf (accessed January 24, 2013).

Flynn, J., P. Slovic, and H. Kunreuther, eds. 2001. *Risk, Media, and Stigma: Understanding Public Challenges to Modern Science and Technology*. Earthscan, London.

Geuter, G. and A. L. Stevens, eds. 1983. *Mental Models*. Lawrence Erlbaum Associates, Hillsdale, New Jersey.

Griffin, R. J., S. Dunwoody, F. Zabala, and M. Kamerick. 1994. "Public Reliance on Risk Communication Channels in the Wake of the Cryptosporidium Outbreak." Paper presented at the Society for Risk Analysis Annual Meeting, December 1994, Baltimore, Maryland.

Kasperson, R. E. 1986. "Hazardous Waste Facility Siting: Community, Firm, and Governmental Perspectives." In R. E., Kasperson, ed., *Hazards: Technology and Fairness*. National Academy of Engineering/National Academy Press, Washington, DC, p. 118–144.

Kasperson, R. E., O. Renn, P. Slovic, H. S. Brown, J. Emel, R. Goble, J. X. Kasperson, and S. Ratick. 1988. "The Social Amplification of Risk: A Conceptual Framework." *Risk Analysis*, 8:177–187.

Leiss, W. and D. Powell. 2005. *Mad Cows and Mother's Milk: The Perils of Poor Risk Communication*, 2nd ed. McGill-Queen's University Press, Montreal, Quebec, Canada.

Morgan, M. G., B. Fischhoff, A. Bostrom, L. Lave, and C. J. Atman. 1992. "Communicating Risk to the Public." *Environmental Science and Technology*, 26(11):2048–2056.

Morgan, M. G., B. Fischhoff, A. Bostrom, and C. J. Atman. 2002. *Risk Communication: A Mental Models Approach*. Cambridge University Press, New York.

NRC (National Research Council). 1989. *Improving Risk Communication*. National Academy Press, Washington, DC.

NRC (National Research Council). 1996. *Understanding Risk: Informing Decisions in a Democratic Society*. National Academy Press, Washington, DC.

Peters, R. G., V. T. Covello, and D. B. McCallum. 1997. "The Determinants of Trust and Credibility in Environmental Risk Communication: An Empirical Study." *Risk Analysis*, 17(1):43–54.

Rogers, E. M. and D. L. Kincaid. 1981. *Communications Networks: Toward a New Paradigm for Research*. The Free Press, New York.

Rowan, K. E. 1991. "Goals, Obstacles, and Strategies in Risk Communication: A Problem-Solving Approach to Improving Communication about Risks." *Journal of Applied Communication Research*, November:300–329.

Sandman, P. M. 1987. "Risk Communication: Facing Public Outrage." *EPA Journal*, November:21–22.

Sandman, P. M. 2003. "Four Kinds of Risk Communication." *The Synergist (Journal of the American Industrial Hygiene Association)*, April:26–27.

Scherer, C. W. and H. Cho. 2003. "A Social Contagion Theory of Risk Perception." *Risk Analysis*, 23(2):261–267.

Shannon, C. E. 1948. "A Mathematical Theory of Communication." *Bell System Technical Journal*, 27:379–425, 623–656; July, October; Bell Labs, Murray Hill, New Jersey.

Tucker, W. T., S. Ferson, A. M. Finkel, and D. Slavin, eds. 2008. *Strategies for Risk Communication: Evolution, Evidence, Experience*. Annals of the New York Academy of Sciences, Vol. 1128, 2008 (April), New York Academy of Sciences, New York.

Waddell, C. 1995. "Defining Sustainable Development: A Case Study in Environmental Communication." *Technical Communication Quarterly*, 4(2):201–216.

Witte, K., G. Meyer, and D. Martell. 2001. *Effective Health Risk Messages*. Sage Publications, Thousand Oaks, California.

ADDITIONAL RESOURCES

Hannigan, J. A. 1995. *Environmental Sociology: A Social Constructionist Perspective*. Routledge Press, London.

Johnson, B. B. 1993. "The Mental Model' Meets 'the Planning Process': Wrestling with Risk Communication Research and Practice." *Risk Analysis*, 13(1):5–8.

Sandman, P. M. 1989. "Hazard versus Outrage in the Public Perception of Risk." In V. T. Covello, D. B. McCallum, and M. T. Pavlova, eds., *Effective Risk Communication: The Role and Responsibility of Government and Nongovernment Organizations*. Plenum Press, New York, pp. 45–49.

3

LAWS THAT MANDATE RISK COMMUNICATION

Although many organizations have realized that it is good business practice to keep communities and interested parties aware of potential risks, risk communication is still often conducted as a result of a law, regulation, or other government inducement. A number of laws and regulations in the United States mandate risk communication as part of the risk assessment and risk management process. New international guides and standards also stipulate risk communication activities. Although these laws run to several volumes, making it difficult for anyone outside of the legal profession to really understand them, those who are communicating risk need to be aware of the laws affecting risk communication efforts and what these laws entail.

> Although many organizations have realized that it is good business practice to keep communities and interested parties aware of potential risks, risk communication is still often conducted as a result of a law, regulation, or other government inducement.

Failing to understand the laws and regulations can have several repercussions:

- Some member of your audience (who knows more than you do about the law) may sue your organization for failing to follow due process. This has happened to a number of federal agencies because they took a law or even their own implementing regulations less seriously than did the audience.
- Your arguments for continued or increased funding for risk communication are weakened if you do not know the law. Organizations are more likely to take notice

Risk Communication: A Handbook for Communicating Environmental, Safety, and Health Risks,
Fifth Edition. Regina E. Lundgren and Andrea H. McMakin.
© 2013 The Institute of Electrical and Electronics Engineers, Inc., and Regina E. Lundgren and
Andrea H. McMakin. Published 2013 by John Wiley & Sons, Inc.

if the risk communication effort is "required" than if it is optional. Optional programs get cut in budget crunches, whereas those required by law usually do not.
- The agency in charge of implementing the regulation may shut down your operations or levy a heavy fine if you are not in compliance.

This chapter highlights some of the major federal laws within the United States and selected international standards. Many states have similar rulings, often with more stringent requirements. In addition, many agencies and organizations have guidelines for implementing the regulations they must comply with most frequently. For example, the National Environmental Policy Act has counterparts in many states (individual state environmental policy acts), and several federal agencies (for example, the U.S. Department of Energy and U.S. Department of Defense) have developed their own policies and procedures to comply with both regulations. Check your state and local laws and how your organization has chosen to implement these laws before beginning your risk communication[1] efforts.

COMPREHENSIVE ENVIRONMENTAL RESPONSE, COMPENSATION, AND LIABILITY ACT

Also known as CERCLA or Superfund, this act and its reauthorization (Superfund Amendment and Reauthorization Act, or SARA) require that specific procedures be implemented to assess the release of hazardous substances at inactive waste sites. Those procedures involve the inclusion of "community relations" in the evaluation process. The term "community relations" refers to developing a working relationship with the public to determine acceptable ways to clean up the site. Figure 3-1 illustrates how the community relations process fits into the technical process for cleanup. Key communication pieces are as follows:

- **Community relations plan.** The community relations plan is very similar to the communication plan described in Chapter 12, "Develop a Communication Plan." The plan incorporates information about the site (for example, history, levels of contamination, and types of contamination), the community interested in the cleanup (demographic information), their concerns and beliefs about the site, and which communication methods will be used to address these concerns and include the public in the cleanup process. Under U.S. Environmental Protection Agency guidelines, in developing the community relations plan, representatives of the organization responsible for cleanup are required to meet with members of the community to listen to their concerns. These community interviews are generally conducted one on one in a location where each member feels comfortable (their home, the local tavern, etc.). The plan is usually updated at least yearly throughout the process, which averages 8 years.
- **Administrative record.** The administrative record is a set of all documents and other information that were used to make a decision about steps in the cleanup process. It is housed in a public library or other location where the audience can have easy access to it. It is updated at each step in the cleanup process.

---

[1] In this chapter, we sometimes use the terms "risk communication" and "public involvement" interchangeably because many of the regulations use the same term to mean both.

- **Information repository.** An information repository is a file containing site information, documents on site activities, and general information about the Superfund program. It too is housed in a library or other location where the audience can have easy access to it. It is updated regularly (the interval depends on how much activity is going on at the site and could range from weekly to quarterly).
- **Advertisement of public involvement opportunities.** Fact sheets, news releases, and proposed plans are some of the devices used to alert the public to opportunities for involvement in how cleanup decisions are made. Figure 3-1 shows the points at which the U.S. Environmental Protection Agency suggests that some of this information be released.

Tools to accomplish community relations (and risk communication) for Superfund can be found on the U.S. Environmental Protection Agency website under Community Involvement Toolkit (http://www.epa.gov/superfund/community/toolkit.htm).

EMERGENCY PLANNING AND COMMUNITY RIGHT-TO-KNOW ACT

This act, a part of SARA, is a freestanding law that requires that the public be provided with information about hazardous chemicals in the community and establishes emergency planning and notification procedures to protect the public from a release of those chemicals. The act also calls for the creation of state emergency response commissions to guide planning for chemical emergencies. Such state commissions have also created local emergency planning committees to ensure community participation and planning. Organizations that generate hazardous chemicals must produce a list each year of the quantities of chemicals stored at each site and make it available to the public and regulatory agencies. In addition, organizations must report any accidental release of hazardous chemicals to the environment and in some cases file reports on routine emissions as well.

EXECUTIVE ORDER 12898, ENVIRONMENTAL JUSTICE IN MINORITY POPULATIONS

Presidential or executive orders provide requirements to federal government agencies and departments. These agencies and departments in turn often pass along similar requirements to civilian organizations that contract to them. This order requires that government agencies and departments consider any potentially disproportionate human health or environmental risks to minority or low-income populations posed by the organization's activities, policies, or programs. As noted in Chapter 5, environmental equity (or environmental justice) has become a rallying cry across the nation as civic organizations have begun to realize that hazardous waste facilities and other industries perceived to be "risky" were apparently more often being sited in minority or low-income areas (for example, see Bullard 1990).

One of the requirements of the order is for agencies to consider translating crucial public documents and public meetings related to human health or environmental risks for those in the audience with limited skills in English. It also requires that information related

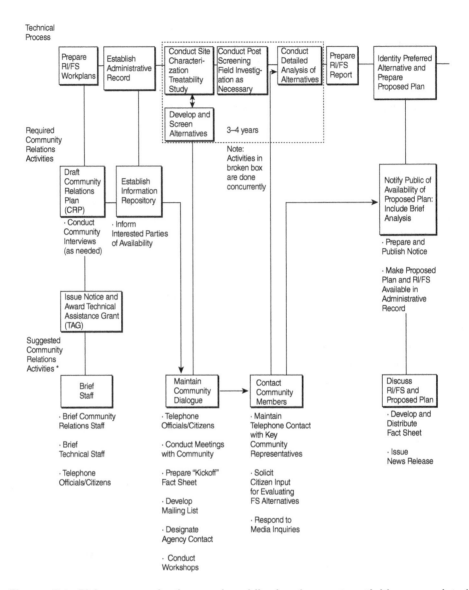

Figure 3-1. Risk communication and public involvement activities associated with the Superfund cleanup process. RI/FS, remedial investigation/feasibility study. *Ongoing throughout RI/FS process.

to human health or environmental risks be concise, understandable, and readily accessible to the public.

EXECUTIVE ORDER 13045, REDUCE ENVIRONMENTAL HEALTH AND SAFETY RISKS TO CHILDREN

This order requires that federal agencies and departments consider the potentially dispro-portionate health and safety impacts to children from the organization's activities, policies,

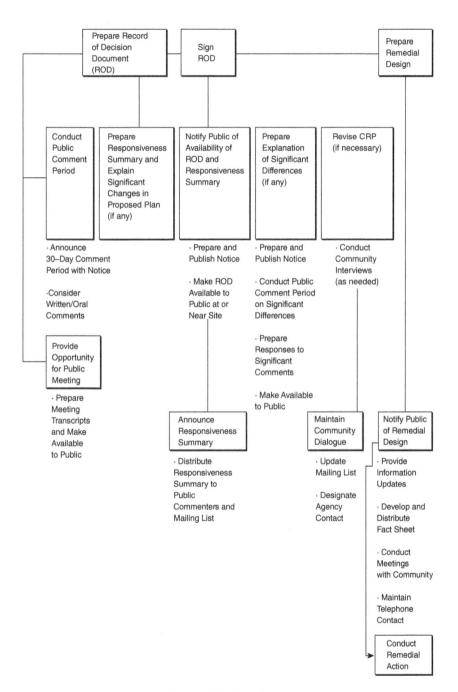

Figure 3-1. *Continued*

> This executive order says to every federal agency and department: put our children first. We Americans owe our largest responsibility to our smallest citizens. From now on, agencies will have to take a hard look at the special risks and disproportionate impact that standards and safeguards have on our children.
> —Former Vice President Albert Gore, national press conference at the Children's National Medical Center, April 21, 1997.

and programs. For example, an agency considering revising a regulation regarding the use of safety belts in automobiles might need to consider the growing body of evidence indicating that people of shorter stature (such as children) are actually in danger using standard over-the-shoulder safety belts. Although the order does not require risk communication per se, it does require that agencies considering enacting regulations submit risk information specifically related to children. Agencies and contractors supporting them need to be aware of this requirement so as to effectively communicate the risks.

FOOD AND DRUG ADMINISTRATION REGULATIONS ON PRESCRIPTION DRUG COMMUNICATION

The U.S. Food and Drug Administration is responsible for overseeing the implementation of a number of regulations on prescription drug use, including labeling and direct-to-consumer communication. For example, the Office of Prescription Drug Promotion has the mission to protect public health by ensuring that prescription drug information is truthful, balanced, and accurately communicated. The office provides a comprehensive surveillance, enforcement, and education program. Direct-to-consumer television ads must be reviewed by the office before showing publically, and other materials may require review as well. Risk communicators charged with sharing information about prescription drugs will want to review Food and Drug Administration guidance early in the planning process to ensure that appropriate time is available for reviews and revision.

NATIONAL ENVIRONMENTAL POLICY ACT

This is the act that mandates environmental impact statements and other environmental assessments. Under this law, any time a federal agency takes some "major" action (whether it be their own action or the granting of a federal permit or right of way), it must consider the impact of that action on the environment. Depending on the potential level of impact, this consideration will be documented in one of the following:

- **Categorical exclusion report.** A categorical exclusion report shows that the agency has considered the action and has determined that it obviously will not have an effect on the environment.
- **Environmental assessment.** An environmental assessment is usually written by an outside organization that has analyzed the action and determined that it will or will not have an effect on the environment.
- **Environmental impact statement.** An environmental impact statement is usually written by an outside organization when the federal agency has determined that the action is likely to have a significant impact on the environment.

Many states have similar documentation requirements for actions taken by state agencies. Each type of assessment has different communication requirements, which become more complex as the assessment becomes more involved. The categorical exclusion report is often simply several standard forms that are filled out and filed. The environmental assessment, although a public document, basically serves to help the agency to decide whether an environmental impact statement is necessary or whether the agency can issue a Finding of No Significant Impact (referred to as a "FONSI"). An environmental assessment should contain information on the need for the proposed action, alternatives to the proposed action, environmental impacts of the proposed action and alternatives, and a listing of agencies and persons consulted in the preparation of the report. Although public involvement is not required for an environmental assessment, some agencies nevertheless issue the report for public comment and consider those comments in determining whether an environmental impact statement will be necessary.

> Under the National Environmental Policy Act, any time a federal agency takes some "major" action, it must consider the impact of that action on the environment.

> The most complex communication requirements are associated with the environmental impact statement.

The most complex communication requirements are associated with the environmental impact statement. Key communication pieces connected with the environmental impact statement are the following:

- **Notice of intent.** The agency must file a notice in the *Federal Register* detailing its intent to prepare an environmental impact statement, the possible content of that statement, and a request for public comments. The notice is usually several paragraphs to several pages long.
- **Scoping meetings.** The agency has the option of conducting formal meetings with the public to determine what should be included in the environmental impact statement.
- **Publication of a draft for comment.** The agency must release a draft environmental impact statement for public comment.
- **Meeting for comments on the draft.** During the comment period, the agency usually holds public meetings to collect comments on the draft environmental impact statement.
- **Publication of final environmental impact statement.** The agency revises the statement based on comments. A discussion of how the comments were used must be included in the environmental impact statement or as a supplement. Another comment period may follow the issuance of the final environmental impact statement, but usually no additional meetings are held.
- **Publication of a record of decision.** The agency publishes a short statement describing its final decision on the action. The record of decision may be posted in the *Federal Register*, mailed to those who commented, and/or placed in a public reading room.

The Council on Environmental Quality issues regulations as to the content and size of environmental impact statements (Council on Environmental Quality, No Date; 40 CFR 1500).

NATURAL RESOURCE DAMAGE ASSESSMENT

A Natural Resource Damage Assessment is the process of determining environmental damage caused by releases of hazardous substances. This process is part of CERCLA. The process determines the condition of a particular part of the environment (including land, fish, wildlife, plants, air, water, groundwater, and drinking water supplies) that is public property (held by federal, state, local, or Native American governments). Natural Resource Trustees (state or federal agencies or Native American nations who act on behalf of the public for these resources) begin the process with a preassessment screen—a review of existing data concerning the resource—to determine whether hazardous substances have been released in sufficient quantity or concentration to damage the resource, whether the resource was damaged, whether data are available or can be obtained to conduct a full assessment, and how quickly it might be possible to restore the resource to its original condition. If the preassessment screen shows that the resource has likely been damaged, an independent organization is usually hired to prepare an assessment plan, more rigorously assess the damage (by studying the resource directly), and report to the trustees. If the trustees find that the resource has been detrimentally changed (chemically, physically, or in its viability as a natural resource; legally called injury), the organization or individual responsible for the injury may have to make restitution to them for loss of use of the resource, the cost to assess the injury, and the cost to restore the resource to its original condition (this money is legally called the damage). The public can also sue the trustees for failing to live up to their responsibilities.

Risk communication, although not mandated specifically for these assessments, can be extremely useful in developing the assessment plan, preparing the final report, and educating the trustees and the public on the potential risks and damages to the resource and the process of assessing those risks and damages.

OCCUPATIONAL SAFETY AND HEALTH ACT

The Occupational Safety and Health Act was passed to ensure that "no employee will suffer material impairment of health or functional capacity" as a result of working at a particular occupation. One provision of the act set up the Occupational Safety and Health Administration (OSHA), which oversees compliance with the other provisions. One of OSHA's duties is to set standards for the limits of exposure to potentially hazardous chemicals and other occupational exposures. A number of public involvement activities surround the setting of standards, but because these activities are conducted only by OSHA, they are not discussed here. Instead, we focus on the communication requirements placed on the organizations employing workers.

Occupational Safety and Health Administration has issued regulations in 29 CFR 1910.1200 (referred to as the Hazard Communication Standard or Hazcom; Occupational Safety and Health Administration, No Date) that require employers to explain chemical and physical risks in the work environment to workers. These regulations are the origin for the use of Material Safety Data Sheets (MSDSs), which are generally one- or two-page documents that explain the properties and risks of various chemicals and mitigation measures to prevent injury. The MSDSs

> Occupational Safety and Health Administration has issued regulations in 29 CFR 1910.1200 that require employers to explain chemical and physical risks to workers.

are prepared by organizations who manufacture, import, or sell the chemical or combination of chemicals. Over 500,000 products have these sheets.

Although no formalized standard exists for the format of these sheets, the same basic information is captured on each:

- Specific chemical identity of the hazardous chemicals involved
- Common names for the chemicals
- Physical and chemical characteristics
- Known acute and chronic health effects and related health information
- Exposure limits
- Whether the chemical can be considered a carcinogen
- Precautionary measures for effective use
- Emergency and first aid procedures
- Identification (name, address, and telephone number) for the organization responsible for preparing the sheet

ANSI Standard No. 2400.1, *Material Safety Data Sheet Preparation*, or the MSDS Form (OSHA 174) can be used as guides to complete such sheets (OSHA 1995). Several commercial firms also sell software to help the user develop these sheets. Under the Occupational Safety and Health Act, the sheets must be made available to workers who could come in contact with the chemical described.

Specific chemicals are subject to additional regulations, some of which stipulate the training of workers in the use of the chemical. The training must include information on the risks, how to prevent possible exposure to the chemical, and what to do if exposed. Additional information, such as the MSDSs, must also be made available.

One portion of the standard that is sometimes overlooked is the requirement for a "written hazard communication program" that describes what steps will be taken to ensure that workers receive information about the chemicals in their work environment. The OSHA publication "Chemical Hazard Communication" (OSHA 1995) describes the types of information to be included in the plan such as the ways in which containers of hazardous chemicals will be labeled, how MSDSs will be made available to employees, a list of the hazardous chemicals in each work area, the means by which the employer will inform employees of the hazards of nonroutine activities, and the hazards associated with chemicals in unlabeled pipes. The written program must be available to employees as well as others such as the Director of the National Institute for Occupational Safety and Health.

In March 2012, OSHA published a final rule to modify its hazard communication policies and practices to align with the Globally Harmonized System of Classification and Labeling of Chemicals, which is being adopted by countries in the United Nations. This system includes criteria for the classification of health, environmental,

> One portion of the standard that is sometimes overlooked is the requirement for a "written hazard communication program" that describes what steps will be taken to ensure that workers receive information about the chemicals in their work environment.

and physical hazards, and it specifies what information must be included on labels of hazardous chemicals and safety data sheets. The rule began implementation in May 2012 and may affect how workplace hazards are communicated in your organization.

All federal agencies have been given authority to administer the OSHA regulations in their areas. In addition, state agencies can administer their own program if it is approved by OSHA. The state programs must be at least as stringent as the federal program and are usually more stringent. Check with your organization and state agencies to determine which programs apply to your risk communication effort and also be sure that you have the latest versions of OSHA's requirements.

RESOURCE CONSERVATION AND RECOVERY ACT

This act establishes regulatory standards for the generation, transportation, storage, treatment, and disposal of hazardous waste. Public involvement in this process is encouraged by the act, but not specified to the extent that it is in the Superfund legislation. Instead, the act discusses ways for the public to take organizations that do not comply to court. To forestall such litigation, many agencies and organizations have developed their own requirements for risk communication and public involvement. Check with your state and local governments, your organization, and regulatory agencies involved before beginning your risk communication efforts.

RISK MANAGEMENT PROGRAM RULE

Under the Clean Air Act Amendments of 1990, the U.S. Environmental Protection Agency, in 1996, issued this rule, which requires facilities that use, make, or store hazardous materials to report accident scenarios for each of their facilities and to make this information available to the public and the agency. These accident scenarios include information on the "worst-case scenario" (the most drastic accident that can be envisioned for a facility), an alternate more likely accident, what the facility has done or is doing to prevent such accidents, and plans for emergency responses should the accident occur.

Because of government and industry concern about the potential criminal and terrorist use of chemical data, the U.S. Environmental Protection Agency limited the types of data available to the public and how that data could be accessed. For example, information on the off-site consequences of chemical releases can only be accessed in person at federal reading rooms after showing appropriate identification, and no copies of the material can be made.

Even this limited accessibility is a benefit to communities trying to determine potential local hazards, and it provides greater incentive for organizations to find ways to communicate risks effectively, as the availability of the information could raise additional questions about local risks.

PRIVACY RULE

The Health Insurance Portability and Accountability Act of 1996 included a provision directing the U.S. Department of Health and Human Services to develop national standards for electronic health care transactions and the security and privacy of health information. The Privacy Rule, signed in April 2001, outlines administrative steps, policies, and procedures to safeguard individuals' personal, private health information, in electronic or other form, including oral communications.

Although this regulation does not mandate risk communication per se, it has affected risk communication in certain cases, particularly where organizations are rigorous in implementing the regulation. Developing strategies to combat certain communicable diseases can be difficult if agency staff refuse to share individual information with the mistaken idea that they will be liable for legal action should they do otherwise. If the risk about which you are communicating must be addressed using private health information, you will want to review this regulation, your state's interpretation or similar regulations, and your organization's policies in this area.

> If the risk about which you are communicating must be addressed using private health information, you will want to review the Privacy Rule, your state's interpretation or similar regulations, and your organization's policies in this area.

OTHER GOVERNMENT INDUCEMENTS

> Besides laws, government agencies and sometimes private organizations can encourage risk communication activities by issuing funds in the form of grants.

Besides laws, government agencies, private organizations, and professional organizations can encourage risk communication activities by issuing standards or funds in the form of grants.

International Standards

Several international agencies and organizations have issued guidelines or standards that can influence risk communication efforts whether in the United States or abroad. For example, in 2010, the European Union published risk assessment and mapping guidelines for disaster management. The International Atomic Energy Agency is also working on guidelines for responding to nuclear disasters. In addition, the International Organization for Standardization (ISO) has issued several standards related to risk communication. Government agencies and private companies around the world apply these standards to their activities. Some industries consider it a competitive advantage to be ISO-compliant. For example, a study of 63 Brazilian companies in the chemical, mechanical, and electronics industries found that one of the four factors in pursuing compliance with ISO environmental management standards had to do with the belief that business would improve as a result (Gavronski et al. 2008).

One of the more far-reaching sets of standards related to risk communication is the ISO 14000 series, which addresses environmental management, including what an organization does to minimize harmful effects it may cause on the environment and continually improve its environmental performance. ISO 14001, first published in 1996 and updated in 2004, describes an organization's approach to developing, maintaining, and evaluating an environmental management system.

ISO 14020 (2000) provides specific tools and guidance related to product labeling and declarations. ISO 14063 (2006) offers additional guidance on communicating environmental information, including risks. Although the standard supports ISO 14001, it can be used in situations in which the organization has no formal environmental management system. This standard discusses principles of environmental communication; development of an environmental policy and strategy; and planning, implementation, and evaluation of environmental communication activities. Although some of the activities (for example, publication of environmental reports with no opportunity for feedback) focus on one-way communication that should not be considered effective risk communication, other guidance is especially useful on developing risk communication programs, whether care, consensus, or crisis communication.

In 2009, ISO issued the beginning of the 31000 series, which addresses a general approach to risk management, applicable to many industries. Part of that standard describes the development of an external communication plan to share information with stakeholders. Again, the focus appears to be more on one-way communication, which can debilitate true risk communication.

Check with your organization to determine which guidelines and standards may apply before beginning risk communication efforts and consider going beyond standards that advocate one-way communication.

North American Standards

Other standards organizations issue standards related to risk communication, particularly in areas of emergency management. The American National Standards Institute, American Society for Testing and Materials, National Council on Radiation Protection and Measurements, and National Fire Protection Association, among others, have developed guides and standards for emergency response. For example, ASTM E2541, *Standard Guide for Stakeholder-Focused, Consensus-Based Disaster Restoration Process for Contaminated Assets* (ASTM 2007), provides a framework to involve stakeholders in the decision-making process when a natural disaster or terrorist attack damages infrastructure (consensus communication). Many of these standards focus on technical aspects such as the assessment of damages, with communication added as a second thought. In some cases, recommendations for good communication practices are a bit dated. However, if your risk communication efforts cover emergency response, check to see if these guides and standards apply, and remember that it is always possible, and often advisable, to improve on the standard.

Grants

When $10 million in grant funding is riding on the development of a risk communication plan, that plan will most likely get written. One of the more recent examples of this tactic is the Public Health Preparedness funding from the U.S. Centers for Disease Control and Prevention (CDC). Each state was allocated a certain amount of money to plan for responding to a public health emergency such as a bioterrorist attack or major disease outbreak, but the money came with the requirement of a grant application that was more of a project plan, divided into areas of focus. Focus Area F constituted risk communication activities, among them a plan on how to communicate during a public health emergency.

To aid states developing such plans, CDC developed a comprehensive training program called CDCynergy, Emergency and Risk Communication. The materials for this program were developed by experts in risk communication, particularly crisis communication, and included videos on specific topics, examples of plans and procedures, and in-depth information on working with the news media and public. CDC has since expanded the CDCynergy program to encompass health communication and social marketing as well.

See the agency website at http://www.cdc.gov/healthcommunication/CDCynergy/ for more information.

SUMMARY

Various laws and regulations can apply to risk communication efforts. Check with federal agencies, state agencies, local governments, national and international standards organizations, and your own organization before beginning a risk communication effort to ensure that you understand and are in compliance with the requirements.

REFERENCES

American Society for Testing and Materials (ASTM). 2007. *Standard Guide for Stakeholder-Focused, Consensus-Based Disaster Restoration Process for Contaminated Assets.* ASTM e2541, ASTM, West Conshohocken, Pennsylvania.

Bullard, R. D. 1990. "In Our Backyards: Minority Communities Get Most of the Dumps." *EPA Journal*, 18(1):11–12.

Council on Environmental Quality. No Date. 40 CFR (*Code of Federal Regulations*) Parts 1500 to 1508. "National Environmental Policy Act Implementing Regulations." Office of the Federal Register, National Archives and Records Administration, Washington, DC.

Gavronski, I., G. Ferrer, and E. Laureano Paiva. 2008. "ISO 14001 Certification in Brazil: Motivations and Benefits." *Journal of Cleaner Production*, 16(1):87–94.

Occupational Safety and Health Administration. No Date. 29 CFR (*Code of Federal Regulations*) Part 1910.1200. "Hazard Communication." Office of the Federal Register, National Archives and Records Administration, Washington, DC.

OSHA (Occupational Safety and Health Administration). 1995. "Chemical Hazard Communication." OSHA 3084, Occupational Safety and Health Administration, Washington, DC.

ADDITIONAL RESOURCES

Baram, M. S. and P. Kenyon. 1986. "Risk Communication and the Law for Chronic Health and Environmental Hazards." *Environmental Professional*, 8(2):165–179.

Carson, J. E. 1992. "On the Preparation of Environmental Impact Statements in the United States of America." *Atmospheric Environment*, 26(15):2759–2769.

EPA (U.S. Environmental Protection Agency). 1989. *Facts about the National Environmental Policy Act.* U.S. Environmental Protection Agency, Enforcement and Compliance Monitoring, Washington, DC.

Hadden, S. G. 1990. *A Citizen's Right to Know: Risk Communication and Public Policy.* Westview Press, Boulder, Colorado.

International Organization for Standardization at http://www.iso.org/iso/home.html (accessed January 18, 2013).

Jia, C. Q., A. diGuardo, and D. Mackay. 1996. "Toxic Release Inventories: Opportunities for Improved Presentation and Interpretation." *Environmental Science and Technology*, 30(2): A86–A91.

4

CONSTRAINTS TO EFFECTIVE RISK COMMUNICATION

A number of factors can place limitations on the risk communication effort. In general, the same factors can limit or constrain care, consensus, and crisis communication. Those who are communicating risk need to be aware of these constraints so that they can recognize and overcome the problems to increase their chances of communicating effectively. This chapter discusses the constraints on those who are communicating risk and the constraints that come from the audience, offering advice on how to recognize and overcome the potential problems.

CONSTRAINTS ON THE COMMUNICATOR

Suppose you are faced with a particularly difficult assignment—explain to a group of mothers that their infants have been inadvertently exposed to a highly toxic chemical that may cause mental retardation, physical disabilities, or even death. You can justifiably expect anger and fear, with people yelling and crying, not to mention the likelihood of lawsuits that could close your organization permanently. You plunge into the task, only to find that these difficulties in communicating this risk information are the least of your worries. What could make them worse? Read on.

Risk Communication: A Handbook for Communicating Environmental, Safety, and Health Risks,
Fifth Edition. Regina E. Lundgren and Andrea H. McMakin.
© 2013 The Institute of Electrical and Electronics Engineers, Inc., and Regina E. Lundgren and
Andrea H. McMakin. Published 2013 by John Wiley & Sons, Inc.

Organizational Constraints

The very organization that has asked you to communicate risks can put any number of roadblocks in your way:

- Inadequate resources
- Management apathy or hostility
- Potential roles dichotomy
- Difficult review and approval procedures
- Corporate protection requirements
- Conflicting organizational requirements
- Insufficient information to adequately plan and set schedules

Inadequate Resources

To effectively communicate risk, you need the funding, staff, equipment, and space to do the job. Unfortunately, in many organizations, the technical aspects of the risk (the risk assessment and risk management) are given the bulk of these key resources. Many organizations that would never refuse a scientist software to calculate dose quite easily turn down a requisition for desktop publishing equipment that would make it easier and less expensive to create and revise risk messages that are far more readable for the intended audience. With so much attention paid to the analysis of risk, the actual communication can be completely forgotten.

The challenge of finding adequate resources for effective risk communication has only been exacerbated in recent years. Government agencies have cut back on all but essential services; private industry is trying to do more with less. Couple this trend with the increased number of communication methods (Internet and social media in particular), staffing resources are quickly overwhelmed, even in care communication.

The empirical data are growing that show how risk communication plays an important part in the risk management process. For example, the *Washington Post* reported in October 2003 that utilities were coming to realize the importance of communicating risk management decisions to their constituencies. One utility took the brunt of public outcry when it failed to provide prompt and reliable estimates of when power would be restored after a devastating hurricane. Concerns escalated to the political level, damaging company credibility and exposing the organization to Congressional inquiry (Davenport 2003).

Those of us who are developing risk communication programs based on laws or other regulations can point to these laws, which mandate involving the audience in how the risk is managed and, therefore, require risk communication. Organizations ignore or neglect these requirements at their own peril. Likewise, when the entire charter of the organization is based on communicating risks, it may not be hard to argue for adequate resources. In the absence of such legal or organizational mandates, those who are communicating risk in more voluntary programs can use case studies and examples of programs that succeeded or failed and how these affected the organization's ability to do business (for example, see Beierle 2002; Hunt and Monaghan 1992; Sanderson and Niles 1992).

In addition, resources of staff and funding can be augmented through partnering. Both government agencies and industry can look to nonprofit organizations and volunteer groups for support where risk interests and organizational cultures align. For example, a large church in Seattle, Washington, offered to supply volunteers to local emergency management agencies to run Internet communications during emergencies. High school and college students may also be willing to trade free hours for work experience or community support.

Management Apathy or Hostility

Even when resources are plentiful, risk communication efforts may still fail because of the lack of support by the organization's managers or other decision makers. This support is necessary to obtain resources; to conduct many of the activities required in care, consensus, and crisis communication; and to evaluate and improve risk communication efforts. Balch and Sutton (1995), Boiarsky (1991), and Dozier et al. (1995) have pointed out that those who are communicating risk must first reach this internal audience before reaching out to an external audience. Only when the managers support a risk communication effort can that effort succeed.

Sometimes, this lack of support may arise from apparent apathy. As in the case of having insufficient resources, with the risk assessment and resulting decision taking the spotlight (often justifiably), managers and other decision makers seem to find it easy to overlook the risk communication effort. Educating managers on the effects poor communication can have on an organization will sometimes be enough to alert them to the need to support your efforts.

> Educating managers on the effects that poor communication can have on an organization will sometimes be enough to alert them to the need to support your efforts.

For example, when the government and beef industry of the United Kingdom ignored both scientific evidence and good risk communication practices during the mad cow outbreak in the 1990s, the costs to subsidize the failing beef industry throughout the European Union and to cull the British herd exceeded $5 billion (Leiss and Powell 2005). In addition, when Jack in the Box Restaurants fumbled their risk communication efforts and were blamed for the death of three children and the hospitalization of 144 customers, stock values plummeted 43% and the company posted a $44 million loss for the year, compared with a $22 million profit the previous year (Henry 2000).

Sometimes, however, the lack of support for risk communication efforts comes from an apparent hostility to the process. Caron Chess, Director of Rutgers University's Center for Environmental Communication, in a presentation at the 1994 Society for Risk Analysis Annual Meeting, described a phenomena called the "Threat Rigidity Response." Simply put, this is how managers of an organization may react when they feel threatened. This feeling can arise from such situations as the threat of a lawsuit, the perceived potential for negative publicity, or the perceived lack of control of the situation. When so threatened, the managers of an organization become more rigid and more controlling. They tighten controls on their staff, the flow of information, and the decision-making process. This kind of rigid response can make successful risk communication virtually impossible.

Chess found that organizations with certain characteristics are more successful at avoiding this rigid response to a perceived threat and, hence, are more successful at communicating risk. These characteristics include the following:

- The organization has a mechanism for the upward flow of information.
- The organization has a diffraction of responsibility (communication is everyone's business as opposed to being within the sole purview of a public affairs function).
- The organization has a permeable boundary—there are numerous ways for the community (or audience) to get information about the organization's activities.

In addition, studies of communication organizations by David Dozier and Larissa and James Grunig of the University of Maryland found that communication effectiveness or excellence was heavily influenced by shared expectations between senior management and those charged with communicating for the organization. Other factors included the core knowledge of the communicators and the level of participation allowed by corporate culture. The more knowledgeable the communicator in communication tactics and strategy, and the more participatory the culture, the better was the chance of effective communication (Dozier et al. 1995).

When the Risk Communication Advisory Committee of the U.S. Food and Drug Administration was asked to review the organization's strategic plan, they also offered recommendations for organizational effectiveness. One recommendation was to develop an organizational structure that ensures the recruitment, retention, and integration of individuals with the needed expertise in risk communication. Another was to develop a workflow system to ensure that communications are integrated into operations, with communicators and other subject matter experts teaming to create, summarize, refine, and deliver the needed information evaluated to a scientific standard (FDA 2009). In addition, the research by the Environmental Agency of the United Kingdom found that if risk communication is not embedded in the organization, it will not be perpetuated beyond certain individuals (UK EA, No Date).

On a routine basis, those who are communicating risk can encourage these types of behaviors and philosophies within their organizations so that hostility to risk communication efforts is less likely to occur. Dozier's work suggests that sometimes "crises" like mergers, change in leadership, new competition, new markets, and new projects open opportunities to convince management to support risk communication activities (Dozier et al. 1995). If hostility does occur, those who are communicating risk can view their management as a hostile audience and use some of the techniques described later in this chapter to deal effectively with such audiences. See also Chapter 5 for more information on satisfying the needs of managers.

Potential Roles Dichotomy

Another constraint that the organization can place on those communicating risk is the role the organization is willing to play in the risk communication process. The organization's perception of this role may come from laws and regulations, community expectations, or corporate policy. Table 4-1 summarizes some of the most common roles organizations can play.

Often, those in care communication situations find themselves in the educator role—trying to provide information to explain a risk so that people will take appropriate steps to protect themselves. In consensus communication, organizations may play the

Table 4-1. Example organizational roles in risk communication

Role	Expectation	Authority
Educator	Explain and inform	To impart knowledge
Facilitator	Encourage and support	To point toward solutions
Partner	Work with others	To jointly solve problems
Manager	Tell them what to do	To prevent or mitigate the risk
Regulator	Mete out justice	To enforce decisions

role of facilitator or partner, working with stakeholders to develop an appropriate risk management approach. In crisis communications, organizations may serve as managers, telling audiences exactly what must be done to stop a risk from escalating. After the crisis, other organizations may act as regulators to ensure that future crises are avoided.

Risk communication efforts are constrained when the audience expects a different role from the one the organization is willing to play. For example, if the audience is expecting the organization to partner with them in developing a risk management approach, but the organization attempts to fill a manager role by telling the audience what must be done, both the organization's credibility and its ability to communicate effectively will suffer. When presented with a roles dichotomy, audiences often react with anger, hostility, and a general unwillingness to help manage the risk.

> When presented with a roles dichotomy, audiences often react with anger, hostility, and general unwillingness to help manage the risk.

For example, a distraught mother called a local agency charged with communicating about air-quality issues. A contractor had just finished remediating her home for asbestos in the ceiling. The mother wanted assurances from the agency that it was safe for her child to return to the home. The agency representative explained the potential risks of asbestos and advised her to talk to her contractor to see what had been done to remediate them.

"You sound like a mother," the woman protested. "What would you do? Would you let your child back in the house?"

In this case, the caller wanted to give the agency representative the role of friend and confidante. Perhaps motivated by anxiety, she might have been trying to put a human face on what she perceived as a bureaucracy. But the agency's role was to inform and educate, not facilitate or regulate. Providing advice as a "mother" could even have resulted in legal action later. The lesson: stay within your role. The representative could have responded something like this:

> I am a mother, but I don't think you called me because of that. I think you may have called because you were hoping my organization had information that could help you. Let me make sure I understand what information you're looking for and then I'll try to help you get the answers you need.

One of the best ways to manage this constraint is to identify the organization's role before the risk communication process starts. Know what your organization expects of you and communicate this consistently to your audiences every time. We sometimes recommend that agency representatives who answer public information lines identify their role in the first few seconds of the conversation. You can do this by using your position title if it communicates what you do to those outside the agency. If it does not, you can add an explanatory phrase when introducing yourself. For written materials, include a sentence on the responsibilities of your organization early on to set expectations.

> One of the best ways to manage a roles dichotomy is to identify the organization's role before the risk communication process starts.

For additional information on developing materials, see Chapter 13. For more information on being a credible spokesperson, see Chapter 15.

Difficult Review and Approval Procedures

Another way an organization can constrain the communication of risk is by requiring review and approval procedures that are either inappropriate or too time-consuming. A good example comes from one of the national laboratories. The laboratory had a detailed review and approval procedure for the release of all technical information. This procedure required no less than nine reviews and signatures, as well as review and approval by the U.S. Department of Energy, which funded the laboratory's work. The system was designed to ensure that preliminary technical information generated in innovative research and development activities was not released prematurely. However, the laboratory was asked to act as an independent agency to determine what past doses the public may have received from radioactive releases from a U.S. Department of Energy facility. To maintain its independence in the eyes of its audience, the laboratory needed to be able to release information, sometimes even preliminary data, as quickly as possible. It also needed to distance itself from the U.S. Department of Energy, which many members of the audience did not trust to produce unbiased results. Recognizing these problems, the laboratory tailored its internal review process (fewer signatures required and no U.S. Department of Energy approval) to allow risk to be communicated effectively for that audience.

To show managers and decision makers that changes are necessary, a complete audience and situation analysis may be all that is required, especially if the audience has already threatened legal action. What reviews does the organization feel are absolutely necessary? Can any be combined? Who does the audience view as credible to review the information? Is this a crisis situation or one in which the audience is particularly hostile? In either of these cases, information will need to be released as quickly as possible. Showing that the audience and the organization's interests are best served by changes and suggesting easy ways to accomplish those changes can result in review and approval procedures that are more appropriate and timely.

> Showing that the audience and the organization's interests are best served by changes and suggesting easy ways to accomplish those changes can result in review and approval procedures that are more appropriate and timely.

Corporate Protection Requirements

Review and approval procedures are only one way an organization seeks to protect itself. Private companies, in particular, can view information as a competitive advantage. Even if some risk information is provided, other information may be closely guarded for the fear of releasing sensitive data that could clue competitors in to company trade secrets.

Since the terror attacks in the United States in 2001, organizations have additional concerns about protecting information. Information on where hazardous materials are stored and in what quantities could be used by those with criminal or terrorist intent. Yet it is often this information that is the key to understanding risk in communities.

Under protest from industry, the U.S. Environmental Protection Agency reconsidered its Risk Management Program Rule activities. Whereas the agency had planned to put such information on the Internet in a publicly accessible database, it agreed to shelter the information in public reading rooms with carefully controlled access. Such access was a compromise between industry fears and public demands. Neither was completely satisfied, but both understood the need.

But corporations are not the only organizations to be concerned about the release of risk information. Government agencies sometimes hesitate to share information because of misplaced optimism or fear of public reaction. For example, during the mad cow disease outbreak in 2000 in Europe, Germany's agriculture minister, hoping to protect the beef market, boldly proclaimed that the country was immune. Germany's regulations would prevent contaminated material from being fed to cattle. One week later, the first sick cow was discovered, and beef sales plummeted. Japan experienced the same pattern in 2001 when its agriculture minister promised no more victims after the first sick cow was found, and two more were found only 2 days later (Green et al. 2007).

Risk communicators must carefully consider what information can be released and look for ways to reach satisfactory compromises. As memories of dangers fade, the public will be less willing to allow companies to protect information in the name of security. For additional information on deciding on what information can be released, see Chapter 5. For more information on choosing appropriate communication mechanisms, see Chapter 10.

> Risk communicators must carefully consider what information can be released and look for ways to reach satisfactory compromises. As memories of dangers fade, the public will be less willing to allow companies to protect information in the name of security.

Conflicting Organizational Requirements

Yet another way an organization can constrain those who are communicating risk is by having policies, however well-intentioned, that conflict with the goals of risk communication. For example, a research and development firm had a standing policy that any information given to the public regarding the organization had to be approved of and released by public affairs representatives. Unfortunately, the public affairs staff had the charter for safeguarding the company's reputation and felt that to fulfill this charter they would allow nothing but completely positive information to be released. When risk communicators wanted to add information about the firm to an environmental impact statement, which analyzed both the negative and the positive aspects of a risk, the public affairs staff responded, "I just can't let you say that." (This type of response is one reason why public affairs staff can be mistrusted by some members of the audience.) However, communicating risk often requires that some of the worst information about an organization be released.

Before any risk communication project begins, those who are communicating the risk should review organizational requirements to see which will conflict and discuss the potential problems with the staff responsible for implementing the requirements. A little planning and explanation of the purposes of risk communication can help avoid a problem.

Insufficient Information to Adequately Plan and Set Schedules

As discussed in Part II, a wealth of information is necessary to create a communication plan and set a schedule. In some organizations, this information is closely guarded. Other organizations seem to simply ignore the planning process. After diligently searching, those who are communicating risk may develop schedules based on inadequate or what turns out to be incorrect information, only to find that their schedules cannot be met. For example, suppose your risk communication effort is scheduled to begin after a risk assessment is finished on October 1, with fact sheets to be issued and public meetings to be held shortly thereafter. Notices of where and when the meetings will be held have already been published. However, unbeknownst to you, the risk assessment has been delayed and will not be finished until November 1. You will have to retract your notices and replan.

Unfortunately, members of your audience may conclude that this is some kind of delaying tactic and that your organization does not want to release the information. When you finally do get the chance to begin communicating the risk, your credibility will already have suffered.

Plans and schedules that are both realistic and flexible must be developed to effectively communicate risk. Some factors to consider include legal requirements, organizational requirements, the scientific process, ongoing activities within the organization and nation, and audience needs. More specific information to be considered in developing schedules is discussed in Chapter 11.

Emotional Constraints

Another type of constraint on those who communicate risk comes from within themselves. Emotions and beliefs can color our attempts to communicate. The three emotional constraints most difficult to overcome are unwillingness to see the public as equal partner in risk decisions, inability to see how stakeholder value systems differ from our own, and the belief that science can never be understood by the public.

Unwillingness to See the Public as an Equal Partner

In a logical world, many feel, decisions about how to manage a risk would be made by those who really understood the situation. Furthermore, such decisions would be based on scientific principles, economic realities, and logic, not emotions, beliefs, and political leanings. Risk communication, however, cannot be effective unless it considers the emotions, beliefs, and political leanings of the audience.

Working with decision makers who show this unwillingness can be difficult. Remind them of their goal: to make a decision that will stand the test of time. As a number of court cases have shown, such a decision can only be reached when all parties agree with it, at least to some extent. It does the organization no good to decide on a perfectly logical, scientifically based, economical strategy, only to be tied up in court for years trying to justify it to people who would likely have accepted it in the first place if they had only been consulted!

Overcoming this belief in yourself can be even more difficult. If you find yourself resisting reaching out to those you know are concerned or whom others have suggested should be contacted, ask yourself why. Is it because of the difficulty involved in developing the contacts, or do you feel that the contacts really should not be made at all? Remind yourself that a number of success stories and research (such as Arvai 2003; Beierle 2002; Hunt and Monaghan 1992; Sanderson and Niles 1992) have shown that making reasonable attempts to consult interested stakeholders results in better, more useful, and more lasting solutions.

Inability to See Differing Value Systems

Everyone has a value system, a method by which they evaluate and choose between alternatives in a given situation. Often, our values are so deeply ingrained that it is difficult to recognize them for what they are. It is even more difficult to recognize that these values differ from the value systems of others. It seems to be a very human tendency to think that everyone else sees things the same way we do. For example, when U.S. and European agencies tried to stop the rise of HIV/AIDS in Namibia with a campaign that featured messages on abstinence and monogamy, the effort flopped. The Namibian culture placed a different value on sexual relations and, in fact, encouraged polygamist relationships (Hillier 2006).

Value systems play a large role in how any member of the audience views a risk. When constructing risk messages, it is easy to focus attention on matters that we feel are important, without thinking through what issues are of importance to the intended audience. To guard against our own biases in framing risk communication messages, those of us who are communicating risk must understand the audience thoroughly. Terryn Barill, a risk communication consultant, recommends interviewing members of the audience and listening to not only the questions they ask but also the words they use, to help you determine what is most important to them and to then include that information in the risk message (Barill 1991). The techniques in Chapter 8 are also designed to help.

Belief that the Public Cannot Understand Science

Many experts who have devoted years to a field of study feel that the public cannot possibly understand scientific or technical information. As one expert stated when asked for a definition of a term used to present risk, "I have a Ph.D. and I've been studying that for 15 years; I can't possibly explain it to you." When this topic was introduced at a lecture at a university, a professor in the audience asked with great skepticism, "Do you honestly believe that you can explain anything to the general public?" Our answer was, and is, a resounding "Yes!"

The Chinese government learned the importance of this belief. During the severe acute respiratory syndrome (SARS) outbreak in 2002/2003, provincial leaders withheld information about the spread of the disease and plans to contain it because

> The principal obstacles to understanding are lay [peoples'] time and attention, not intelligence.
> —Granger Morgan et al. (2002, p. 8).

they felt that their citizens were not educated enough to understand it. But when the villagers learned that their area might be used as a quarantine station, they panicked. The lack of information led them to riot and block roads to keep strangers away (Green et al. 2007).

Like risk communicators all over the world, we have spent our careers explaining scientific information to nontechnical audiences—how a nuclear reactor works, how hydrocarbon contamination can "float" on aquifers, and what a risk of 10^{-6} means. Presenting each piece of information was a challenge, requiring that we knew our audience and had a good command of language. We often had to borrow or develop innovative ways to present the information. But that just proves that it can be done. The techniques described later in Part III are designed to help you present risk information in a variety of ways.

CONSTRAINTS FROM THE AUDIENCE

Constraints do not just come from within those who are communicating risk or the organization. The audience itself often brings constraints that must be overcome if risks are to be communicated. These constraints include hostility and outrage, panic and denial, apathy, mistrust of risk assessment, disagreements on the acceptable magnitude of risk, lack of faith in science and institutions, and learning difficulties.

Hostility and Outrage

Many audiences react with hostility toward risk messages. By hostility, we mean anxiety, anger, frustration, and contempt. This can be true in care and crisis communication efforts; however, it seems to particularly plague consensus communication efforts. Peter Sandman, among others, has presented factors that may affect the audience's hostility level. As mentioned in Chapter 2, in the discussion of the Hazard Plus Outrage Approach, he sees risk as having two components: hazard, or the scientific aspect of the actual danger, and outrage, or the audience's other concerns about the danger. Factors that influence the level of outrage include the voluntariness of the risk (Do they have a choice?), level of dread (How scary is it?), issue of fairness (Do they see it as being fair?), and the moral relevance (the more relevant it is to their morality, the more hostile they become).

In addition to Sandman's outrage factors, there are other reasons that the audience may be hostile (Table 4-2), reasons that have nothing to do with the risk itself. Those who are communicating risk need to be aware of these issues because the higher the level of

Table 4-2. Reasons the audience can be hostile

Factors	More hostility	Less hostility
Catastrophic potential	Grouped in time and space	Scattered in time and space
Level of familiarity	Uncommon	Very common
Level of understanding	Not well understood by science	Well understood by science
Level of personal control	Controlled by a distrusted individual	Controlled by the source
Voluntariness	Involuntary	Voluntary
Effects on children	More likely to affect children	Less likely to affect children
Moral relevance	Relevant	Not relevant
Timing of effects	Immediate effects	Delayed effects
Identity of victims	Closely related	Statistical victims
Level of dread	Greatly feared	Apathetic
Level of trust in institutions	Distrust	Trust
Amount of media attention	Highly popularized	Seldom mentioned
History of accidents	Well-known accidents	No accidents
Equity (fairness)	Viewed as unfair	Viewed as fair
Distribution of benefits	Benefits distributed unequally with risk	Benefits distributed equally with risk
Reversibility	Damage irreversible	Damage reversible
Personal stake	Strong	Weak
Origin	Artificial	Natural
Level of uncertainty	Unknown to science	Known to science
Tone of message	Too positive	Objective
Organizational attitude	Organization ignores	Organization seeks out concerns and acknowledges concerns
Degree of change in lifestyle	Sharp change from normal	Little change from normal
Degree of understanding of process/data	Process/data presentation too complex	Process/data presentation aimed at audience

Adapted from various works by Paul Slovic, Vincent T. Covello, and Peter Sandman, for example, Covello et al. (1988).

hostility, the less chance that the audience will hear the risk message, and the less chance that any real communication will take place.

One reason an audience may be hostile is that the organization communicating the risk is seen as not being credible. This was the situation for the U.S. Environmental Protection Agency during the Reagan Administration. The agency was viewed by many as being more often on the side of the polluter than on the side of the environment; hence, anything agency officials said was viewed with great suspicion. Those who are communicating risk may not be able to do much once credibility has been lost, except to recognize it and try to keep future risk communication efforts as credible as possible. The trust between organization and audience can be rebuilt slowly if the relationship is built on trustworthy efforts.

Another reason for audience hostility has to do with the message being viewed as too positive. Most audiences react with hostility if they feel that they are merely being placated and if the message is full of trite phrases or facile reassurances, especially in the face of negative information in the press. They do not want to hear, "Trust us; everything is fine." They will often only be satisfied if they or someone they trust has reviewed all the data and reached the same conclusion.

> Audiences may react with hostility if the message is viewed as too positive. Most audiences react with hostility if they feel they are merely being placated or if the message is full of trite phrases or facile reassurances, especially in the face of negative information in the press.

Audiences can also be hostile if their concerns have been ignored. One of the principles of risk communication is to listen and deal with specific concerns (see Chapter 6). Hance et al. (1988) cite the case of an office building that was contaminated with dioxin (a cancer-causing chemical) after a fire. Although the risk assessment showed the building to be safe for the employees to return to work, and this information was communicated to the employees, the hostility level continued to rise. It turned out that the employees' main concern was where to park. The parking garage had also been damaged, and parking was not available in the downtown area near the building. The employees were not willing to listen to the message that the workplace was safe and they could return to work until their concerns over parking were addressed. Thus, those who are communicating risk must make sure that risk messages address audience concerns, even if the concerns seem to relate to peripheral issues.

Another reason audiences can be hostile is the very human perception that change is bad. For example, in a case where the groundwater in a rural area had been contaminated by chemicals leaching from a landfill, the response from many area residents was, "My grandfather drank from this well, my father drank from this well, and I don't see why I can't." A possible way to overcome this type of attitude is to show not only the dangers of continuing risky behavior but also the benefits of changing behavior to something less risky. However, information alone does not generally lead to changes in behavior, as shown by some of the research related to the mental models approach to risk communication (see Chapter 2).

Another reason audiences can be hostile is that they do not understand either the process or the data being communicated. The information may be too technical (full of difficult concepts or laced with acronyms and jargon) or the presentation may not meet their needs (for example, a presentation in English for an audience whose primary language is Spanish). The obvious way to overcome this reason for hostility is to use language and a method that meet the needs of your audience. The information in Part III is designed to help.

Panic and Denial

Sometimes, the nature of the risk and the audience's situation raise an even deadlier response than hostility: panic and denial. Such cases are, we are thankful to say, rare. Panic is the extreme response to a combination of fear, dread, and a lack of perceived control. When audiences panic, it is because they perceive themselves or those they love to be in imminent, life-threatening danger that they cannot influence. Panic can stop all action to prevent a risk, effectively freezing a person, or lead to unproductive actions that can actually make the damage worse. Panic also tends to interfere with cognitive processing, as chemicals flood the brain, preventing risk communication messages from being heard and acted upon.

> Panic is the extreme response to a combination of fear, dread, and a lack of perceived control. Beyond panic is denial, when the perceived result of a risk is so horrific that other coping mechanisms break down.

Beyond panic, however, is denial. Denial occurs when the perceived result of a risk is so horrific that other coping mechanisms break down. We cannot accept that something so awful has happened or could happen, so we simply refuse to think about it. An audience in denial is deceptively calm. Risk communication messages attempting to increase concern only push those at risk into a deeper denial.

Panic and denial are beyond the skills of most risk communicators to manage. However, those who are communicating risk need to realize that denial and panic are one end of the response spectrum for risks. A thorough understanding of audience needs can help prevent the introduction of risk messages or use of mechanisms (such as a graphic picture or live footage) that could thrust an audience into this difficult psychological area. Chapter 8 provides additional information on understanding those with whom you are communicating.

Apathy

At times it seems that those who are communicating risk simply cannot succeed. Audiences faced with crises may panic. Audiences in consensus communication situations often exhibit hostility and outrage. Audiences in care communication (and to a lesser degree some crisis communication efforts) are often apathetic.

When an audience is apathetic toward a risk, it is often because what has been deemed a high risk by scientists and government agencies seems impossible or trivial to the audience in question. A good example is the early efforts of the U.S. Environmental Protection Agency to warn of the dangers of radon in homes. Despite a major risk communication effort, the intended audience still did not change to less risky behavior, in this case, testing their homes for radon. Only when risk communication researchers did a more thorough audience analysis, using the mental models approach discussed in Chapter 2, did the risk communication effort begin to show some success (Morgan et al. 1992).

Even those thoroughly familiar with the risk can grow apathetic. For example, in planning for pandemic influenza, the U.S. Occupational Safety and Health Administration found that health care workers refused to wash their hands, even though hand washing is cited as one of the most effective methods for stemming a pandemic. In general, people need to feel sufficient concern about the risk to take action, and risks that are familiar and perceived to be low are less likely to engender that concern (OSHA 2007).

When faced with apathy, those who are communicating risk need to conduct as thorough an audience analysis as time and other resources allow. Only by understanding the audience can we hope to identify matters of importance to them and then to link the risk communication efforts to those important issues. First consider, however, the ethical implications of manipulating your audience (see Chapter 5 for additional information).

> When faced with apathy, those who are communicating risk need to conduct as thorough an audience analysis as time and other resources allow.

Mistrust of Risk Assessment

As mentioned in Chapter 1, risk communication efforts are built around a process that starts with a risk assessment and ends with a risk management decision and subsequent action. Unfortunately, a number of health, environmental, and civic organizations have over the last few years presented their opinions that any form of risk assessment is so seriously flawed as to prevent its use in all but the most specific of circumstances. Researcher Alon Tal of the Arawa Institute for Environmental Studies in Israel conducted a study of 17 national and 16 local environmental groups to determine their attitudes toward risk assessment. Over 75% of those surveyed felt that risk assessment was a disempowering process (undemocratic), led to regulatory delays, and was used as a ruse for deregulation. Another telling observation was that 58% felt that risk assessments are inescapably biased to underestimate risk, when most risk assessors would agree that assessments are grossly over conservative (Tal 1997).

Faced with such negative perceptions of risk assessment, what can those who are communicating risk hope to accomplish? The key appears to be in understanding what is driving the difficulties. For example, many environmental organizations in Tal's study felt that one of the major flaws in the risk assessment process was the characterization or definition of risk. One way to address this flaw is to involve the audience in the risk characterization process to identify the aspects of the risk to be studied and the study methods. Another area of concern was the ethical dimension of risk. According to Tal (1997), many environmental organizations felt that "risk assessment is fundamentally immoral, consigning people (and in some critiques, ecosystems) to intolerable environmental fates without their consent" (p. 473A). As noted earlier in this chapter, addressing the value systems of your audience can help in this case by adapting the process when possible. For example, risk assessors found a way to show via icons the various steps and assumptions in the risk assessment process when the audience expressed concerns that risks were calculated by a "black box."

> Until these issues are addressed, environmentalists will remain extremely suspicious of risk numbers. Without enhanced scientific validity, they will oppose the growing influence of risk numbers over environmental health decisions. 'Garbage in garbage out' is a frequently heard aphorism among interviewees.
> —Alon Tal (1997, p. 473A).

Disagreements on the Acceptable Magnitude of Risk

Scientists, regulators, and risk managers often disagree with the public on the acceptable magnitude of a certain risk. Industry and government standards deem a certain level of risk to be safe, that is, unlikely to cause harm to most people who follow some lifestyle. Unfortunately, the public sometimes sees risk as an all-or-nothing proposition. Any level of risk may seem to be too much. For example, some people do not want to hear that the level of trichloroethylene in their drinking water is below the U.S. Environmental Protection Agency standards; they want the level to be zero. Given this attitude, it can be very difficult to communicate risk.

In addition, scientists and the public often approach risks from two very different viewpoints. Scientists and engineers tend to reject suggestions of a cause with no positive evidence. The public demands strong negative evidence against a cause that looks intuitive. In other words, scientists and engineers come from a "there's no proof it is" standpoint, while the public says, "there's no proof it isn't!" Scientists and engineers also tend to look at population risks when the general public wants to know how the risk affects individuals. Even if the audience agrees with the risk assessors on the magnitude of the risk, they may still oppose risk management efforts or refuse to take action because of personal values (for example, a desire for a pristine environment or the need for personal autonomy; Bennett and Calman 1999).

Providing more technical information will not necessarily eliminate this constraint.

However, letting the audience see all the data and helping them understand how trade-offs might be made can help many members of the audience come to terms with the risk. They still may not accept that the risk is necessary and may go on fighting for cleaner water and cleaner air, but they may come to accept the risk communication efforts.

Lack of Faith in Science and Institutions

A number of researchers have pointed out the decreasing credibility of scientists and government institutions with regard to communicating risk information. More than 25 years ago, Roger Kasperson (1986) warned that public opinion surveys were showing a steady decrease in confidence in industry and government officials, with a decline of up to 50% in some areas from 1966 to 1980. Another study found that confidence in government and industry has been declining for the last 30 years (Harris poll as cited by Peters et al. 1997). Unfortunately, many of these same industries and government agencies are faced with the challenge of communicating risk information today.

Those who are communicating risk need to be aware of their audience's attitudes toward the organization in charge of the risk communication effort and any associated organizations. In some cases, it may be necessary to partner with someone the audience finds credible to convey the risk message. In a survey to measure how audiences determined trust and credibility of organizations, Peters et al. (1997) found that defying a negative stereotype was the most important factor to improving trust and credibility for government and industry. The researchers cited the example of Johnson and Johnson's strong response in the wake of the Tylenol tampering in 1982,

> Be aware of your audience's attitudes toward the organization in charge of the risk communication effort and any associated organizations. In some cases, it may be necessary to partner with someone the audience finds credible to convey the risk message.

exhibiting more openness than is usually attributed to big industry. In any case, those who are communicating risk need to continue to strive for credible and open risk assessment, management, and communication efforts, for only these will fully rebuild the faith that has been lost through past practices.

Learning Difficulties

Another difficulty inherent to the audience is the fact that many people reach adulthood lacking a number of learning skills. Much has been written about the growing problem of illiteracy. Even when people are highly literate, words and phrases can be misunderstood. Australian researchers Janet Carey and Mark Burgman (University of Melbourne) propose four areas of linguistic uncertainty:

1. **Ambiguity.** A word may have more than one meaning, and synonyms are often used interchangeably.
2. **Vagueness.** The same word may mean different things to different people (for example, the word "significant").
3. **Underspecificity.** Some words are too general and result in differing interpretations.
4. **Context dependence.** The meaning of words can change when they are placed near other words (Tucker et al. 2008).

These kinds of uncertainties challenge the risk communication effort as much as uncertainties in the risk assessment itself.

Additional studies focus on numeracy, the ability of people to understand and use numeric information. People have a limited ability to process large amounts of data, and that ability decreases with age and with stress. Highly numerate people gravitate to numbers, understand them, put them into practice, and use them to make decisions about risks. People who are not very numerate tend to base decisions on emotions and trust (Tucker et al. 2008).

In addition, the future will see more children moving through the education system with problems associated with having parents who were addicted to drugs or alcohol. Psychologists also note that today's youth relies heavily on television and the Internet for their information, entertainment, and role models. All these factors can conspire to make the job of communicating the scientific concepts of risk extremely difficult.

One way around this is to know your audience. If you are communicating the dangers of reusing unsterile needles to a community who never finished high school, is seldom sober, and watches television 15 hours a day, you will need to tailor your message differently than if you are explaining the need to maintain cleanliness to restaurant workers in a Midwestern town where everyone graduated from high school. The information presented in Parts II and III should help you determine which techniques are best suited for your audience's needs.

CONSTRAINTS FOR BOTH COMMUNICATOR AND AUDIENCE

As if these constraints were not enough, other constraints affect both those who are communicating risk and their audiences: stigma and stability of the knowledge base.

Stigma

Both the audience and the risk communicator can be affected by the stigma of being at risk. In many cases, the actual physical risk can be exacerbated by impacts on the economy and society. For example, when a small town in Nevada found itself the focus of a cancer cluster investigation, property values plummeted, people moved to other nearby towns, and tourism took a heavy hit, all because of the perception that something dangerous must be happening there.

Stigma can also affect the psychological well-being of the audience. For example, researchers Robin Gregory and Theresa Satterfield found that dairy farmers in the Tilla-mook River estuary, long admired for its pristine beauty, felt themselves unwelcome by the growing retirement community because their work had resulted in increased fecal coliform bacteria counts in the local rivers. Because the farmers were associated with this increased health risk, they felt that they were being accused of moral deficiency by simply earning a living (Gregory and Satterfield 2002).

When those at risk are stigmatized, their level of outrage and hostility rises. This increase in emotion can be seen as a destabilizing influence on risk communication efforts. For example, researcher Theresa Satterfield found that some groups were denied access to public meetings because it was assumed they would be disruptive (Satterfield 1996). However, for risk communication efforts to succeed, such exclusionary attitudes must be overcome.

Stigma experts James Flynn et al. (2001) suggest two ways that risk communicators can help to overcome stigma:

1. **Reduce perceived risk.** It is the perception of the danger that drives stigma. This perception can be managed, at least in part, by creating and maintaining trust in those charged with managing the risk, informing and educating the public about the risk, and educating scientific experts on how to more effectively present risk infor-mation without increasing stigma.
2. **Reduce the amplification of stigma.** Once begun, stigma has a tendency to grow. Educate the media as well as government regulators on the potential effects of overstating or sensationalizing a risk.

To avoid creating or amplifying stigma, those communicating risk must factor audience concerns into risk communication efforts. As mentioned previously, these include even those concerns that might be seen as tangential to the risk itself. Again, a thorough under-standing of your audience may be your best approach. See Chapter 8 for additional information.

Stability of the Knowledge Base

Both science and the information your audience has been exposed to change daily. Today's scientific "facts" may be derided in years to come as tales of superstition and ignorance. Once even learned people thought that the earth was flat and the sun revolved around the earth. Given the knowledge at the time, this was a logical assumption; our knowledge may be similarly outdated in the future. In addition, experts on a particular risk often disagree on the magnitude or effects of the risk. The study that you quote extensively can come under fire as being too general or too specific. The experts you bring forth to discuss the risk may be confronted with results of a study by colleagues who directly contradict their findings. These kinds of problems can make the public dubious about any scientific procedure.

Your audience's knowledge base also changes. Two weeks ago when you finished your audience survey, the local news station had not aired "complete" coverage of the risk you were trying to communicate. We once watched in dismay while a news broadcast explained why the waste tanks at a nuclear facility (our customer) were "burping" dangerous gases, all the while showing a picture of radiation-suited individuals loading plutonium pellets into a nuclear reactor. The spoken story was about the risks associated with storing hazardous and radioactive waste; the visual story was about what a reactor looks like. At no time were the tanks or procedures there shown. Those of us who were communicating the risks now had to contend with a number of audience misperceptions.

> Both science and the information your audience has been exposed to change daily. Today's scientific "facts" may be derided in years to come as tales of superstition and ignorance.

This constraint can be difficult to overcome because it is often the least within the control of the communicator. Planning ahead, keeping abreast of what is happening in science and the community, and keeping a sense of humor can all help. Planning is discussed further in Part II. Keeping abreast can be enhanced by subscribing to those information sources most respected by your audience and by whatever branch of science you are communicating about. Read the local newspaper, watch local newscasts, and develop relationships with local print and television news media representatives (see Chapter 8 for more details).

SUMMARY

A number of constraints can hinder the effective communication of risk, including organizational and emotional constraints affecting those who are communicating the risk, hostility and outrage, panic and denial, apathy, mistrust of the risk assessment process, disagreement on the acceptable magnitude of risk, lack of faith in science and institutions, learning difficulties for the audience, and stigma and the changing knowledge base for both communicators and audience members. Those who are communicating risk need to recognize potential constraints and act to overcome them for risk communication efforts to succeed.

REFERENCES

Arvai, J. L. 2003. "Using Risk Communication to Disclose the Outcome of a Participatory Decision-Making Process: Effects on the Perceived Acceptability of Risk-Policy Decisions." *Risk Analysis*, 23(2):281–290.

Balch, G. I. and S. M. Sutton. 1995. "Putting the First Audience First: Conducting Useful Evaluation for a Risk-Related Government Agency." *Risk Analysis*, 15(2):163–168.

Barill, T. 1991. "Communicating Risk to Communities." In *Superfund '90, Proceedings of the 11th National Conference*, pp. 98–100. Hazardous Materials Control Research Institute, Washington, DC.

Beierle, T. C. 2002. "The Quality of Stakeholder-Based Decisions." *Risk Analysis*, 22(4): 739–750.

Bennett, P. and K. Calman. 1999. *Risk Communication and Public Health*. Oxford University Press, New York.

Boiarsky, C. 1991. "Writing for Multiple Readers with Conflicting Needs: An Opportunity for Improving Communication with Regulatory Agencies." In *The Engineered Communication: Designs for Continued Improvement, Proceedings of the 1991 International Professional Communication Conference*, Volume 2, pp. 313–317. 91CH3067-4, Institute of Electronics and Electrical Engineers, Washington, DC.

Covello, V. T., P. M. Sandman, and P. Slovic. 1988. *Risk Communication, Risk Statistics, and Risk Comparisons: A Manual for Plant Managers*. Chemical Manufacturers Association, Washington, DC.

Davenport, C. 2003. "Utilities Discover Message Matters." *Washington Post*, Monday, October 6, p. B01.

Dozier, D. M., L. A. Grunig, and J. E. Grunig. 1995. *Manager's Guide to Excellence in Public Relations and Communication Management*. Lawrence Erlbaum Associates, Mahwah, New Jersey.

FDA (U.S. Food and Drug Administration). 2009. *Minutes of the Risk Communication Advisory Committee, FDA*. http://www.fda.gov/downloads/AdvisoryCommittees/CommitteesMeeting Materials/RiskCommunicationAdvisoryCommittee/UCM190625.pdf (accessed January 18, 2013).

Flynn, J., P. Slovic, and H. Kunreuther, eds. 2001. *Risk, Media, and Stigma: Understanding Challenges to Modern Science and Technology*. Earthscan, London.

Green, M., J. Zenilman, D. Cohen, I. Wiser, and R. Balicer. 2007. *Risk Assessment and Risk Communication Strategies in Bioterrorism Preparedness*. NATO Security Through Science Series-A: Chemistry and Biology, Springer, Dordrecht, Netherlands.

Gregory, R. S. and T. A. Satterfield. 2002. "Beyond Perception: The Experience of Risk and Stigma in Community Contexts." *Risk Analysis*, 22(2):347–358.

Hance, B. J., C. Chess, and P. M. Sandman. 1988. *Improving Dialogue with Communities: A Risk Communication Manual for Government*. New Jersey Department of Environmental Protection, Division of Science and Research, Trenton, New Jersey.

Henry, R. 2000. *You'd Better Have a Hose if You Want to Put Out the Fire: The Complete Guide to Crisis and Risk Communications*. Gollywobbler Productions, Windsor, California.

Hillier, D. 2006. *Communicating Health Risks to the Public: A Global Perspective*. Gower Publishing, Farnham, United Kingdom.

Hunt, B. and J. Monaghan. 1992. "How Public Issues Shape Environmental Restoration Plans Experiences with Colorado UMTRA Projects." In *ER '91: Proceedings of the Environmental Restoration Conference for the U.S. Department of Energy*. U.S. Department of Energy, Washington, DC.

Kasperson, R. E. 1986. "Six Propositions on Public Participation and Their Relevance for Risk Communication." *Risk Analysis*, 6(3):275–281.

Leiss, W. and D. Powell. 2005. *Mad Cows and Mother's Milk: The Perils of Poor Risk Communication*, 2nd ed. McGill-Queen's University Press, Montreal, Quebec, Canada.

Morgan, G., B. Fischhoff, A. Bostrom, L. Lave, and C. J. Atman. 1992. "Communicating Risk to the Public." *Environmental Science and Technology*, 26(11):2048–2056.

Morgan, G., B. Fischhoff, A. Bostrom, and C. J. Atman. 2002. *Risk Communication: A Mental Models Approach*. Cambridge University Press, Cambridge, United Kingdom.

Occupational Safety and Health Administration (OSHA). 2007. *Pandemic Influenza Preparedness and Response Guidance for Healthcare Workers and Healthcare Employers*. OSHA 3328-05, OSHA, Washington, DC.

Peters, R. G., V. T. Covello, and D. B. McCallum. 1997. "The Determinants of Trust and Credibility in Environmental Risk Communication: An Empirical Study." *Risk Analysis*, 17(1):43–54.

Sanderson, W. and K. Niles. 1992. "Effective Outreach Is Good Public Policy." In *ER '91: Proceedings of the Environmental Restoration Conference for the U.S. Department of Energy*. U.S. Department of Energy, Washington, DC.

Satterfield, T. A. 1996. "Pawns, Victims, or Heroes: The Negotiation of Stigma and the Plight of Oregon's Loggers." *Journal of Social Issues*, 52(1):71–83.

Tal, A. 1997. "Assessing the Environmental Movement's Attitudes Toward Risk Assessment." *Environmental Science and Technology*, 31(10):470A–476A.

Tucker, W. T., S. Ferson, A. M. Finkel, and D. Slavin, eds. 2008. *Strategies for Risk Communication: Evolution, Evidence, Experience*. Annals of the New York Academy of Sciences, Vol. 1128, 2008 (April). New York Academy of Sciences, New York.

UK EA (Environmental Agency of the UK). No Date. "Improving Institutional and Social Responses to Flooding." Science Summary SCHO0509BQBM-E-P, Environmental Agency of the UK, London.

ADDITIONAL RESOURCES

Chess, C., P. M. Sandman, and M. R. Greenberg. 1990. *Empowering Agencies to Communicate about Environmental Risk: Suggestions for Overcoming Organizational Barriers*. Rutgers University, Cook College, Environmental Communication Research Program, New Brunswick, New Jersey.

Hadden, S. G. 1990. "Institutional Barriers to Risk Communication." *Risk Analysis*, 9:301–308.

Hance, B. J., C. Chess, and P. M. Sandman. 1988. *Improving Dialogue with Communities: A Risk Communication Manual for Government*. New Jersey Department of Environmental Protection, Division of Science and Research, Trenton, New Jersey.

Sandman, P. M. 1989. "Hazard versus Outrage in the Public Perception of Risk." In V. T. Covello, D. B. McCallum, and M. T. Pavlova, eds., *Effective Risk Communication: The Role and Responsibility of Government and Nongovernment Organizations*. Plenum Press, New York, pp. 45–49.

5

ETHICAL ISSUES

Ethics can be a difficult subject. One reason for this is that each of us has our own ethical code, our own morality, that has been shaped by our experiences and beliefs. This personal code makes ethical issues subjective; what is an issue for me may not be one for you. Another reason is that ethics is a philosophical study with its own language and concepts. Although we are not ethicists, we wanted to provide a general overview of some of the ethical issues often faced, either directly or indirectly, by those who are communicating risk. Therefore, this chapter discusses potential ethical issues and associated decisions. It does not tell you what decision to make but rather helps you weigh the potential outcomes of the possible choices.

Producing any form of technical communication can result in ethical difficulties. Questions such as how much information to release, to whom should it be released, and who makes those decisions are connected with the dissemination of any type of information. The Society for Technical Communication, the largest professional society for those who communicate scientific or technical information, publishes "Ethical Principles for Technical Communicators" (STC 1998) that recognize such ethical issues as complying with regulations, honesty, confidentiality, and fairness.

The communication of risk carries additional potential ethical issues, such as when and how much information should be released, whether the use of persuasion

> The Society for Technical Communication, the largest professional society for those who communicate scientific or technical information, publishes principles that recognize such ethical issues as complying with regulations, honesty, confidentiality, and fairness.

Risk Communication: A Handbook for Communicating Environmental, Safety, and Health Risks, Fifth Edition. Regina E. Lundgren and Andrea H. McMakin.
© 2013 The Institute of Electrical and Electronics Engineers, Inc., and Regina E. Lundgren and Andrea H. McMakin. Published 2013 by John Wiley & Sons, Inc.

is appropriate, and the relationship between public danger versus private interests. Such ethical issues can arise in care, consensus, and crisis communication. Ethical issues in risk communication can be divided into three areas: social ethics, organizational ethics, and personal ethics.

SOCIAL ETHICS

Social ethics comprise the code of conduct by which society judges our behavior. As a society evolves and changes, so will its ethics and the resultant behavior. Not surprisingly, social ethics can differ between countries and between cultures within that country (see Chapter 22 for more information). Risk communication has also evolved to meet changing societal demands. However, the influence of society on the communication of risk can itself be an ethical issue. Other social ethical issues related to risk communication are how the risk idiom is used, by whom, and when; whether the risk is being applied equally to all ethnic and social groups; who should bear the consequences if messages are misunderstood; and the stigma.

The Sociopolitical Environment's Influence

Society has changed over the years, especially in the way in which it views risks and how risk decisions are made. Joseph Beck, social scientist and former congressional and presidential advisor, attributes these changes in public involvement to changes in governance ideologies. Until the 1950s, he says, the United States insisted on strong national governance, or federalism. The threat of Communism and the Cold War in the 1950s led to a change in the education system, in which the "virtues of the national government being controlled by its peoples was (sic) firmly entrenched in the minds of the largest mass of school children ever taught in any school system in the world" (Beck 1991, p. 1). As this group moved through the educational system and into the larger society, they made sure that this view of government by the people was upheld. A good example, according to Beck, is the way this group mounted an effort that resulted in the American abandonment of the Vietnam War.

James Creighton of Creighton and Creighton, the consulting firm responsible for many of the Bonneville Power Administration's successes in the area of risk communication and public involvement, has also observed changes in society in the areas of public involvement and risk communication. His focus was on how the public decided to accept any particular decision, such as how a risk will be managed. According to Creighton, in the 1950s, the public felt that a decision was acceptable if they had been informed about it. All an organization had to do was to produce the proper information and disseminate it widely. In the 1960s and 1970s, society's views changed: Now, the public felt that it must be heard first before a decision could be made. Agencies moved into the age of public involvement, but mostly in the form of written testimony and formal hearings. Beginning in the 1980s and continuing today, Creighton says, society feels that it must actually influence the decision. This has led to a form of public involvement that is referred to as consensus building, trying to get all parties to agree on a decision before it is implemented and involving the public in all aspects of risk assessment, management, and communication (Creighton 1992).

More recently, Granger Morgan and his associates from Carnegie-Mellon University laid out historical stages through which organizations have viewed their charter to assess, communicate, and manage risks:

- All we have to do is get the numbers right.
- All we have to do is tell our audience the numbers.
- All we have to do is explain what we mean by the numbers.
- All we have to do is show our audience that they have accepted similar risks in the past.
- All we have to do is show our audience that it is a good deal for them.
- All we have to do is treat our audience nicely.
- All we have to do is make our audience our partners (Morgan et al. 2002).

If these researchers are correct, and we believe they are, these changes in society can pose a number of ethical issues. If the public demands the right to be involved in risk decisions throughout the cycle of assessment, management, and communication, then is it appropriate for organizations to stick to the old ways of making decisions and informing the public afterward? Is it right to take public testimony but simply go on with whatever decision the organization hoped for to begin with? Is it prudent to exclude public participation in a crisis? Should the mandates of such agencies as the U.S. Environmental Protection Agency and local health departments be changed to allow and encourage them to bring the public more fully into the decision-making process?

The choice for those who are communicating risk is whether or not to involve interested participants in the full cycle of assessment, management, and communication of risk and to what extent. Admittedly, this involvement will be different in care, consensus, and crisis communication; however, each type can carry some kind of involvement. Table 5-1 shows

Table 5-1. Advantages and disadvantages of involving the public in risk assessment, management, and communication

	Involve the public	Do not involve the public
Advantages	Because public participates in risk decision, decision is likely to last.	Organization will not have to change the way it does business.
	Project schedules and budgets less likely to be affected later by lawsuits.	No chance of loss of control.
	Can increase organization's credibility.	
	Provides organization with broader information net.	
Disadvantages	Risk managers may resist because of fear of loss of control.	Risk analysis, decision, and communication can be held up in court indefinitely, delaying project schedules and increasing budgets.
	Lack of organizational commitment can result in loss of credibility.	Organization's credibility decreases.
	Requires more time at the beginning of the process.	Loss of potential information critical to understanding risk.

the advantages and disadvantages of each choice. For a more in-depth discussion of public involvement, see Chapter 17.

The Use of the Risk Idiom

On a slightly smaller scale is the ethical issue of who determines what is a risk and to whom the risk will be communicated. Audiences often take umbrage at the terms "insignificant" or "negligible" when applied to the magnitude of risk, yet these are terms experts often use to generalize complicated risk assessment calculations. Who decided that the risk was insignificant? Was the judgment based on scientific principles alone or were the audience's values considered? Who decided what part of the information was to be disseminated if not the entire set of assessment calculations? Who decided who would receive the information? These issues derive from the ethical questions of power and fairness.

> Who decided what part of the information was to be disseminated if not the entire set of assessment calculations? Who decided who would receive the information? These issues derive from the ethical questions of power and fairness.

The choice for those who are communicating risk is to determine how soon to become involved in the risk assessment process as the audience's representative. Although the expert who is assessing the risk (and who may be communicating it) is involved from the beginning, the technical communicator, risk manager, or public affairs person is sometimes not involved until the risk assessment has been completed. In either case, those who are communicating risk must determine audience concerns and perceptions of the risk and help decide how to factor these concerns and perceptions into the risk assessment process. Concerns and perceptions can be factored in at several stages in the process (Table 5-2), or not at all,

Table 5-2. Stages when audience concerns and perceptions can be factored into the risk assessment or risk communication efforts*

Stage	Advantages of consideration	Disadvantages of consideration
Risk assessment	Less likely to delay schedule or increase budget. Concerns integrated throughout process.	Because risk is not fully understood, integration in planning may be difficult and require changes later.
Scenario development	Suggest additional or different scenarios.	Potentially increases time and cost to explore more options.
Data collection	Can suggest additional data sources.	Potentially increases time and cost to explore more options.
Analysis	Provide audience-specific review of data and results.	Potentially increases time and cost to explore more options.
Risk communication	Communication more likely to be understood.	Assessment is less likely to be accepted if this is the only place of involvement. May require more resources to communicate effectively.

*Note that some agencies advise that risk assessment, risk management, and risk communication will be most successful if the audience is involved in all stages of the process (NRC 1996).

leaving the risk communication efforts to compensate. In many situations, the best stage to factor in audience concerns and perceptions is during the planning of the risk assessment because concerns and perceptions can be more easily integrated without a delay in schedule or an increase in budget. Another stage is when the scenarios (that is, the lifestyle characteristics and other factors to be considered in the calculation of risk) are being developed (for example, the risk to children from ingesting apple products coated with a pesticide). At this stage, the audience concerns and perceptions may point to different scenarios than those being considered by the experts, and revision may give the risk assessment more credibility for the audience.

Another stage at which audience concerns and perceptions can be factored into the risk assessment is in the choice of data to be collected. Using the example of the children ingesting apple products, perhaps the expert had decided not to collect data on preschool children because of the potential dangers associated with testing chemicals on humans and the greater expense of other data collection methods. If an analysis of audience concerns and perceptions shows that the danger to preschool children is the main concern, innovative methods may have to be developed so that these data can be collected and this concern can be addressed.

Another stage in which concerns and perceptions can be factored into the risk assessment and management process is how the scenarios and data are analyzed. Is there a particular method that will be more credible with the audience? The latest generation of environmental and health risk software include graphical interfaces that allow stakeholders to manipulate data and set up analyses. Also, many audiences find the assessment more believable if more than one group of experts analyzed the information and reached similar conclusions. On the other hand, some members of the audience will be satisfied only if their chosen expert has conducted the analysis.

Audience concerns and perceptions must be factored into the risk communication effort, if nowhere else. Note, however, that extensive experience taught U.K. public health experts Peter Bennett, David Coles, and Ann McDonald that "risk management that overlooks stakeholders' basic concerns cannot usually be saved by good communication techniques late in the day" (Bennett and Calman 1999, p. 207).

Those who are communicating risk must know to whom they are communicating and what information that audience requires or risk communication efforts will likely fail. However, if audience concerns and perceptions have not been considered earlier in the process, those who are communicating the risk will likely have a much more difficult job of explaining (and justifying) the risk information.

Fairness of the Risk

One social ethical issue of concern is whether the risk is spread equitably over all ethnic and social groups. According to Dr. Robert Bullard, Director of the Environmental Justice Resource Center, Clark Atlanta University, and author of *Dumping in Dixie: Race, Class, and Environmental Quality* (Bullard 1990), "People of color (African Americans, Latinos, Asians, and Native Americans) have borne a disproportionate burden in the siting of municipal landfills, incinerators, and hazardous waste treatment,

> People of color (African Americans, Latinos, Asians, and Native Americans) have borne a disproportionate burden in the siting of municipal landfills, incinerators, and hazardous waste treatment, storage, and disposal facilities.
> —Robert Bullard (1992, p. 11).

storage, and disposal facilities" (Bullard 1992, p. 11). In the early 1990s, Congress passed legislation to ensure that future efforts to site potential environmental risks such as factories and landfills considered the equity of the risk. The term coined was "environmental equity" or "environmental justice" (see Chapter 3 for more information on the executive order that mandates environmental justice).

Because of recent laws, environmental equity is often considered long before any risk communication effort begins. On the other hand, poor risk communication, accompanied by unresponsive risk management, may cause those at risk to perceive a lack of equity. In either case, those who are communicating risk need to be aware of the potential issue because of its possible effect on risk communication efforts. If the audience's perception is that the risk is being shared inequitably, the level of anger and hostility will rise, making any risk communication effort (care, consensus, or crisis) extremely difficult. (Angry people do not listen.) One way to lessen the anger and hostility is to give the audience a role in how the risk is being assessed and/or managed. For example, if the concern is that the fumes from the siting of a sewage plant will pass over a particular neighborhood, possibly endangering the inhabitants, those inhabitants could be involved in assessing the risk of breathing the fumes (perhaps by determining scenarios or data to be collected) or in operating community air monitoring stations to ensure that the releases do not go over an agreed-upon level.

Consequences of Multiple Meanings

Another ethical issue within society is what happens when messages are misunderstood. Who is to blame when a worker misunderstands a safety procedure and is injured? Is it the fault of the worker for not reading properly, the organization for failing to properly train the worker, the manufacturer of the equipment involved, or the communicator who wrote a message that could be misunderstood?

In any given situation, any message can be misunderstood. No matter how much we analyze our audience, there will always be someone within it who misinterprets the risk message. For example, at a public meeting on the proposed siting of a wind energy project (windmills to generate electricity), the representative of the company making the proposal spoke at length about how the windmills were constructed and how they would be arranged on the site. At one point he made the statement, "We have to site the windmills so far apart because when the wind hits them, they take 25% out of the wind and the wind needs time to recover before it hits the next windmill for maximum efficiency." A member of the audience later expressed a concern: "But if you take away some of the wind, won't there be less air for the rest of us?" The spokesperson was talking about velocity, but the member of the audience was thinking in terms of volume.

> In any given situation, any message can be misunderstood. No matter how much we analyze our audience, there will always be someone within it who misinterprets the risk message.

The choice for those who are communicating risk is how much information about the audience to collect. What information is necessary to understand how the audience thinks? A principle of technical communication related to gathering information about the audience is "audience, purpose, and use." For risk communication, this equates to know your audience, know why you are communicating with them (purpose), and know how they plan to use the information. In

Chapter 8, we rank gathering information on the audience into three tiers: baseline, midline, and comprehensive. In choosing the appropriate tier, consider your resources (time, money, and staff), the purpose of your risk communication effort (care, consensus, or crisis), and your specific objectives. Also consider pretesting your messages to identify and correct as many potential misunderstandings as possible before wide dissemination.

The Issue of Stigma

As mentioned in Chapter 4, communities or individuals facing certain risks can also face societal disapproval or stigmatizing behavior. A child born to a mother with AIDS may be avoided by neighbors uneducated in the transmission mechanisms. A scenic coastal area may lose tourist trade after an oil spill blackens beaches farther north.

Those at risk and organizations charged with managing the risks are often quick to blame the news media for such stigma. While sensationalized stories in the public view can increase stigma, so can other factors. For example, researchers Theresa Satterfield and Robin Gregory found that risk mitigation measures, such as protective barriers around a hazardous waste site, actually increased perceptions of risk and thus stigma (Gregory and Satterfield 2002).

More pointedly, a major contributor to stigma can be the careless use of risk information by otherwise conscientious scientists. Even civic-minded organizations have been known to manipulate risk information to make their point about the perceived dangers of a particular substance. For example, when the National Resources Defense Fund protested the use of the chemical Alar on apples and their story was prominently featured on national television, prices for the fruit plummeted.

The question for those communicating risk is how to present information appropriately for various segments of the audience without raising undue concern. Those at risk must understand their situation if they are to take appropriate action to eliminate or minimize the potential impacts. Those charged with managing the risk must understand the magnitude of the risk and potential management mechanisms. James Flynn, expert on the topic of stigma, advises educating risk assessment scientists and the news media on how stigma can impact communities (Flynn et al. 2001). Even if this education is impossible given the particular circumstances, risk communicators must understand their audiences. See Chapter 8 for additional information.

> Manipulation of framing [in public health information] is arguably unethical and probably impractical.
> —Peter Bennett and Sir Kenneth Calman (1999, p. 216).

ORGANIZATIONAL ETHICS

Besides the societal scheme of ethics, agencies and corporations have their own organizational ethics. In many organizations, it is now standard practice for a newly hired employee to be required to read the organization's code of ethics and to sign a formal statement promising to abide by it. In others, the code is less formal and may even take some deciphering by the new employee. Nevertheless, some such code exists, formal or informal, within every organization. Organizational ethical issues relate to how that code handles such things as the legitimacy of representation, designation of primary audience, release of information, and attitude toward compliance with regulations.

Legitimacy of Representation

Legitimacy of representation refers to who is trusted to speak for whom—for the organization or for the audience—and whether the information being presented on the risk actually represents the risk. An organization usually has rules as to who is allowed to represent it before external audiences, how that person is prepared, and how the information to be presented is tailored for release. In some organizations, only staff in the public affairs department are allowed to speak for the organization. In others, managers or staff knowledgeable in a particular area are allowed to present information, as long as they have received training in such areas as public speaking and media relations. In rare cases, anyone who happens to be in the right (or wrong?) place at the right time will end up as the organization's spokesperson.

Usually, the information presented is passed through some sort of organizational screening review first. This review can range from a grammar check or presentation dry run to ensure that the information is in standard language, to a complex system in which a written version of the planned speech or a draft of an article is reviewed by the legal office, line or project managers, peer reviewers, and communications specialists.

The choices for those who are communicating risk are who to send to a particular audience, how to train that person, and how to ensure that the information presented meets the audience's needs. The choice of spokesperson is among an expert on the risk, a person who has a high level of accountability in managing the risk (risk manager), a communications specialist (technical communicator, public relations specialist, or public information officer), or a celebrity (someone the audience knows and trusts). Table 5-3 shows which audiences are likely to accept each spokesperson and areas in which each spokesperson may need training. For more information on choosing a spokesperson, see Chapter 16. For more information on determining audience needs, see Chapter 8.

A related ethical issue is who the organization accepts as having the right to speak for the audience. For example, a group or individual may step forward and claim to be representatives of "the public." They may act as if they are the *only* representatives and their

Table 5-3. Matching spokespersons to audience characteristics

Audience characteristics	Spokesperson	Potential training needed
Are interested in technical details	Expert	Public speaking
Are not particularly hostile		Media relations
Have at least a basic understanding of the risk		
Are interested in who is accountable	Risk manager	Public speaking
Are hostile		Media relations
Have a basic understanding of the risk		Risk assessment
Are interested in passing information on to others (for example, news media)	Communications specialist	Risk assessment*
Are not particularly hostile		
Have little understanding of the risk		
Have some general knowledge of risk area	Celebrity	Risk assessment
Are apathetic to or unaware of specific risk		

*Assumes that communications specialists have backgrounds in public speaking and media relations.

views are the only legitimate ones. However, seldom does any one group or individual encompass all aspects of an audience.

The choice for those who are communicating risk is how to determine who represents the audience. The way to make this choice is to know all the components of your audience. For example, in assessing the risk of developing carpal tunnel syndrome among those who use computers, components of the audience could be divided into job categories related to computer use: computer programmers and database managers, secretaries and clerks, other nonclerical professionals, and managers. The members of each of these components view computer use differently (and may stand at greater or lesser risk). In choosing who represents the audience, those who are communicating risk may want to choose at least one representative from each of these components. More information on audience analysis can be found in Chapter 8.

> A group or individual may step forward and claim to be representatives of "the public." They may act as if they are the *only* representatives, and their views are the only legitimate ones. However, seldom does any one group or individual encompass all aspects of an audience.

Designation of Primary Audience

Another potential ethical issue within the organization is the designation, either formally or informally, of the "primary audience." The primary audience is the segment of the audience with the highest priority in the risk communication effort. In many cases, the needs of this segment of the audience are considered above the needs of other audience segments. When resources are scarce, the needs of the primary audience are often the only needs considered.

The choice of primary audience should be based on several considerations, such as the following:

- Which segment of the audience is most at risk
- Which segment of the audience has the least information with which to make choices on how to manage the risk
- Which segment of the audience will be most involved in making choices on how to manage the risk (including those that must be involved for legal reasons)

In many situations, the first aspect, which segment of the audience is most at risk, would be the primary consideration. However, in some organizations, the third aspect becomes the most, and sometimes the only, aspect of importance. In other words, those who are communicating risk are forced to communicate in such a way as to meet only the needs of internal senior management who will make a final decision concerning the risk, an audience that is likely not part of the audience at risk. When only this segment of the audience has its needs met, the risk communication effort will fail. However, if this audience is ignored, risk communication efforts will lack the needed support to succeed (see Chapter 4 for more information). For example, in an organization responding to an external audit of a safety program, appeasing internal management can come before fixing some of the problems identified.

The dilemma for those who are communicating risk is how to meet the needs of all audiences. Carolyn Boiarsky, a communication consultant for industry and government, suggests holding a meeting of those who are communicating risk, subject matter experts, and senior management to understand the context of the risk communication effort and to align internal approaches. She also suggests designing an appropriate format for documentation of risk in which information for particular audiences is separated into sections or placed in an appendix that can be skimmed or ignored by less interested readers (Boiarsky 1991).

Releasing Information

Another ethical issue within organizations is the release of information. This issue has two aspects. First, audiences generally want as much information as they can get as early in the process as possible. However, organizations often release as little as possible as late as possible, for several good reasons. One is that early risk information—information gathered shortly after the risk has been identified—has not been subjected to the kind of peer review necessary to ensure the validity of the scientific results. Another reason is that much risk information is either classified in some way (so that releasing it before a certain time would be detrimental to national security) or proprietary (early release would damage the organization's financial or competitive standing). However, the question arises of whether it is right to put scientific, national, or organizational interests before those of the people at risk. For example, when the government of Japan chose to wait to issue updated information on the risks associated with the Fukushima incident in 2011, the citizens of one city evacuated north, thinking that the winds would be blowing to the south. Instead, they walked directly into the plume of radioactive contamination (Center for Biosecurity 2012).

> Audiences generally want as much information as they can get as early in the process as possible. However, organizations often release as little as possible as late as possible, for several good reasons.

The choice for those who are communicating risk is to determine when and how much information to release. The choice should be based on organizational, legal, and audience requirements. The risk communication literature advocates releasing as much information as possible as soon as possible. However, if this is your choice, remember to release that information with sufficient caveats as to its stage of scientific uncertainty. One caveat of our own, however, is that if you are releasing information associated with some legal proceedings (a lawsuit, compliance with the National Environmental Policy Act, or meeting some other legal commitment), be mindful of your legal responsibilities. One federal agency released an environmental impact statement for a nuclear plant that would use an innovative method of producing isotopes. To protect themselves (and because this was the standard procedure for the organization preparing the statement), they put a lengthy disclaimer on the inside cover of the document. The disclaimer basically said that the information in the statement was so preliminary that no one in the organization or the agency was willing to be accountable for anything in the environmental impact statement! Based largely on that disclaimer, an environmental activist group promptly sued the agency for failing to live up to the due process of law. Because of this lawsuit, the project was scrapped entirely.

The second aspect of releasing information that can be an ethical issue is the archiving and possible release of draft information. This is the information that led up to the preliminary results: draft input on scenarios to be considered, hand calculations of various types, and even the first drafts of the risk communication messages. Those who are communicating risk (as well as others in the risk assessment process) often file a number of early "drafts," most of which are never intended to be read by anyone outside the group assessing and communicating the risk and which may contain proprietary or even embarrassing information (some reviewers can be quite scathing in their comments on drafts). However, when a lawsuit has been filed against an organization, these drafts may be some of the first information to be requested for use in court. In addition, the Privacy Rule for individual health information (see Chapter 3) describes ways to manage some risk information. So, the question is, should an organization limit the amount of this type of information that can be kept on file, or would this limitation be a form of censorship?

The choice for those who are communicating risk is how many drafts should be archived, based on the needs of the audience and the needs of others within the organization who are using the information. If the organization has a policy that is too limited for audience and risk assessment purposes (for example, a policy that only final versions of documents are kept), those who are communicating risk may need to work to advocate a change in policy. For example, an external panel of experts overseeing an effort to determine radiation doses received by a population near a government nuclear facility had trouble finding unclassified information. When it became apparent that it was inappropriate to declassify all the necessary information, they requested the government to grant certain members high-level security clearances so they could review the information and assure the public that no pertinent data were being left out of the assessment. See Chapter 4 for more information on dealing with restrictive review procedures or lack of management support.

Attitude toward Compliance with Regulations

Another ethical issue that may arise in organizations is the attitude toward compliance with regulations. Most organizations have wisely chosen to follow regulations, and those that have decided otherwise face legal as well as ethical difficulties that are beyond the scope of this book. Almost as important as following a regulation, however, is the way in which the regulation is followed. When compliance is viewed as an onerous duty at best or a way to subvert democracy at the very worst, risk communication, and all other efforts, can only suffer. For example, the staff for one government contractor refers to their response to Freedom of Information Act requests as "malicious compliance." That is, when a public group has to resort to threatening to sue to get necessary risk information, the organization dumps every piece of unanalyzed, unfiltered raw data onto the group in hopes that this truckload of information will take the group so long to decipher that they will no longer have the resources to "bother" the organization. This approach further frustrates and alienates the audience.

> Communicator's Serenity Prayer: "Grant us the serenity to compromise with the publics we cannot change, the courage to persuade the publics we can change (when it is socially responsible to do so), and the wisdom to know the difference."
> —David Dozier et al. (1995, p. 14).

The choice for those who are communicating risk is how to comply with regulations in such a way as to assist the risk communication effort. As with many of the ethical questions faced by communicators, the issue is to balance the needs of the organization with those of the audience. One key to this is understanding and respecting the audience. The information in Chapter 8 should provide some useful guidelines.

PERSONAL ETHICS

Another area that must be considered is your own personal ethics. What do you believe is the right way to present risk information? What do you believe your role as the communicator of risk should be? What is your personal code of honor, and what would you do if following it conflicted with following your organization's code of ethics?

Using Persuasion

> . . . [T]he power to change behavior carries the immense ethical responsibility to use this power wisely.
> —David B. McCallum (1995, p. 65).

One of the ways to present risk information is by using persuasion, which purposefully presents risk information with the intent of forcing an opinion on the audience. Persuasive arguments may be used to alarm the audience and motivate them to action for fear of loss of life or livelihood. Those who communicate risk in this way often justify it by saying that in some situations, such as a crisis, time is limited and risks are high; therefore, the risk communicator should use every communication tool to get people to do what is best for them. But even in a crisis, does any organization have the right to tell others what is best?

The choice for those who are communicating risk is whether persuasion is justified in their situations. Situations in which persuasion has been justified by risk communicators are those that have one or more of the following characteristics:

- At least some component of the audience is in immediate danger of injury or death (as in crisis situations).
- Those at risk are not the same as those engaged in the behavior and have little control over those engaged in the behavior (for example, unborn babies of alcoholic mothers).
- The audience consists of fewer than 10 people who all feel that they are social equals of the risk communicator (in a small group of equal standing, there is more likelihood that the audience will consider the arguments and not feel as if they are being coerced).
- The audience has specifically asked to be persuaded (for example, by inviting a speaker in for a lively debate).

The Role of the Communicator

Another personal ethical question is, "What should my role be in the communication of risk information?" Should we be disseminators of information, the conduit through which technical information flows and the audience's needs are communicated back to the decision maker? Should we "sell" the risk decision? Or should our knowledge of the audience and communication methods help the decision maker determine what the ultimate decision should be?

> Those charged with making and implementing policies governing our affairs must understand the limits of science as well as its promise, and scientists must learn the limits of policy, the processes by which policies are established in a democratic society, and how to communicate scientific information to the policy process.
> —American Institute of Biological Sciences.

Your choice of role will depend on organizational and personal factors. If the organization recognizes that those who are communicating risk are a vital part of the risk assessment team, the range of roles available will be wider than in an organization that views risk communication as a necessary evil. Some audiences will want to interact only with decision makers and, hence, limit interactions with others who might communicate risk. In addition, some risk communicators may not have the skills necessary to fill a larger role or have not realized that a larger role can exist.

Organizational Ethics or Personal Ethics?

Perhaps the most difficult ethical dilemma comes when personal ethics conflict with organizational ethics. The organization has asked you to downplay, ignore, or, worse, cover up some risk information that, if released, could prevent the injury or death of a number of people. Yet you have a certain loyalty to the organization that issues your paycheck. Which do you listen to, your organization's need to protect itself (and perhaps your job) or your conscience's need to ensure that no one is hurt? Examples of this can be seen in some classic disasters: the engineering staff who warned about the integrity of the O-rings before the explosion on the space shuttle *Challenger*, and the medical staff who warned of potential problems with silicone-gel breast implants.

Oftentimes, the dilemma is not as clear-cut as this. Sometimes the organization simply wishes to limit the amount of information provided to an audience concerning a particular risk. How much information is the right amount? Which pieces will the audience feel are necessary? When faced with such dilemmas, the communicator really has three choices: follow organizational dictates, step down from the work in question, or find someone who will recognize the problem and give it the attention it deserves. Sometimes this person is a higher level manager within the same organization (and we encourage you to try this avenue first if at all possible). Sometimes it is an outside agency with oversight over your organization. If all else fails, you can turn "whistleblower" and tell your story to the media. This may make you a celebrity in the short run; however, despite laws that stipulate no harassment of whistleblowers, telling your story to the press may have disastrous results for your career. Organizations take a dim view of those who break corporate dictates, however justifiably.

SUMMARY

This short discussion has covered only a few of the ethical issues that can face those who are communicating risk. Many more might be added. Being aware of these issues and how they might be resolved can help those who are communicating risk meet the challenges of communicating in an ethical manner.

REFERENCES

Beck, J. E. 1991. "Public Involvement through Negotiation, Mediation, and Arbitration." Seminar presented to staff, November 25, 1991, Pacific Northwest National Laboratory, Richland, Washington.

Bennett, P. and K. Calman. 1999. *Risk Communication and Public Health*. Oxford University Press, New York.

Boiarsky, C. 1991. "Writing for Multiple Readers with Conflicting Needs: An Opportunity for Improving Communications with Regulatory Agencies." In *The Engineered Communication: Designs for Continued Improvement, Proceedings of the 1991 International Professional Communication Conference*. Institute of Electronic and Electrical Engineers, Washington, DC, pp. 313–317. 91CH3067-4.

Bullard, R. D. 1990. *Dumping in Dixie: Race, Class, and Environmental Quality*. Westview Press, Boulder, Colorado.

Bullard, R. D. 1992. "In Our Backyards: Minority Communities Get Most of the Dumps." *EPA Journal*, 18(1):11–12.

Center for Biosecurity. 2012. *After Fukushima: Managing the Consequences of Radiological Release*. University of Pittsburgh Medical Center, Baltimore, Maryland.

Creighton, J. 1992. "What Does It Take for a Decision to 'Count'?" Presentation to U.S. Department of Energy, Richland Operations Office, Richland, Washington. Creighton and Creighton, Palo Alto, California.

Dozier, D. M., L. A. Grunig, and J. E. Grunig. 1995. *Manager's Guide to Excellence in Public Relations and Communication Management*. Lawrence Erlbaum Associates, Mahwah, New Jersey.

Flynn, J., P. Slovic, and H. Kunreuther, eds. 2001. *Risk, Media, and Stigma: Understanding Challenges to Modern Science and Technology*. Earthscan, London.

Gregory, R. S. and T. A. Satterfield. 2002. "Beyond Perception: The Experience of Risk and Stigma in Community Contexts." *Risk Analysis*, 22(2):347–358.

McCallum, D. B. 1995. "Risk Communication: A Tool for Behavior Change." *NIDA Research Monograph*, 155:65–89.

Morgan, M. G., B. Fischhoff, A. Bostrom, and C. J. Atman. 2002. *Risk Communication: A Mental Models Approach*. Cambridge University Press, New York.

NRC (National Research Council). 1996. *Understanding Risk: Informing Decisions in a Democratic Society*. National Academy Press, Washington, DC.

STC (Society for Technical Communication). 1998. *Ethical Principles for Technical Communicators*. http://www.stc.org/about-stc/the-profession-all-about-technical-communication/ethical-principles (accessed January 31, 2013).

ADDITIONAL RESOURCES

Chess, C., P. M. Sandman, and M. R. Greenberg. 1990. *Empowering Agencies to Communicate about Environmental Risk: Suggestions for Overcoming Organizational Barriers*. Rutgers University, Cook College, Environmental Communication Research Program, New Brunswick, New Jersey.

Covello, V. T., D. B. McCallum, and M. T. Pavlova. 1989. "Principles and Guidelines for Improving Risk Communication." In V. T. Covello, D. B. McCallum, and M. T. Pavlova, eds., *Effective Risk Communication: The Role and Responsibility of Government and Nongovernment Organizations*. Plenum Press, New York, pp. 3–16.

Gelobter, M. 1992. "Expanding the Dialogue: Have Minorities Benefited . . . ? A Forum." *EPA Journal*, 18(1):32.

Kasperson, R. E. 1986. "Hazardous Waste Facility Siting: Community, Firm, and Governmental Perspectives." In R.E. Kasperson, ed., *Hazards: Technology and Fairness*. National Academy of Engineering/National Academy Press, Washington, DC, pp. 118–144.

Morgan, M. G. and L. B. Lave. 1990. "Ethical Considerations in Risk Communication Practice and Research." *Risk Analysis*, 10(3):355–358.

6

PRINCIPLES OF RISK COMMUNICATION

The risk communication literature discusses a number of principles regarding how best to communicate risk. Two that have been covered extensively are the fact that the audience must find the communicating organization credible and trustworthy and the fact that the audience must be allowed to participate in the risk management decision. Because following both of these principles is often outside the control of those who are communicating the risk, we will not discuss them here.

Another important principle that is often outside the control of the risk communicator is that actions, policies, and language must be congruent for risk communication to succeed. One of the most disastrous examples of a mismatch in this area is the British government's handling of the mad cow disease outbreak in the 1990s. Various ministers attempted to curb public concerns by expressing supreme confidence in the beef industry, yet at the same time failed to ensure that policies were implemented to reduce the risk (Leiss and Powell 2005). Another lesson learned from this incident is that an organization that only appears to act under pressure will increase public hostility levels and make risk communication much more difficult (Bennett and Calman 1999).

The following principles, then, focus on those aspects of risk communication that

> The risk communication literature discusses a number of principles regarding how best to communicate risk. Two that have been covered extensively are the fact that the audience must find the communicating organization credible and trustworthy and the fact that the audience must be allowed to participate in the risk management decision.

Risk Communication: A Handbook for Communicating Environmental, Safety, and Health Risks, Fifth Edition. Regina E. Lundgren and Andrea H. McMakin.
© 2013 The Institute of Electrical and Electronics Engineers, Inc., and Regina E. Lundgren and Andrea H. McMakin. Published 2013 by John Wiley & Sons, Inc.

are within the purview of those who are communicating risk: the principles related to the risk communication process, risk communication presentation, and risk comparison. Unless specifically noted, these principles apply equally to care, consensus, and crisis communication. For more information on risk communication principles, consult Resources at the back of the book.

PRINCIPLES OF PROCESS

Principles of process relate to the process of planning and conducting a risk communication effort. They are ways of setting up the risk communication process that help ensure that the effort achieves its objectives.

Know Your Communication Limits and Purpose

To effectively communicate risk, you must know why you are communicating and any limitations to your ability to communicate risk. Your communication limits may be defined by:

- **Regulatory requirements.** For example, the U.S. Environmental Protection Agency specifies what community relations activities are to be conducted for Superfund cleanup sites (see Chapter 3).
- **Organizational requirements.** For example, some organizations cannot allow preliminary risk data to be released for proprietary reasons (see Chapter 4).
- **Audience requirements.** For example, some members of the audience may have difficulty in reading or processing information (see Chapter 4).

These kinds of limits affect how you can communicate risk.

Another way to think about this principle is "do not promise what you cannot deliver." Define the audience's role at the beginning of the process and frequently thereafter so that both the audience and your organization know what to expect. If you and your audience understand why you are communicating about the risk and the limits to that communication, you will be less likely to promise what you cannot deliver, and they will be less likely to demand a bigger role than they can legally have. See Chapter 7 for more information on setting purposes and determining limits.

> Define the audience's role at the beginning of the process and frequently thereafter so that both the audience and your organization know what to expect.

Violating this principle can increase hostility in the audience, making it more difficult for the organization to communicate credibly and effectively. An example comes from one of the U.S. Department of Energy's defense production laboratories (those laboratories whose research has focused on better ways to produce nuclear weapons). A group of citizens who lived in the area near the laboratory was concerned about the risk of having nuclear materials so close and distressed that their local economy was driven by making better bombs. They petitioned for and received funding from a philanthropical organization to study alternative uses for the laboratory. The communication process between the laboratory and the concerned citizens had never been good; suspicion was strong on both sides. However, even as the laboratory had begun to communicate risk more effectively, opening its doors to tours and inviting comments on activities, the citizens group was trying to identify ways to change the laboratory, with no input from the laboratory or the U.S. Department of Energy. When the citizens group comes up with its recommendations, will anyone listen? And if the recommendations fail to get the hearing the group feels is deserved, what kind of communication will be possible then? In a situation like this, it would have been best if the laboratory and the U.S. Department of Energy had gone to the citizens group as soon as the grant was received and discussed what options were possible given the department funding and mandate. With expectations set, the group could have developed strategies within existing constraints.

Whenever Possible, Pretest Your Message

Audience analysis should be part of every effort to communicate risk. Factors such as reading level, knowledge of the subject, and level of hostility must be considered if risk is to be communicated effectively. Whenever possible, however, the message should also be pretested, reviewed by a group representing the intended audience, before dissemination, to determine that the audience analysis information was correct and that the risk message achieves the desired results.

Pretest even before the message is designed by asking your potential audience about issues to be covered, concerns to be addressed, and levels of information needed. Pretest the risk communication message prototype before dissemination to make sure that you have addressed the concerns and are not alienating anyone with the presentation. Test between communication rounds so that you can build on your efforts and refine them (Arkin 1989). More information on pretesting is presented in Chapter 8.

Communicate Early, Often, and Fully

This principle has two aspects: timing of communication and amount of information released. Risk communication must be timed to involve the audience throughout the process, not only during a crisis or once in the life of a project. As mentioned previously (see Chapter 5), many members of the audience will expect to be involved from the beginning. In fact, many will consider such involvement as their right. Denying them this opportunity will increase hostility and make risk communication more difficult.

> If no new information becomes available in a timely manner, let the audience know that the risk is still being studied and that they have not been forgotten.

Therefore, risk communication should begin as soon as a risk has been identified and continue as new information becomes available. If no new information becomes available in a timely manner, let the audience know that the risk is still being studied and that they have not been forgotten. The length of time between communications will vary by risk and level of interest from the audience. A risk that seriously threatens entire communities and has the audience extremely concerned will require more frequent communications (hourly to daily, as in a flood) than will a risk that results in less immediate danger to an audience that is unconcerned (quarterly to yearly).

The second aspect involves the amount of information released. As noted in Chapter 4, many organizations find it difficult to release information about a risk because of national security or proprietary concerns. However, withholding information, even to confirm the accuracy of the data, can make the audience suspect that the organization is trying to hide something, eroding its credibility, increasing hostility, and generally making risk communication more difficult. Therefore, do not restrict information. When in doubt, ask your audience what level and type of information they want and provide as much of it as you can within organization and resource constraints.

An example of how this principle has been implemented comes from the New York City Department of Health officials. When the West Nile virus outbreak hit in 1999, they developed a detailed response plan that allowed them to communicate early. To allow them to communicate often, they also used multiple communication channels (including TV and radio public service announcements), extensive media outreach and announcements during daily mayoral press conferences, brochures and fact sheets prepared in 10 languages, posters placed throughout the city, bill inserts mailed with the cooperation of city utilities, hotlines staffed around the clock, a website, and town-hall public meetings. To ensure that they were communicating fully, they provided information that answered peoples' questions, explained protective measures that people could take to reduce their risk, and described what the city was doing. As a result, communication efforts went more smoothly than what might otherwise have been the case (Covello et al. 2001).

Remember That Perception Is Reality

This principle can be difficult for some technical experts to apply. To them, reality is built on carefully constructed, tested, scientific truths, not someone's possibly uninformed perceptions. However, such regulatory agencies as the U.S. Environmental Protection Agency sometimes make decisions based heavily on the audience's perception, not just on the technical aspects.

For example, at a garbage dump near Spokane, Washington, dangerous chemicals had the potential to leak into the groundwater and from there to the drinking water supply. A scientific study of alternative treatment methods recommended the use of capping, that is, pouring cement over the dump and monitoring it to ensure that nothing leaked out as the most cost-effective treatment. However, during the public review of the study and recommendations, the public overwhelmingly preferred the alternative of pumping out some waste, treating the remaining waste, and then capping. Although this alternative did not significantly lessen the risk and significantly increased the cost, it was the alternative chosen by the U.S. Environmental Protection Agency. The moral: risk assessments and subsequent decisions are not based on the technical aspects of the risk alone. Audience perceptions and concerns must be considered if risk decisions, and their communication, are to be successful.

> The moral: risk assessments and subsequent decisions are not based on the technical aspects of the risk alone. Audience perceptions and concerns must be considered if risk decisions, and their communication, are to be successful.

PRINCIPLES OF PRESENTATION

Another set of principles addresses how to present the risk information in ways that best communicate the risk to the intended audience.

Know Your Audience

You cannot communicate unless you know to whom you are communicating. This is the one principle you should always follow; if you follow it, you will be in a position to know how to apply any of the other principles. In fact, knowing your audience is crucial to knowing what risk communication methods to use. For example, if you know that your audience wants to see the risk information immediately, you might forego the principle of pretesting so that you can release the information quicker. And you might use the method of informing the news media rather than using a method that could take longer (such as holding a public meeting 30 days after announcing it in the paper). For more information on this issue, see Chapter 8.

> You cannot communicate unless you know to whom you are communicating. This is the one principle you should always follow; if you follow it, you will be in a position to know how to apply any of the other principles.

Do Not Limit Yourself to One Form or One Method

Any audience for a risk communication message will be made up of a variety of segments, each with different levels of knowledge about the risk, of interest in the risk, and of being at risk. Because of these factors, no single method of communication is likely to meet the needs of your entire audience. You will need to find methods that best meet the needs of each segment. For example, in a community near a Superfund site in Alaska, we found that the written word was the best approach for a large segment of the population because other forms of communication (radio and television) were subject to outages during inclement weather, and the library was the one place almost everyone eventually visited. For those in more rural areas, who came into town only for major community events such as fairs, we also developed a traveling exhibit. See Chapter 10 for more information on which methods best serve which segments of the audience.

Simplify Language and Presentation, Not Content

When trying to communicate the complex issues behind a risk, it is easy to leave out information that seems to be overly technical. Unfortunately, by simplifying the content of a risk message, you may leave out key information that your audience would need to make a decision. Your audience will understand the concepts better, and be better informed about the risk, if you simplify the way you present the content instead of the content itself. Any technical subject can be understood by the public if it is presented properly. Technical communicators have made careers out of this fact. The audience does not have to understand it at the same level as the risk expert, but they can understand it well enough to make an informed decision.

Be Objective, Not Subjective

Quantify information whenever possible. Avoid words like "significant," "negligible," and "minor." They beg the questions, "Significant to whom? Under what conditions? Based on what evidence?" Whenever possible, give examples, numbers that can be put in perspective, and concrete information.

Communicate Honestly, Clearly, and Compassionately

To communicate honestly, you must differentiate between opinions and facts. Any risk communication message, whether spoken, written in a report, or printed on a bulletin board, can be questioned by the audience. Responding credibly to a question about a fact is much easier than substantiating an opinion.

To communicate clearly, you must present information at your audience's level of understanding. Audiences reject information that is too difficult for them, either by refusing to acknowledge it or by becoming hostile. On the other hand, audiences may become hostile when information is too easy for them because they feel patronized.

To communicate compassionately, do not ignore audience concerns, even ones that seem to be based on information about something other than the risk itself. A scientist we once worked with was wonderful about listening to his audience. He attended every public meeting, pored over letters and comments, and carefully categorized every comment. Unfortunately, he then identified groups of comments that he felt were ridiculous: "That's a stupid comment, I won't sink to that level to answer it. This one is clearly out of scope, it doesn't have anything to do with the risk. This comment is purely emotional."

The audience's concerns will not go away simply because you refuse to deal with them. On the contrary, they will likely keep coming up until you are forced to deal with them, perhaps under less favorable conditions such as a lawsuit. It is better to deal with them as soon as they are aired, show your audience that you are listening, and allow them to move on to other questions that may more directly involve the risk itself. How you deal with them depends on the risk communication method. In print messages and technology-based applications, use a question-and-answer format with the questions being their concerns and the answers your responses, or incorporate your responses into a graphic. In face-to-face methods and stakeholder participation, respond directly to questions and concerns as you hear them.

Listen and Deal with Specific Concerns

Besides dealing with the emotions behind concerns, listen to what people are saying about the risk itself. Then deal with each specific concern you hear. Do not discount concerns that seem to be based on faulty scientific information or are peripheral to the situation. A good example comes from a scientist who was asked to speak about atmospheric fallout at a public meeting. After explaining the process and associated risks, the scientist asked if there were any questions. A man from the back of the room rose and identified himself as a local farmer: "You tell me that I have this plutonium stuff all over my crops. What exactly does that mean? Can I still sell my crops? Can I eat them myself? Should I let my children play outside? I don't know who to trust anymore!" The scientist responded, "Excuse me, sir, but it's strontium-90, not plutonium."

> Besides dealing with the emotions behind concerns, listen to what people are saying about the risk itself. Then deal with each specific concern you hear. Do not discount concerns that seem to be based on faulty scientific information or are peripheral to the situation.

Now, the scientist was just trying to correct a technical mistake. However, as you can imagine, the level of hostility in that room skyrocketed. No one listened to anything else the scientist had to say. It would have been far more effective for the risk communication process if the scientist had recognized the underlying confusion and fear and said something like, "I understand why you might not know where to turn. There's a lot of information out there, but there's also a lot of misinformation. Let's see what I can do to clarify a few points. For one thing, it's not plutonium, it's strontium-90, and what that means is. . . ."

Convey the Same Information to All Segments of Your Audience

As mentioned previously, different segments of your audience will have different needs—for information, for involvement, and for responding to the risk. To communicate effectively, you must communicate with each segment in a way that meets those needs. However, as Callaghan (1989) found, you can change the method and amount of detail, but you cannot change the basic information or you will lose trust and your efforts will be useless. You must provide the same information to each segment to retain credibility.

Deal with Uncertainty

In communicating about risks, you can never present results as definitive; no study is ever the final word. Instead, you must discuss sources of uncertainty, such as how the data were gathered, how they were analyzed, and how the results were interpreted. The sources of uncertainty and how you communicate about them vary among care, consensus, and crisis risk assessments.

In care communication situations, the risk assessment has been conducted and found to be credible by most of the audience. Sources of uncertainty, then, are less important and may be discussed very little if at all. In consensus communication, the audience will probably be involved in determining how the risk is analyzed and may help determine which types of uncertainty are acceptable. Therefore, sources of uncertainty become well understood and may be discussed less and less as time goes on. In some types of crisis communication, the risk is obvious, and the areas of uncertainty can be left out of messages unless a particular audience has requested them. In other types of crisis communication, for example, in a terrorist attack, the risk and final consequences may be less well defined.

> The research on risk communication provides insights into the utility of risk comparisons. They can be useful but only when they are a part of an overall communication strategy. This strategy requires that the communicator: understand the nature of the risk—both the hazard that it presents and the qualitative attributes that influence perception by the target audience; understand the audiences that are being addressed and their relationship to the hazard; understand how the risk comparison interacts with other components of the message; and have a way to evaluate the audiences' response.
> —David B. McCallum and Susan Santos, special paper to support the Commission (1997, p. 212).

In such cases, the uncertainty may lie in our response to the crisis. Risk communicators must be open with what they do not know and stress what they do know and what they are doing to resolve the uncertainties.

In cases where uncertainty is to be discussed, the process of assessing the risk should be discussed first to frame where the sources of uncertainty can be found. In data gathering for an environmental risk assessment, for example, were the data gathered over the same time period, or are you comparing between years? Were the same collection methods used each time? Were the data always gathered in the same location? In data analysis for a safety risk assessment, which methods were used? How reliable are they? Are they new methods or ones that have stood the test of time? For interpreting the results of a health assessment, what was the basis for determining significance? How certain are the standards or limits used? Who determined whether these standards actually protected human health? Your audience will often want the answer to one particular question, "Is it safe?" Although they may not like answers with caveats attached, they will be even less happy if they are given an answer that later turns out to be wrong because of the uncertainties involved. See Chapter 14 concerning the visual representation of risk for more information on portraying uncertainty and probability.

PRINCIPLES FOR COMPARING RISKS

In risk communication efforts, comparing risks can be helpful but challenging. As you try to present information at a level that your audience understands, comparisons look like an easy answer. Unfortunately, what little empirical research has been devoted to this area is complicated, confusing, and contradictory (see, for example, Johnson 2002, who provides a frank discussion of compounding influences). Much of the advice that follows is based on informed opinions of risk communication researchers and practitioners. The best way to choose which of these principles or methods to apply is to know your audience and pretest messages.

Risk communication researchers and practitioners agree that some types of comparisons can alienate certain segments of the audience. Unfortunately, which comparisons are best for which audiences has yet to be determined. Covello et al. (1988) listed ways to compare risks according to which methods might be most acceptable with most audiences (Table 6-1). However, studies at Carnegie-Mellon University and elsewhere (for example, Roth et al. 1990) have shown that some of these methods may in fact be more acceptable to certain audiences than Covello et al. originally thought.

Use Analogies, but Do Not Trivialize

According to the Commission (1997) and our own experience, most people, including physicians and some risk assessment experts, do not easily relate to terms like 10^{-6} or

Table 6-1. Acceptability of risk comparisons*

Most acceptable	Less desirable	Even less desirable	Rarely acceptable
Same risk at different times	Doing something versus not doing it	Average risk versus peak risk at a particular time or location	Risk versus cost
Risk versus standard	Alternative ways of lessening risk	Risk from one source of harm versus risk from all sources of that harm	Risk versus benefit
Different estimates of the same risk	Risk in one place versus risks in another place	Occupational risk versus environmental risk	Risk versus other specific causes of same harm

*Adapted from Covello et al. (1988).

E-6 when describing risk. Analogies can help put these risks in perspective. For example, a risk of one in a million (10^{-6}) is equivalent to 30 seconds in a year, 1 inch in 16 miles, or 1 drop in 16 gallons (Commission 1997).

> Analogies can help put risks in perspective. For example, a risk of one in a million is equivalent to 30 seconds in a year, 1 inch in 16 miles, or 1 drop in 16 gallons (Commission 1997).

Using analogies, however, can be problematic for several reasons. One reason is that risk is such a multifaceted situation that it is difficult to find something completely analogous with which to compare it. For example, the analogies discussed in the previous paragraph, while endorsed by the Presidential/Congressional Commission, deal with volume or distance, not toxicity, and may in fact confuse some of the audience.

Another reason analogies can be problematic is that they can come across as trivial. As you try to compare a technical concept of risk to something that is more familiar to your audience, it is easy to find something that may look too simplistic. For example, you might say that the risk of contracting cancer from being exposed to a certain chemical is as small as one piece of toilet paper in a roll stretched from New York to San Francisco. This does present the concept that the risk is very small, but you seem to have compared someone's life to a roll of toilet paper, a comparison that will offend at least some if not all members of your audience.

Use Ranges

You can express a risk using a range of numbers. (Unfortunately, some studies show that this can be problematic; again, know your audience.) One end of the range may represent a level that has been determined to be "safe," the other end may represent a level that has been determined to be risky, and another number may represent the findings of your risk assessment. Audiences can then compare your risk assessment findings with those in the range. This method of risk comparison is especially good for hostile audiences in that your audience can determine for themselves where the risk falls on a hazard scale, and you do not have to decide "significance" for them, a practice that can lead to increased hostility. However, be careful to explain the ranges—why they were chosen, by whom, and what they mean to your audience. This further explanation helps put the risk in perspective.

Compare with Standards

A number of standards have been developed by regulatory agencies and interested groups to describe levels at which certain risks can cause certain levels of harm. One of the most often used at Superfund sites, for example, is the National Primary Drinking Water Regulations mandated by the U.S. Environmental Protection Agency. These standards show what level of contaminant is considered safe in drinking water supplies. You could compare the results of your risk assessment with such a standard. If your results are higher than the standard, you are showing your audience that they should be concerned. If your results are lower than the standard, you are showing your audience that they probably have no reason for concern. One caution, however, is that recent studies have shown that this rule is only as good as the standard used. If your audience already feels that the standard is too low or high, comparing with it may not be your best choice.

Compare with Other Estimates of the Same Risk

For any particular risk, a number of assessments are usually performed. For example, government researchers, university researchers, and independent researchers hired by concerned citizen groups may all study a particular risk. In addition, some risks have been studied by the same organization over many years. You can compare the results of each of the studies. If the results are similar, you are reinforcing the risk assessment. If the results vary widely, you are reinforcing recognition of the uncertainties involved. This type of comparison has been called "dueling Ph.D.s," which points to a possible problem. If having too many studies or contradictory studies will confuse or alienate your audience ("I knew it—these scientists will say anything!"), you may want to try another way of comparing the risk.

Compare Traits

> Another way to compare risks is to base the comparison on different traits of segments of your audience. Use age groups (risk to infants vs. risk to senior citizens), geographic regions (risk on the East Coast vs. risk on the West Coast), or lifestyles (risk to the avid sportsman vs. risk to the farmer; risk to the farmer vs. risk to the city dweller). This personalizes the risk for each member of your audience by allowing them to determine which trait best applies.

Another way to compare risks is to base the comparison on different traits of segments of your audience. Use age groups (risk to infants vs. risk to senior citizens), geographic regions (risk on the East Coast vs. risk on the West Coast), or lifestyles (risk to the avid sportsman vs. risk to the farmer; risk to the farmer vs. risk to the city dweller). This personalizes the risk for each member of your audience by allowing them to determine which trait best applies. Those at lesser risk may find this determination comforting. Those at greater risk may be motivated to find ways to lessen their risk.

Do Not Compare Risks with Different Levels of Associated Outrage

The term outrage refers to the feelings of anger and frustration often associated with certain risks. (For a more in-depth discussion, see Chapter 4.) Risks that are high in outrage include those that are imposed, government controlled, seen as unfair by the audience, from an untrustworthy source, artificial, exotic, associated with disasters, dreaded, undetectable, or not scientifically well understood. Nuclear power is a prime example of a high-outrage risk, being imposed, government controlled, artificial, exotic, associated with disasters like those at Chornobyl and Fukushima, dreaded, undetectable (the radiation is), and relatively new. In contrast, smoking cigarettes is voluntary, familiar, detectable, and scientifically well understood, so outrage is generally lower. The reasoning behind this principle is that when you compare something like the high-outrage risk of contracting cancer from radiation exposure from nuclear power to the low-outrage risk of contracting cancer from smoking cigarettes, you alienate your audience. Your audience will see the two as totally different risks, having nothing to do with each other.

However, studies at Carnegie-Mellon University (Roth et al. 1990) indicate that some audiences will not be alienated by such comparisons. They will, in fact, learn from such comparisons and even find them comforting, if the high-outrage risk they were concerned about is actually lower than the low-outrage risk. Unfortunately, the research has not delineated exactly which audiences these are. So, for now, it is probably best to avoid comparing high-outrage and low-outrage risks.

Explain Reductions in Magnitude

One of the common misperceptions among those at risk is that a reduction in magnitude (one chance in a thousand to one chance in a million) is the same as a linear reduction (10,000 to 9999). Although research in this area is quite complicated, one suggestion is to illustrate this concept graphically. The Commission (1997) recommends bar charts to show the radical differences between these scales. See Chapter 14 for additional information on visual representation of risk.

SUMMARY

Risk communication is still far from being a science. The principles that have been developed through years of study can be distilled into two maxims: know your audience and know your situation. Understand what your audience needs to know (what they want to know and what you need to tell them to help deal with the risk), how they want to receive that information, and what you can do within certain constraints. The other principles can then be followed as they apply to your audience and situation. Part II deals with the issues of audience and situation.

REFERENCES

Arkin, E. B. 1989. "Translation of Risk Information for the Public: Message Development." In V. T. Covello, D. B. McCallum, and M. T. Pavlova, eds., *Effective Risk Communication: The Role and Responsibility of Government and Nongovernment Organizations*. Plenum Press, New York, pp. 127–135.

Bennett, P. and K. Calman. 1999. *Risk Communication and Public Health*. Oxford University Press, New York.

Callaghan, J. D. 1989. "Reaching Target Audiences with Risk Information." In V. T. Covello, D. B. McCallum, and M. T. Pavlova, eds., *Effective Risk Communication: The Role and Responsibility of Government and Nongovernment Organizations*. Plenum Press, New York, pp. 137–142.

Covello, V. T., P. M. Sandman, and P. Slovic. 1988. *Risk Communication, Risk Statistics, and Risk Comparisons: A Manual for Plant Managers*. Chemical Manufacturers Association, Washington, DC.

Covello, V. T., R. G. Peters, J. G. Wojtecki, and R. C. Hyde. 2001. "Risk Communication, the West Nile Virus Epidemic, and Bioterrorism: Responding to the Communication Challenges Posed by the Intentional or Unintentional Release of a Pathogen in an Urban Setting." *Journal of Urban Health: Bulletin of the New York Academy of Medicine*, 78(2):382–391.

Johnson, B. B. 2002. "Stability and Inoculation of Risk Comparisons' Effects under Conflict: Replicating and Extending the 'Asbestos Jury' Study by Slovic et al." *Risk Analysis*, 22(4):777–788.

Leiss, W. and D. Powell. 2005. *Mad Cows and Mother's Milk: The Perils of Poor Risk Communication*, 2nd ed. McGill-Queen's University Press, Montreal, Quebec, Canada.

Commission (Presidential/Congressional Commission on Risk Assessment and Risk Management). 1997. *Risk Assessment and Risk Management in Regulatory Decision-Making*, Volume 2. Commission on Risk Assessment and Risk Management, Washington, DC.

Roth, E., M. G. Morgan, B. Fischhoff, L. Lave, and A. Bostrom. 1990. "What Do We Know about Making Risk Comparisons?" *Risk Analysis*, 10(3):375–387.

ADDITIONAL RESOURCES

Covello, V. T. and F. W. Allen. 1988. *Seven Cardinal Rules of Risk Communication, OPA-87-020.* U.S. Environmental Protection Agency, Washington, DC.

Covello, V. T., D. B. McCallum, and M. T. Pavlova. 1989. "Principles and Guidelines for Improving Risk Communication." In V. T. Covello, D. B. McCallum, and M. T. Pavlova, eds., *Effective Risk Communication: The Role and Responsibility of Government and Nongovernment Organizations*. Plenum Press, New York, pp. 3–16.

Hance, B. J., C. Chess, and P. M. Sandman. 1988. *Improving Dialogue with Communities: A Risk Communication Manual for Government*. New Jersey Department of Environmental Protection, Division of Science and Research, Trenton, New Jersey.

Hance, B. J., C. Chess, and P. M. Sandman. 1990. *Industry Risk Communication Manual*. CRC Press/Lewis Publishers, Boca Raton, Florida.

NRC (National Research Council). 1989. *Improving Risk Communication*. National Academy Press, Washington, DC.

II

PLANNING THE RISK COMMUNICATION EFFORT

For any effort to be effective, a certain amount of planning is needed. To plan a risk communication effort, whether a one-time message or a variety of messages for a variety of audiences over a longer time, you need to determine your purpose and objectives, analyze your audience, develop your message, determine the proper method, set a schedule, and pull all these pieces together into a comprehensive plan. Having a risk communication plan can help you focus your efforts and keep all those involved in assessing, communicating, and managing the risk informed so that they can work together as a team.

> Attention to details (e.g., dress, language used in risk communication materials, meeting location, and room arrangement) is often critical to effective risk communication.
> —Vincent T. Covello, David B. McCallum, and Maria Pavlova, (1989, p. 9).

7

DETERMINE PURPOSE AND OBJECTIVES

Two variables you have to consider in communicating risk are why you are communicating the risk (purpose) and what you hope to gain by it (objective). The purpose is a general statement. It answers "why" questions: Why are we communicating? Why are we educating this group? Why are we trying to build consensus? Objectives are statements of specific, measurable details to be accomplished. Objectives often answer "how" questions: How will we communicate? How often will we communicate? How many messages will we use? For example, if your purpose is to decrease teenage smoking, one of your objectives might be to have 50% of your audience stop smoking by June 15.

When you have determined your purpose and objectives, formalize them by writing them down and by getting concurrence from all those involved in the project, as high up in the organization as you can. This formal agreement can help you communicate risk more effectively because it:

> Two variables you have to consider in communicating risk are why you are communicating the risk (purpose) and what you hope to gain by it (objective).

- Gives everyone a common ground on which to build
- Lets upper management know why you are doing what you are doing
- Gives you a yardstick for measuring success

When you are determining purpose and objectives, a number of factors should be considered.

Risk Communication: A Handbook for Communicating Environmental, Safety, and Health Risks,
Fifth Edition. Regina E. Lundgren and Andrea H. McMakin.
© 2013 The Institute of Electrical and Electronics Engineers, Inc., and Regina E. Lundgren and
Andrea H. McMakin. Published 2013 by John Wiley & Sons, Inc.

FACTORS THAT INFLUENCE PURPOSE AND OBJECTIVES

Your purpose and objectives may at first seem obvious. You are communicating to provide the audience with information they need to make a decision about a risk to their health or safety or the environment. However, your purpose and objectives are necessarily influenced by a number of factors, including legal issues, organizational requirements, the risk itself, and audience requirements. These factors must be consciously considered in determining your purpose and objectives or you may find yourself in conflict with something that could seriously impair, if not cancel, your entire effort. For

> Your purpose and objectives are necessarily influenced by a number of factors, including legal issues, organizational requirements, the risk itself, and audience requirements.

example, if the purpose and objectives of your risk communication effort at a Superfund site conflict with the communication requirements for these sites mandated by the U.S. Environmental Protection Agency, your organization will be liable for strict financial penalties.

Legal Issues

Several legal issues may influence your choice of purpose and objectives. Chief among these are the laws that may dictate your risk communication efforts. We describe some of the major U.S. laws and their requirements in Chapter 3. They include the Comprehensive Environmental Response, Compensation, and Liability Act (CERCLA); the National Environmental Policy Act (NEPA); and the requirements set forth by the Occupational Safety and Health Administration (OSHA). In addition, many government agencies have policies regarding how risk communication will be conducted or how the public will be involved in risk management decisions.

For example, in late 1992, the U.S. Department of Energy's Office of Environmental Management first issued a policy on public involvement/communication that specified involving the audience fully in decisions related to the environmental cleanup of the department's sites. In 2001, the U.S. Centers for Disease Control and Prevention issued guidance for risk communication activities as part of multimillion dollar grants to the states. These kinds of requirements must be considered when planning a risk communication effort because they generally provide guidance and sometimes specify activities that must be conducted. If the purpose and objectives of risk communication efforts that fall under these legal requirements differ from what is specified, special arrangements such as legal waivers or exemptions may be necessary.

Organizational Requirements

After considering the legal issues that may affect your efforts, you also need to consider your own organization's requirements and policies regarding the communication of risk, the involvement of the public, the release of information, and the development of communication materials and processes in general. The policies may be formal, or they may be in the form of tradition ("we always do it this way").

Check with those in charge of communication in your organization to determine what they expect from risk communication efforts. For example, do they feel that the only proper purpose for communication efforts is to advertise the organization or otherwise make it look good? Is one of their objectives the dissemination of information to key political figures who might be able to influence the future of the organization? You will need to be aware of these issues before you determine exactly what you should do to communicate risk. How you handle their expectations depends on your philosophy concerning the communication

> Also consider your own organization's requirements and policies regarding the communication of risk, the involvement of the public, the release of information, and the development of communication materials and processes in general.

of risk. For example, if you agree that promoting the organization should be part of the risk communication effort, then you will include that as a purpose with associated objectives. If you disagree, then you may try to convince those in charge that your audience will want more from a risk communication effort, perhaps by showing how risk communication has helped speed the implementation of projects by avoiding costly and time-consuming legal battles (see Chapter 4 for more information on how this might be accomplished).

The Risk Itself

As mentioned in Part I, your risk communication efforts will probably fall into one of the three categories: care communication (for situations in which the risk is no longer in doubt), consensus communication (for situations in which the audience will help assess and/or manage the risk), or crisis communication (for an immediate risk). Each of these has its own requirements for communication.

Some examples of risks that fall under care communication include the risks from using tobacco, contracting the AIDS virus, and failing to wear protective clothing when handling hazardous materials in the workplace. Most experts agree that these risks are dangerous to human health (although they may disagree about the magnitude or specific exposure routes). When communicating about these risks, it may not be necessary to review the possible dangers, which are usually recognized. The purpose, then, is to alert your audience and provide information that will encourage them to change to less risky behaviors.

Consensus communication involves risks in which the audience and the decision maker must reach an agreement over how the risk will be assessed or managed. Examples include operation of a hazardous waste incinerator, siting of electrical power transmission lines, and cleanup of a Superfund site. Often, no consensus has been reached about what constitutes safe or dangerous levels of exposure or about the acceptability of the risk to those affected by it. The purpose of risk communication in such cases is to build consensus as a basis for making a risk management decision.

Crisis communication relates to those risks brought about by an emergency: a chemical plant fire, an earthquake, or a train derailment. Again, the danger is clearly recognizable. In these situations, there is no time to develop advisory groups to assess ways of dealing with the risk (although such groups may have been involved in the emergency planning process long before the actual crisis). The purpose of crisis communication is to alert your audience to the danger and provide alternatives to minimize the risk.

In addition to your communication type, another consideration in determining your purpose and objectives is the relative newness of the risk and its visibility to your audience (how risky it seems; Figure 7-1). If the risk is relatively new and not very visible,

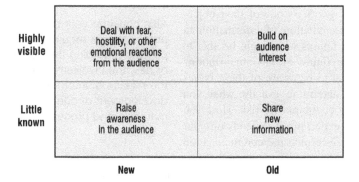

Figure 7-1. Primary purpose of risk communication related to newness and visibility of the risk.

Another consideration in determining your purpose and objectives is the relative newness of the risk and its visibility to your audience.

you will have to first raise awareness before you can communicate more technical information, encourage behavior changes, or build consensus. If this risk is something that has been discussed for years and has been visible for some time, the audience may be apathetic, and you may have to find new ways to awaken audience interest and concern. You might use new information to pique audience interest or present the older information in relation to something your audience is now concerned about. If this risk is relatively new and highly visible, you may have to deal with fear and hostility before effective risk communication can take place. Begin by acknowledging and addressing audience concerns so that the audience can move beyond the fear and hostility to understanding the risk itself.

Audience Requirements

The fact that audience requirements are last on the list does not imply that they are the least important. They may in fact be most important because the needs and concerns heavily affect any type of risk communication. Therefore, what the audience wants from you should be among the first things you consider in determining your purpose and objectives.

What does your audience want from the risk communication effort? The specific answer to this question will differ for each effort; however, the question can be answered generally for each category of risk communication. For care communication, your audience generally wants to know about any risks that would prevent their living a good life (enjoyable, long, and worry free) and about how they can minimize those risks. For consensus communication, your audience generally wants to contribute to a decision about how to assess and manage the risk. Depending on how concerned your audience is, they may want to make a larger or smaller contribution. For crisis communication, your audience generally wants to learn how to minimize their risk as quickly as possible. However, after the crisis is over, they may want to know more about how it began.

Another way to look at audience requirements is to consider the relationships between those who are communicating the risk and those who are at risk. Kasperson and Palmlund (1989) present purposes and objectives of risk communication based on relationships such as doctor to patient, employer to employee, and government or private organization to community. Consider the information in Table 7-1 if your risk communication effort is based on one of these relationships. Also see Chapter 8 and Chapter 9 for more information on what the audience likely wants to know about a particular risk.

Table 7-1. Risk communication purposes based on job relationships*

Doctor to patient	Employer to employee	Organization to community
Change behavior	Inform	Encourage involvement in decision making
Increase responsibility for living a healthy life	Motivate to action	Motivate to action

*Adapted from Kasperson and Palmlund (1989).

CHECKLIST FOR DETERMINING PURPOSE AND OBJECTIVES

The purpose and objectives of my risk communication effort are based on:

- ☐ Associated legal requirements
- ☐ Organizational requirements for
  - ☐ Public involvement
  - ☐ Release of information
  - ☐ Development of communication materials and processes
- ☐ Type of risk communication
  - ☐ Care
  - ☐ Consensus
  - ☐ Crisis
- ☐ Newness of the risk
- ☐ Visibility of the risk
- ☐ Audience requirements
- ☐ Functional relationships between my organization and the audience
  - ☐ Doctor to patient
  - ☐ Employer to employee
  - ☐ Organization to community
- ☐ The purpose and objectives are agreed upon by those involved in the risk communication effort:
  - ☐ Those assessing the risk
  - ☐ Those managing the risk
  - ☐ Those communicating the risk
- ☐ The purpose, objectives, and agreement have been documented

REFERENCE

Kasperson, R. E. and I. Palmlund. 1989. "Evaluating Risk Communications." In V. T. Covello, D. B. McCallum, and M. T. Pavlova, eds., *Effective Risk Communication: The Role and Responsibility of Government and Nongovernment Organizations*. Plenum Press, New York, pp. 143–158.

ADDITIONAL RESOURCES

Rowan, K. E. 1991. "Goals, Obstacles, and Strategies in Risk Communication: A Problem-Solving Approach to Improving Communication about Risks." *Journal of Applied Communication Research*, 19:300–329.

Santos, S. L. 1990. "Developing a Risk Communication Strategy." *Management and Operations*, 82:45–49.

8

ANALYZE YOUR AUDIENCE

Whenever we communicate, we make assumptions about our audience. When we talk to our spouses about our children, we can use nicknames and allusions to past behavior because we can assume that our spouses will know the children and situations. When we discuss our jobs with our managers, we assume that they know what we do. When we communicate risk to a broad audience, however, we cannot make such assumptions because the audience can be divided into segments, each with its own characteristics and needs.

With whom are we communicating? Although answering this question is a must in communicating risk (or any information for that matter), some risk communication efforts are still conducted with a total lack of information about the audience. Instead, communication efforts are based on one of two assumptions: either the people we are communicating with are just like us or they are somehow different.

We do not do too badly if the audience really is the same as we are. Scientists generally communicate well with other scientists. Managers generally communicate well with other managers. Within each group, there is a common language, a shared way of viewing the world. In addition, some scientists have found effective ways of communicating with managers and vice versa. However, once we step very far outside this shared vision, effective communication becomes much more difficult.

Suppose you are vacationing in a country where English is seldom spoken and you know little of the local language. You notice that the elevator door in your hotel is stuck open on the fifth floor. Being a good risk communicator, you try to alert the first hotel employee that you see to the potential danger. To your chagrin, you find that the helpful chambermaid does not speak a word of English.

Risk Communication: A Handbook for Communicating Environmental, Safety, and Health Risks,
Fifth Edition. Regina E. Lundgren and Andrea H. McMakin.
© 2013 The Institute of Electrical and Electronics Engineers, Inc., and Regina E. Lundgren and
Andrea H. McMakin. Published 2013 by John Wiley & Sons, Inc.

Obviously, your usual mode of communication will not work. You cannot explain to her that the risk of people falling to their deaths is approximately 3E-6. You cannot hand her Poole's latest paper discussing all the possible ways in which open elevator doors can cause increased morbidity and reduced life spans. You cannot hold an open public forum and take depositions and written comments for a 45-day period.

You have to try something different. You could draw a picture, you could take her out and show her the door, or you could wander in search of someone to translate. Doubtless, there are also other ways that you could convey your message. The most effective ways, of course, will be those that address her needs. For example, if she has been told *not* to fraternize with guests and to do her work more quickly, she will probably avoid or ignore you as much as she can to get on with her work, a behavior for which you will have to compensate if you are to succeed in your communication efforts.

Audience analysis (that is, determining the audience's characteristics and needs) is a tool too often neglected in risk communication. In almost every case in which communication efforts failed, inadequate or faulty audience analysis is at least partly to blame. *You have to know to whom you are communicating if you are to communicate with any hope of success!*

BEGIN WITH PURPOSE AND OBJECTIVES

You can begin to develop an understanding of your audience by looking at your purpose and objectives. Why are you communicating risk? If you are conducting care communication—risk communication in which the risk is not in doubt—you are generally communicating to increase awareness and change behavior. Whose behavior are you trying to change: a group of workers, a specific group in a community, an entire community, or a specific group across the country? If you are conducting consensus communication—risk communication to reach agreement about the way a risk is assessed or managed—you are generally communicating to encourage consensus building. Who must reach that consensus: a federal agency, its contractors, concerned citizens groups, or industry representatives? If you are conducting crisis communication—risk communication relative to an immediate threat—you are generally communicating to alert the audience and provide ways to minimize the risk. Who is at risk: several communities, one community, or specific people within that community? Answering these questions will give you a basic knowledge about who you are trying to reach.

Be aware, however, that sometimes the audience is broader than you might think. For example, Dr. Jeremy Green at Brigham and Women's Hospital and Harvard Medical School found that family members and friends were equally important members of the audience when trying to communicate with patients with diabetes through social media (Green et al. 2010). When it comes to consensus communication, the U.S. Nuclear Regulatory Commission's Guidelines for External Risk Communication recognizes several types of audiences, including stakeholders impacted because they belong to an organization (for example, licensees), people impacted because their personal lives will be affected (perhaps by citing

> Crisis responses have to be considered in terms of the needs of local communities and citizens immediately affected by the crisis, but also of the diverse needs of those who feel connected to or part of the crisis events even though they may live thousands of miles away.
> —Blankson et al. (2012, p. 220).

of a new facility), the generally concerned public, and the news media (Persensky et al. 2004).

CHOOSE A LEVEL OF ANALYSIS

After examining your purpose and objectives, you have limited knowledge of your audience. How much you more fully analyze your audience will depend on several factors within and without your organization. Factors within your organization include funding, schedule, availability of staff and information sources, and approvals required. The first three factors are fairly self-explanatory. The more funding available, the more you may be able to use for audience analysis, and the more detailed your analysis can be. The more time available, the more detailed your analysis can be. More staff and resources will also help you achieve a more detailed analysis. However, the fourth factor, approvals required, merits additional discussion.

> How much you fully analyze your audience will depend on several factors. Factors within your organization include funding, schedule, availability of staff and information sources, and approvals required.

Approvals for some of the audience analysis techniques discussed in this chapter can be difficult to obtain. If you are a federal agency or government contractor, for example, any survey with more than 10 participants will require approval by the U.S. Office of Management and Budget. This approval can take some time. In addition, because of liability and privacy issues, many organizations also have approval requirements for "human subjects research," which may include any form of questioning. Legal offices and public affairs departments may also have to approve any discussions with people outside the organization. Before you start any audience analysis, find out which approvals will be necessary, how long they will take, and who makes the final decision.

> Before you start any audience analysis, find out which approvals will be necessary, how long they will take, and who makes the final decision.

Also find out if anyone else has been through this maze and whether you can follow their path. Then draw yourself a map of that path with all its turnings so that you can not only make it through the maze but also plan for future efforts.

Factors outside your organization to consider when choosing an appropriate level of audience analysis are those that derive from your purpose and objectives and from the audience itself. What you have to communicate and what you hope to accomplish affect the amount of information you need about your audience. More information is always better than less; however, if your purpose is merely to raise awareness, you may need less information than if your purpose is to change behavior patterns. To raise awareness, you may need to know only reading or education levels and preferred ways of communicating. To change behavior, you need a more complete psychological profile, including why the members of your audience are practicing the behavior to begin with, their feelings about the risk, and what would motivate them to change.

> Lessons learned from devastating wildfires in Australia show that failure to reach the audience contributed to failure to protect the community.
> —Galloway and Kwansah-Aidoo (2012).

Not surprisingly, your audience also influences your choice of levels of audience analysis. Are the members of the audience spread over wide areas? If so, do you have the time and funding to reach each area? Do they differ greatly in other ways (for example, are some highly educated while others failed to finish high school)? Do you have the time and funding to fully analyze each segment or will you have to do one in depth or both more shallowly? Is the audience already so hostile to your organization or efforts that they might refuse to be interviewed? In that case, you will have to use less-direct methods to gather information. Do you already know that they have reading difficulties that require you to use methods that would not require them to read surveys or write? Some of these questions are a little like the question of whether the chicken or the egg came first—you need to know the information before you can properly analyze their needs, but you can only get the information after doing some sort of analysis.

Audience analysis efforts can generally be divided into three levels:

1. **Baseline audience analysis.** This includes information largely related to the audience's ability to comprehend the communication, such as reading ability, preferred methods of communication, and level of hostility. At least a baseline audience analysis should be conducted for any risk communication effort. However, for crisis communication, it may be the only level needed.

2. **Midline audience analysis.** This includes baseline information plus information about socioeconomic status, demographics, and cultural information, such as age, gender, and occupations. A midline audience analysis will usually suffice for care communication in which the purpose is to increase awareness.

3. **Comprehensive audience analysis.** This includes baseline and midline information plus psychological factors, such as motivations and mental models of risk. A comprehensive audience analysis is usually necessary in consensus communication and care communication in which the purpose is to change behavior.

The cost and time associated with any of these levels depends largely on how the information is gathered, but both generally increase with each level of effort. A baseline audience analysis will usually take one person working full time between 4 hours and 2 weeks, a midline 1 week to 1 month, and a comprehensive 3 weeks to 2 months.

The U.S. Environmental Protection Agency (EPA) has issued an excellent resource for audience analysis: *Community Culture and the Environment: A Guide to Understanding a Sense of Place* (EPA 2002). It is designed to provide "a process and set of tools for defining and understanding the human dimension of an environmental issue" (p. 3). Although it is focused on helping the EPA foster community-based environmental protection, the guide's process could be applied to any health intervention at a community level. The approach is flexible, but if it is used to its full extent, the resulting analysis would be beyond comprehensive. Some of the characteristics suggested for analysis include the following:

- **Community boundaries.** Natural, physical, administrative, social, and economic characteristics that distinguish one community from another
- **Community capacity and activism.** How local leaders and citizens influence local decisions
- **Communication interaction and information flow.** How people interact and exchange information
- **Demographic information.** Population description

- **Economic conditions and employment.** History, present, and future of the local economy
- **Education.** The level of schooling achieved and the role of education in the community
- **Environmental (or health) awareness and values.** Knowledge, concerns, and perceptions and how these influence daily life
- **Governance.** How decisions are made from the local to federal levels
- **Infrastructure and public services.** Roads, schools, police, fire, and so on
- **Local identity.** Quality-of-life issues, history, art, and local traditions
- **Local leisure and recreation.** How community members spend their free time
- **Natural resources and landscapes.** Natural features of the area
- **Property ownership, management, and planning.** Who owns land and who is responsible for land-use planning
- **Public safety and health.** Personal safety and health issues
- **Religious and spiritual practices.** Importance, variety, and beliefs associated with local religions and spiritual practices

In addition, consider where your audience stands on the perception of the risk you are communicating. Studies in Norwich, United Kingdom, showed that audience's perceptions move through three distinct phases:

1. Development, in which people first discover the risk and begin making associations based on social and cultural structures as well as personal considerations
2. Maintenance, in which people selectively take in new information, with a bias toward enforcing previously held perceptions
3. Transition, in which a new event changes perceptions (usually rapidly) (Bennett and Calman 1999)

Many of these characteristics are also noted in Table 8-1, Table 8-2, and Table 8-3 for the baseline, midline, and comprehensive audience analysis, respectively.

DETERMINE KEY AUDIENCE CHARACTERISTICS

Once you have determined the appropriate level of analysis (considering organizational and audience factors), make a detailed list of the characteristics you need to know. Refer to Table 8-1, Table 8-2, and Table 8-3 for suggestions. Depending on your specific situation, you may want to add characteristics to these lists or delete some. For example, if you will not be developing written materials, then reading level is not a characteristic you will need to determine.

Each level of analysis builds on the last; that is, the midline includes everything in the baseline, and the comprehensive includes everything in the baseline and midline.

DETERMINE HOW TO FIND AUDIENCE ANALYSIS INFORMATION

There are many ways to gather the audience analysis information listed previously. The best way is to go out and talk to your audience, actually meet them face to face. This can be done by conducting interviews, conducting surveys (face to face or less directly

Table 8-1. Key audience characteristics for the baseline audience analysis

Characteristics	Questions to ask yourself	How answers affect risk communication
Experience with the risk	Is risk new to the audience or something they have been living with for a long time?	If new, build awareness first. If familiar, build on known concepts.
Experience with organization communicating risk	Are they familiar with the organization? Do they find it credible?	If unfamiliar, explain organization's role. If familiar and credible, build on goodwill. If familiar but not credible, use an outside spokesperson.
Background in risk subject matter	How much do they understand about the risk scientifically?	If little, provide an explanation. If a great deal, build on concepts.
Reading level	At what level do they read?	If lower level, simplify language, organization, and sentence and paragraph structure. At higher level, can use more complex language.
People they trust and believe	Whom do they trust and believe?	Choose that person as the spokesperson.
Information sources	Where do they get information (television news, newspapers, radio, Internet, family networks, personal experience)?	Use that source to disseminate risk messages.
Education level	What is the highest level of education completed? What is the range?	If higher levels, use more complex concepts. If lower levels, provide basic information.
Group size	How many people are in the audience?	If larger audience, use a method like television. If smaller audience, use a more intimate method like meetings.
Goals/expectations for risk communication process	What do they expect to happen when you communicate risk?	Whenever possible, match or exceed their goals and expectations.
Their role in the risk communication process	Based on laws and organization's requirements, what can the audience's role be?	Whenever possible, involve them in the way they want to be involved.
Their "hot buttons"	Are there words and concepts that infuriate them?	Avoid those words. Find other ways to discuss concepts.

Table 8-2. Key audience characteristics for the midline audience analysis

Characteristics	Questions to ask yourself	How answers affect risk communication
Age	What age range do they fall into? What 5-year range has the most people in it?	Consider possible concerns: families, careers, retirement.
Culture	How many different cultures make up the audience? How does each view the world?	Address different views.
Gender	Are they mostly male or female?	Consider how gender affects risk probability.
Turnover within the community (yearly)	Is this a transient community or one with deep local ties?	If transient, use self-contained messages. If more stable, build on previous messages.
Preferred social institutions	Where do they go to relax? To play? To worship?	Determine possible concerns. Use preferred locations to hold meetings.
Length/history of involvement	How long have they been involved with the risk? Has the involvement been one of passive listening, consensus building, or reactive argument?	If passive, build on concepts and encourage activity. If consensus building, provide information and encourage involvement. If reactive, acknowledge concerns to lower hostility and encourage activity.
Jobs/occupations	Where do they work? What do they do there? Is the risk part of their workplace?	If jobs relate to risk, focus on how to mitigate. If not, determine possible concerns.
Geographic areas	How near is the risk?	If close to risk, provide information to mitigate. In general, determine possible concerns. If large geographic area, use method like television. If smaller area, use more intimate method like meetings.

through the mail), sponsoring members of the audience as advisors, hosting focus groups, or pretesting prototype risk messages with audience members. These direct methods often provide the most current, risk- and situation-specific information. Unfortunately, these methods are not often used in risk communication efforts for several reasons:

> There are many ways to gather audience analysis information. The best way is to go out and talk to your audience, actually meet them face to face.

- The members of the audience are so dispersed or in such large numbers that it is impossible to meet them all or even representatives.
- The costs and time involved are prohibitive in a given situation.
- The audience is hostile and refuses to associate with anyone connected with the risk or its assessment, management, or communication.

Table 8-3. Key audience characteristics for the comprehensive audience analysis

Characteristics	Questions to ask yourself	How answers affect risk communication
Concerns and feelings about risk	What kinds of concerns do they have? How do they feel about the risk (angry, frustrated, apathetic)?	Address concerns and feelings in risk messages.
Experience with other risks	Have they had good examples you can build on? Bad examples to overcome?	If good, build on them. If bad, acknowledge them and begin with basic risk information.
Exposure to news media or other coverage	Have they seen comprehensive coverage or tabloid-style journalism?	If comprehensive, build on it. If tabloid, acknowledge it and begin with basic risk information.
Effect of the risk on them	How do the experts think the risk can affect them? How does the audience think it can affect them?	If two views differ, address misconceptions to correct. If two views are identical, build on concepts.
Their control over the risk	Can they mitigate the risk or must they live with it?	If mitigatable, give ways. If no control, try to empower with knowledge.
Goals of organized groups	What are they trying to accomplish?	Determine possible concerns and feelings.

- The very idea of actually meeting with the audience terrifies some members of the organization that is conducting the risk communication effort.

If these reasons apply in your situation, you may have to choose other methods of gathering audience information. Less-direct methods of gathering audience information include using surrogate audiences and consulting existing sources of information. Surrogate audiences are people to whom you have easy access who seem to approximate the general audience you are trying to reach. If, for example, you are trying to construct a risk message for a distant community, but have no time to fly there and conduct a more direct analysis, you might find a local community that seems similar and interview or survey the people there. The danger in using this method is that you are forced to make assumptions based on very little information. These assumptions may prove wrong and cause your risk communication efforts to be less than successful. However, some audience information is generally better than none.

> Surrogate audiences are people to whom you have easy access who seem to approximate the general audience you are trying to reach.

Another way to use surrogate audiences is to look to the risk communication literature to find examples of audiences that have been studied in detail and seem to match your audience (based on your purpose and objectives). J. D. Callaghan (1989), for example, divided audiences into the scientific/medical community, government agencies, a specific community, and the public (which would include a number of communities). He determined characteristics for each of these audiences and used that information to tailor risk messages. His advice for each was:

- To present risk effectively **to the scientific/medical community**, work through professional societies and conferences, personal communication, and impartial databases. Use simple English (as opposed to the scholarly writing often seen in professional journals), with references for additional information. Target mailings to society members and develop special interest groups within societies. Begin a newsletter on your risk that includes both scientific information and clearly labeled opinions, and develop traveling seminars that could be given to society groups.
- To present risk effectively **to government agencies**, provide in-depth information at the beginning. Use information packages that consist of a position statement in nonspecialist language, backed up by research information. Have a scientist or other expert in the field present the information to the group. Identify the decision makers and those who will offer advice and concentrate on key individuals who will be able to influence others.
- **For communities**, use more personal communication. Use open meetings with a skillful moderator. Follow up with stories in union and company publications, and with video or classroom instructional materials.
- **For the public** in which there is little or no hostility, use the news media (press conferences, news releases, interviews, video and audio tapes, and media hotlines; see Chapter 16 for more information). If there is hostility, send credible experts to radio and television talk shows, radio call-in shows, and major public forums. Also, submit articles to the newspapers for the opinion-editorial page and stories to magazines, which can be reprinted and sent to specific segments of the audience.

Another indirect method of gathering audience analysis information is to use existing sources of information:

- **Staff sociologists.** Many large organizations have on staff a group of social scientists who compile information about the local community and other people with whom the organization comes into contact regularly. They can provide a wealth of information for audience analysis. If your organization does not have such a staff, and you have time and money, you can hire social scientists (from colleges and universities or independent consulting firms) to develop surveys, interview key audience members, and compile information to meet your needs. Sometimes graduate or undergraduate classes will do the research for you for the experience and/or a nominal fee as a class project.
- **Environmental impact statements.** Many environmental impact statements incorporate information about the local communities and economy. Although they will not give you sufficient information for a comprehensive audience analysis, they can be a source of information for the baseline and midline analyses. Sources include local libraries and government document repositories. Some documents are now available online as well.
- **Documentation of work for the Comprehensive Environmental Response, Compensation, and Liability Act (CERCLA or Superfund).** By law, a Superfund site must have a community relations plan. This plan should include information about the community (and, thus, the audience with whom the environmental cleanup contractor must communicate). Sources include local libraries, government document repositories, the U.S. EPA, and the Internet.
- **The Internet.** Nearly every community and group has a website that can provide a wealth of information, often in the language and style preferred in the area.

- **Social media.** Many organizations and communities have their own sites and groups as well. If you are charged with communicating risks over the long term to a specific community, consider joining the conversation.
- **Census data.** These data cover demographics, economic trends, and education levels. Sources include local libraries, government repositories, and the Internet.
- **Local television station and radio station advertising profiles.** Stations need to know to whom they are broadcasting to attract advertising and viewers or listeners. They may be willing to share this information with you if you explain why you need it. They may charge a fee.
- **Local newspapers and magazines advertising profiles.** These media also need to know their readership to attract advertising and increase subscriptions. They may be willing to share this information with you if you explain why you need it. They may also charge a fee.
- **State or local political groups.** Again, these organizations need to know their constituencies to be reelected. They may be willing to share this information with you if you explain why you need it.
- **Health care agencies and cancer centers.** Public affairs or communications groups within larger health care agencies often have to communicate to a large group and will have developed community profiles. They may share these if you explain why you need the information.
- **Chambers of Commerce or other community economic development organizations.** These organizations also conduct community research to provide information that might attract new businesses to the area. They may be willing to share if the purpose of your risk communication effort will assist the community.
- **Letters to the editor in the local newspaper.** These will tell you what the local concerns are and which groups are most vocal.
- **Market analysis information.** Firms such as Gallup, Harris, and Opinion Research conduct surveys and interviews to gather information for marketing purposes. They may be willing to share existing information for a fee and can develop specific surveys and interviews to meet your needs. These tailored services can be more expensive than using your own staff, but may be cost- and time-effective for large risk communication efforts.
- **Related information materials.** In consensus communication and sometimes in care communication, other organizations besides yours will also be communicating about the risk. These organizations may in fact be part of the audience. Look at how they frame questions, what factors seem important to them, and the language they use. This information can give you important clues to answer your audience analysis questions.
- **Job descriptions.** For communication focused on specific types of workers, job descriptions can indicate education levels preferred, years of experience, and other factors useful in analyzing that audience.

Which sources you use to gather information concerning your audience depends not only on your situation (time, funding, and organizational support) but also on whether you are conducting care, consensus, or crisis communication. For care communication, useful sources differ depending on whether you are conducting health care communication (in which human health in general is at risk) or industrial risk communication (in which worker health or safety is at risk). Industrial risk communication can be divided into industrial hygiene and individual worker notification. The less-direct sources of audience analysis information most useful for each of these types of risk communication are shown in Table 8-4.

A possible issue with using these indirect methods is the validity of the information. The less direct, that is, the farther the source is from the original (your audience itself), the more likely that the information is actually an interpretation or at least partly assumed. To compensate for this problem, use multiple sources of information to confirm audience characteristics and needs.

> Which sources you use to gather information concerning your audience depends not only on your situation but also on whether you are conducting care, consensus, or crisis communication.

INCORPORATE AUDIENCE ANALYSIS INFORMATION INTO RISK COMMUNICATION EFFORTS

Once you get all this information, what do you do with it? Audience analysis information is used to tailor risk messages to meet specific audience and situational needs. The information can tell you what media to use, how much audience interaction is needed, and what concerns must be addressed, among other factors (Table 8-5).

Audience analysis information is also often used to determine the proper "style" for written messages. Many organizations use computer software such as "style checkers," or apply readability formulas. Although these tools will give you written messages that

Table 8-4. Less-direct sources most useful for audience analysis for various types of risk communication

Type of risk communication	Most useful source
Care communication	
Health care (heterogeneous audience)	Market analysis information Health care agencies
Health care (homogeneous audience)	Surrogate audience Advertising profiles Internet Census data
Industrial risk	
Industrial hygiene	Job descriptions
Worker notification	Surrogate audience
Consensus communication	None—direct contact required for this type of communication to succeed
*Crisis communication**	Staff sociologists Environmental impact statements CERCLA documentation Political groups Letters to the editor

*Crisis communication sources also depend on the timing of the analysis. Sources listed are most useful before the crisis, during emergency planning. During a crisis, use whatever sources are immediately available, including surrogate audiences.

Table 8-5. Using audience analysis information to tailor risk messages

Information learned	How to tailor the message
Audience unaware	Use graphic method—high color, compelling visuals, and theme.
Audience apathetic (or feels like victims)	Open risk assessment and management process to stakeholder participation. Show where past interactions have made a difference. Provide choices.
Audience well informed	Build on past information.
Audience hostile	Acknowledge concerns and feelings. Identify common ground. Open risk assessment and management process to stakeholder participation.
Audience highly educated	Use more sophisticated language and structure.
Audience not highly educated	Use less sophisticated language and structure. Make structure highly visible, not subtle.
Who the audience trusts	Use that person to present risk information.
Where the audience feels comfortable	Hold meetings in that location.
The method by which the audience gets most of its information	Use that method to convey your message.
Who makes up the audience	Ensure that the message reaches each member.
How the audience wants to be involved in risk assessment or management	If at all possible, given time, funding, and organizational constraints, involve the audience in the way they want to be involved.
Misconceptions of risk or process	Acknowledge misconceptions. Provide facts to fill gaps in knowledge and correct false impressions.
Audience concerns	Acknowledge concerns and provide relevant facts.

conform to a particular grade or reading level, they may also give you messages that are flat, boring, and monotonic.

Studies have shown that the concept loading, or number of concepts, and the placement of those concepts in the sentence, is far more important to readability and comprehension than the length of words or their number in a sentence. Take, for example, the following sentence:

> Because of differences in lifestyles, certain members of this group, particularly infants and the elderly, are more likely to be affected.

This sentence would be rated at about the tenth grade reading level according to the Fog Index, a readability formula developed by Robert Gunning and Douglas Mueller. After all, it contains 21 words, several with three or more syllables. However, according to the theories of concept loading and placement, this sentence should be readable at a much lower level. There are three main concepts, with one clarifier, each contained in its own phrase or component of the sentence. With any skill, the communicator can stack such phrases so that each sentence and paragraph builds on the next. An outstanding user's guide developed for the U.S. Food and Drug Administration provides additional advice on using readability formulas and other tools to meet audience needs (Fischhoff et al. 2011).

CHECKLIST FOR ANALYZING YOUR AUDIENCE

Based on the purpose and objectives, general audience, and organizational constraints, the most appropriate level of audience analysis is:

- ☐ Baseline
- ☐ Midline
- ☐ Comprehensive

The list of key characteristics to be analyzed was based on the:

- ☐ Purpose and objectives
- ☐ General audience
- ☐ Level of analysis

Based on:

- ☐ Time
- ☐ Funding
- ☐ Availability of staff
- ☐ Approvals required

Audience analysis information will be gathered by:

- ☐ Direct methods
  - ☐ Interviews
  - ☐ Surveys
  - ☐ Focus groups
- ☐ Less-direct methods
  - ☐ Surrogate audiences
  - ☐ Existing sources of information

For direct methods:

- ☐ Necessary approvals have been determined
- ☐ Necessary approvals have been received

For less-direct methods:

- ☐ Multiple sources were used to confirm audience information

Audience analysis information was used to:

- ☐ Tailor risk communication strategies
- ☐ Determine appropriate
  - ☐ Language
  - ☐ Sentence structure
  - ☐ Organization

REFERENCES

Bennett, P. and K. Calman. 1999. *Risk Communication and Public Health*. Oxford University Press, New York.

Blankson, I. A., S. Natasia, and M. Liu. 2012. "A Triple Disaster in One Fell Swoop: Rethinking Crisis Communication in Japan after March 11." In A. M. George and C. B. Pratt, eds., *Case Studies in Crisis Communication: International Perspectives and Hits and Misses.* Routledge, New York, pp. 196–226.

Callaghan, J. D. 1989. "Reaching Target Audiences with Risk Information." In V. T. Covello, D. B. McCallum, and M. T. Pavlova, eds., *Effective Risk Communication: The Role and Responsibility of Government and Nongovernment Organizations.* Plenum Press, New York, pp. 137–142.

EPA (U.S. Environmental Protection Agency). 2002. *Community Culture and the Environment: A Guide to Understanding a Sense of Place.* Office of Water, Washington, DC. EPA 842-B-01-003.

Fischhoff, B., N. T. Brewer, and J. S. Downs, eds. 2011. *Communicating Risks and Benefits: An Evidence-Based User's Guide.* U.S. Food and Drug Administration, Washington, DC.

Galloway, C. and K. Kwansah-Aidoo. 2012. "Victoria Burning: Confronting the 2009 Catastrophic Bushfires in Australia." In A. M. George and C. B. Pratt, eds., *Case Studies in Crisis Communication: International Perspectives and Hits and Misses.* Routledge, New York, pp. 279–292.

Green, J. A., N. K. Choudry, E. Kilabuk, and W. H. Sharnk. 2010. "Online Social Networking by Patients with Diabetes: A Qualitative Evaluation of Communication with Facebook." *Journal of General Internal Medicine*, 26(3):287–292.

Persensky, J., S. Browde, A. Szabo, L. Peterson, E. Specht, and E. Wright. 2004. *Effective Risk Communication: The Nuclear Regulatory Commission's Guidelines for External Risk Communication.* U.S. Nuclear Regulatory Commission, Washington, DC. NUREG/BR-0308.

ADDITIONAL RESOURCES

Arkin, E. B. 1989. "Translation of Risk Information for the Public: Message Development." In V. T. Covello, D. B. McCallum, and M. T. Pavlova, eds., *Effective Risk Communication: The Role and Responsibility of Government and Nongovernment Organizations.* Plenum Press, New York, pp. 127–135.

Babbie, E. 1973. *Survey Research Methods.* Wadsworth Publishing Company, Belmont, California.

Barke, R. P. No Date. "Surveys on How Specific Audiences See Risk." School of Public Policy, Georgia Institute of Technology, Atlanta, Georgia.

Butler, L. M. 1995. *The "Sondeo" A Rapid Reconnaissance Approach for Situational Assessment.* Community Ventures, A Western Regional Extension Publication, WREP0127. Washington State University, Pullman, Washington.

Hodges, M. 1992. "How Scientists See Risk." *Research Horizons*, Summer, pp. 22–24. Georgia Institute of Technology, Atlanta, Georgia.

International Association for Public Participation. http://www.pin.org/iap2.htm (accessed January 23, 2013).

McDonough, M. H. 1984. "Audience Analysis Techniques." *Supplements to a Guide to Cultural and Environmental Interpretation.* U.S. Army Corps of Engineers, Waterways Experiment Station, Vicksburg, Mississippi.

Pearsall, T. E. 1969. *Audience Analysis for Technical Writing.* Glencoe Press, Beverly Hills, California.

Santos, S. L. 1990. "Developing a Risk Communication Strategy." *Management and Operations*, November, pp. 45–49.

Warren, T. L. 1993. "Three Approaches to Reader Analysis." *Technical Communication*, 40(1):81–88.

Weinstein, N. D. and P. M. Sandman. 1993. "Some Criteria for Evaluating Risk Messages." *Risk Analysis*, 13(1):103–114.

9

DEVELOP YOUR MESSAGE

When conveying risk-related information, it often helps to develop key messages as part of the planning process. Messages help focus all communication participants on the most important information and how to convey it. In care communication, messages usually convey the essential nature of a risk and what people can do to avoid or reduce it. In consensus communication, a stakeholder group may want to develop its own messages when recommending policy or actions to be taken by decision makers. In crisis communication, making all organizational participants aware of the key messages (even if they evolve during the crisis) can make recovery actions more effective, reduce confusion, and increase organizational credibility. In a crisis, key messages are especially important for media spokespeople and those who staff phone hotlines.

Message development in risk communication is not the same as developing a catchy slogan in an advertising campaign. Message development is not manipulative, nor is it a substitute for audience analysis or public participation. The point is not to try to bombard people with what you think they ought to know but to understand what they want and need to know and addressing those things in a clear, concise way. As public health researchers in the United Kingdom found, the idea is to state the risk information in a way that supports continued sharing on all sides (Bennett and Calman 1999).

> Message development is not manipulative, nor is it a substitute for audience analysis or public participation.

The following sections discuss some common pitfalls in developing messages, what people want to know about risks, and how to craft messages for various risk situations.

Risk Communication: A Handbook for Communicating Environmental, Safety, and Health Risks, Fifth Edition. Regina E. Lundgren and Andrea H. McMakin.
© 2013 The Institute of Electrical and Electronics Engineers, Inc., and Regina E. Lundgren and Andrea H. McMakin. Published 2013 by John Wiley & Sons, Inc.

COMMON PITFALLS

Message development is not as easy as it sometimes appears. Everything from the words that are used to the order they are used to the way they are expressed can hinder sharing information between the risk communicator and the intended audience and from the audience to the communicating organization. One of the most common pitfalls is framing.

Framing deals with how risk information is presented. Studies have shown that how information is framed can affect people's reaction to it or the conclusions they draw from it. For example, two groups were shown the same risk information on a new cancer treatment. In the first group, the information was framed as to how the treatment affected the chance of dying. In the second group, the information was framed as to how it affected the chance of survival. Choices to accept the treatment more than doubled in the second group, even when the participants were physicians who might have been expected to understand the difference in the framing (Bennett and Calman 1999).

Another pitfall is the concept of "no risk." When exhaustive studies have shown that a particular public concern is unlikely to pose a risk, it is tempting to craft messages that claim there is nothing to fear. However, no risk assessment is without uncertainty, and no situation is ever completely without risk. After seeing the billions of dollars lost and thousands of cattle killed following the mad cow disease outbreak and subsequent risk communication failure in the United Kingdom, risk communication experts William Leiss and Douglas Powell advocated banishing "no risk" messages (Leiss and Powell 2004).

Another challenge in crafting messages is how to share numerical information. As noted earlier in the book (see Chapter 4), people generally have a limited ability to process numerical data. The National Adult Literacy Survey says that nearly half of all people have some trouble understanding simple math. Ellen Peters of the noted research firm Decision Research recommends the following to help convey numerical messages:

- Highlight the most important information.
- Pretest symbols and graphics.
- Align data with general thinking (for example, in a choice of one to five, the highest number would be best).
- If you state probabilities as 1 chance in X, keep X consistent.
- Give visual clues as to the importance of information (for example, use larger fonts or bold items).
- Consider expressing risks as absolute risks (1 in 10) as opposed to relative risk (10%) and do not use decimals (Tucker et al. 2008).

See Chapter 14 for more guidance on conveying numerical information visually.

The idea of benefits associated with a risk can also prove a stumbling block for risk communicators. Benefits are the so-called good that can come from taking a particular risk. A nuclear power plant, for example, might bring new jobs, an increased tax base, and additional electricity to an area, as well as the potential risks from industrial and environmental accidents and the long-term problem of nuclear waste. Some industries advocate cost–benefit analysis, where the environmental, safety, and health risks are allocated some dollar value and compared with the dollar value of the benefits.

But benefits can be a tricky component of risk communication messages if not used with care. When explaining potential health care risks in care communication, for example, a patient would expect a doctor to provide the benefits as well as the risks so that the patient can make an informed decision. In consensus communication, all those working together to determine the most appropriate outcome should weigh both the risks and the benefits. In crisis and emergency communication, however, a discussion of benefits would seem out of place. As the flood waters rise, who wants to hear that plummeting property values will allow more first-time homeowners into the market? In many cases, the benefits of implementing the risk advice also seem obvious (evacuate or risk drowning, shelter in place to avoid exposure).

An example of the potential dangers in including benefit information along with risk information is the U.S. Food and Drug Administration's (FDA) announcement in 2007 of a Risk Communication Advisory Committee. The FDA created this committee on the Institute of Medicine's recommendation to address how the agency communicates information about the efficacy, safety, and use of drugs and other regulated medical products. The FDA decided to broaden the committee's scope to include all products and benefits along with the risks. Coming so soon as it did after the deaths of many pets from tainted food, the idea that the committee would consider benefits ignited an online firestorm of criticism among stakeholders that took months to calm.

To determine whether to include benefits in risk communication information, consider the following:

- Would the person at risk expect to hear about the risk's benefits from your organization? When a regulatory agency discusses the benefits of a risk, the audience often reacts with hostility, assuming that the agency has forgotten its charter to protect the public.
- Would including benefits further the risk communication dialogue? For example, a stakeholder group evaluating a new product might need benefit information.
- Would having benefit information allow the person at risk to make a more informed decision? For example, knowing the benefits of an operation as well as the risks allows for better patient decision making.

If you cannot answer yes to at least one of these three questions, think twice before mixing risk and benefit information. Remember the cardinal rule of risk communication: Know your audience!

INFORMATION PEOPLE WANT

What people want to know about a particular risk seems to vary, depending on several factors including their familiarity with the risk and the expectation for the method being used (for example, static information materials vs. interactive social media). When the audience is faced with an unfamiliar risk, for example, research has pointed to some key pieces of information

> [Our research] seems to indicate that people first want to know whether the risk is relevant to them: what it is, what are the consequences, and/or is it likely that one will be exposed to it.
> —Lion et al. (2002, p. 775).

people are most likely to want to know. Though some people do not seek information out of fear or inability to process the risk, many others want to know everything (Lion et al. 2002). In general, people are interested in the following types of information:

- **Description of the risk.** People want to go beyond technical descriptions to familiar analogies. Thus, risk communicators may want to provide example analogies to aid risk understanding.
- **Risk consequences.** This includes effects and the level of danger associated with the risk.
- **Level of control about the risk and its consequences.** People want to know the answers to questions such as "What should I do?" and "What are agencies doing?"
- **Exposure information.** This includes risk intensity, duration, acceptable risk levels and how they are measured, how long the exposing agent is dangerous, how long it persists, and how it accumulates in the body.

People with higher levels of education also wanted to know how research on the risk was conducted; those with lower education were interested only in the results of the research. The advantages of the risk were uniformly ranked as among the least important information (Lion et al. 2002).

Additional research has focused on how people process risk messages depending on their level of familiarity. A group of university researchers led by LeeAnn Kahlor of the Center for Health Systems Research and Analysis in Wisconsin found that the less people knew about a risk, the more likely that they were to process information in a systematic matter. Such systematic processing was methodical and evaluative. This type of processing tends to lead to attitudes that are more stable and resistant to change, which can help in crisis preparedness and consensus communication efforts. It may also lead to behavior changes, which can help in crisis response and care communication efforts. To encourage systematic processing, risk communicators need to include what audience members *perceive* they need to know, or audiences will not attend to the information (Kahlor et al. 2003).

When it comes to social media, audiences appear to want emotional support as well as specific guidance and feedback on appropriate actions to take. For example, Dr. Jeremy A. Green and his colleagues at Brigham and Women's Hospital and Harvard Medical School found that patients with diabetes divided their messages between information about disease management strategies (approximately two-thirds) and requests for emotional support (approximately one-third; Green et al. 2010).

MENTAL MODELS

The mental models approach described in Chapter 2 can be used to develop accurate risk messages. Because the time involved can be extensive, this approach is most appropriate for care and consensus communication and for the planning phase of crisis communication.

Messages are developed after conducting interviews that show how people understand and view the risks associated with various phenomena. The approach is not designed to persuade people that risks are small and under control, but rather to supply laypeople with the accurate information they need to make informed, independent judgments about risks to health, safety, and the environment. Morgan et al. (2002) describe a six-step process to learn what people believe and what information they need to make the decisions they face:

1. **Create an expert model of the risk, using an influence diagram.** This is a scientifically accurate model of the processes that determine the nature and magnitude

of the risk. Figure 9-1 and Figure 9-2 show two examples, for Lyme disease and HIV/AIDS.

2. **Conduct mental models interviews.** Using the expert model, ask questions to learn people's beliefs about the hazard, expressed in their own terms.
3. **Conduct structured interviews.** Interview larger groups of people using a questionnaire that captures the beliefs in the open-ended interviews, to determine the prevalence of the beliefs in a larger population.
4. **Compare the responses with the experts' understanding of the risk.**
5. **Draft the risk communication.** Use the results from the interviews, along with an analysis of the decisions people face, to address the most significant incorrect beliefs and knowledge gaps.
6. **Evaluate the communication.** Test and refine the communication with various groups until the communication is understood as intended.

The goal of this process is neither to persuade people that they should make decisions the way scientists do nor to transform people into scientists. It is to make sure that people have the scientifically, technically, and medically accurate information they need to make their own decisions.

Invariably, this process reveals inaccurate (and often unexpected) beliefs that must be addressed in risk communication. For example, interviewers found that some people believe that houses containing radon are permanently contaminated and that radon remediation in homes is impossible or prohibitively expensive. Thus, argues Granger Morgan et al. (2002), the central message in the U.S. Environmental Protection Agency's 1986 *Citizens' Guide to Radon*—"You should test your house for radon"—may have been

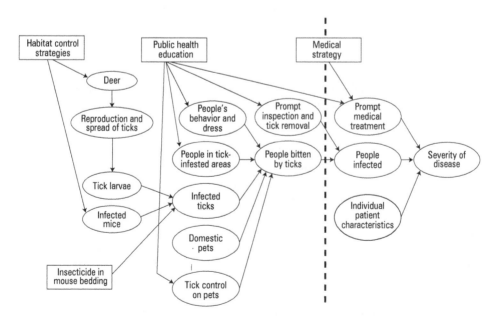

Figure 9-1. Simplified expert model of the risk of infection from Lyme disease. The part to the left of the vertical dashed line deals with exposure processes. The part to the right of the vertical dashed line deals with effects processes. (Source: Morgan et al. 2002; used with permission).

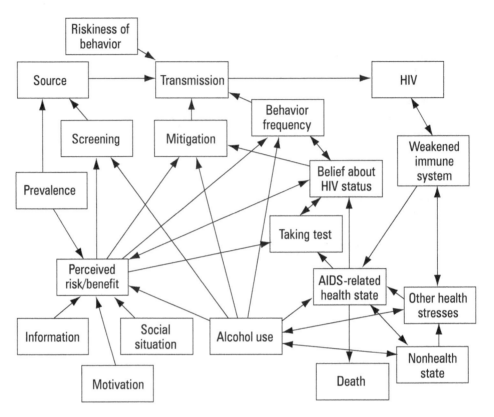

Figure 9-2. Expert model of the risks associated with HIV and AIDS. (Source: Fischhoff and Downs 1997; used with permission).

undermined by the public belief that testing was futile. Better messages would have been that radon decays over time and that inexpensive devices are available to monitor radon levels in homes.

When the mental models process was used with teens for HIV/AIDS awareness, it showed that teens already understood the big picture about HIV/AIDS, but they needed to have information gaps filled and misconceptions corrected. Thus, rather than repeating familiar HIV/AIDS facts, communication materials were addressed with messages such as "Knowing your partner is not enough to prevent HIV," "The more times you have sex with an HIV-infected person, the more likely you are to get HIV," and "The only way to know if you have HIV is to be tested."

A mental models process about climate change risks led to communication materials that directly addressed common misconceptions. The materials used messages such as "Climate change and the loss of the ozone layer are two different problems that are not very closely connected," "Using aerosol spray cans has almost no effect on climate change," and "Nuclear power does not contribute to climate change."

A mental models process about health risks from electric power fields led to messages such as "The strength of electric fields decreases rapidly as you move away from them." Figure 14-1 later in this book shows an illustration that was created to convey this message.

As with any other technique, the mental models approach has its limitations. In one study, mental models interviews were used to design materials to explain electroplating risks. Consequently, electroplating workers showed a better understanding of long-term health effects and were more aware of precautions to take in the plating shop that were not based on direct experience (Petts et al. 2002). However, the researchers also raised several cautions about the mental models approach. For example, the approach may incorrectly assume that all expert beliefs are consistent, may undervalue layperson knowledge and belief, and may skip over behavioral and organizational culture issues, especially in the workplace. The researchers endorsed the user-centered approach for designing risk information, but with these added guidelines:

- Focus not only on what people believe but also on why and how they reduce risks.
- Include divergent views related to demographics, length of work experience, and disability.
- Realize that there may be more than one expert model, for example, for occupational health experts and factory inspectors, and that the expert model of hazards and risk reduction may not be better than those held by an industry worker.
- In presenting risk information, answer key questions such as the following: How might this happen? What can it to do me? How can I protect myself?
- Test whether new risk messages affect behavior, not just improve understanding.

MESSAGE MAPPING AND MESSAGE DEVELOPMENT TEMPLATES

Risk communication consultant and Columbia University professor Vince Covello developed the concept of a message map. It is a template for displaying detailed, hierarchically organized responses to anticipated questions or concerns. Message maps are one way to make sure that everyone understands the organization's messages for high-concern or controversial issues.

> A message map is a template for displaying detailed, hierarchically organized responses to anticipated questions or concerns.

Message maps should be created with interested and affected parties, including scientists, communication specialists, and policy experts as applicable. Broad involvement will reveal a diversity of viewpoints, but the end goal is to come up with messages and supporting points with which everyone agrees.

Start by developing a list of anticipated questions, to identify common sets of underlying concerns. Develop key messages in response to the underlying concerns and questions. Each key message should have up to three supporting facts associated with it. To reduce confusion and increase comprehension and retention, Covello recommends limiting the number of key messages and keeping them at a middle-school level of readability. Table 9-1 shows an example of a message map to answer the question, "How contagious is smallpox?"

HEALTH RISK COMMUNICATION

Public-health-related messages are similar to other risk messages, but with the added goal of human behavioral change. Though health messages often contain an element of

Table 9-1. Sample message map. Stakeholder: General Public. Question: How contagious is smallpox?

Key message 1	Key message 2	Key message 3
Smallpox spreads slowly compared with measles or the flu.	This allows time for us to trace contacts and vaccinate those people who have come in contact.	Vaccination within 3–4 days of contact will generally prevent the disease.
Supporting fact 1-1	Supporting fact 2-1	Supporting fact 3-1
People are only infectious when the rash appears and they are ill.	The incubation period for the disease is 10–14 days.	People who have never been vaccinated are the most important ones to vaccinate.
Supporting fact 1-2	Supporting fact 2-2	Supporting fact 3-2
It requires hours of face-to-face contact.	Resources for finding people are available.	Adults who were vaccinated as children may still have some immunity to smallpox.
Supporting fact 1-3	Supporting fact 2-3	Supporting fact 3-3
There are no asymptomatic carriers.	Finding people who have been exposed and vaccinating them is the successful approach.	Adequate vaccine is on hand, and the supply is increasing.

Source: Centers for Disease Control and Prevention et al. (2003).

persuasion, they should still be based on understanding the audience's concerns, needs, and incentives to act. Health communication researchers Kim Witte et al. (2001) recommend a detailed process for audience analysis and message development, based on communication research:

- Identify the health threat, recommended response, and target audience.
- Conduct formative research about the target audience's beliefs about the threat, including beliefs to change, reinforce, or introduce. Develop one or more audience profiles of "typical" members of your target audience, including lifestyle practices, cultural beliefs, religious values, and so on.
- List the source, channel, and message preferences, making sure that they fit with the audience's values, demographic characteristics, and needs.
- Determine the stage of change readiness (unaware of or apathetic about the health threat, considering change, preparing to change, action, and behavior maintenance) for your profiled audience members and describe ideas for moving them to the next stage.
- Develop and test messages using the above research and using communication theories that address persuasion and behavior change.
- Deliver the message and measure belief and/or behavior change results.

The University of Toronto's Health Communication Unit promotes the following tips for health message development, also based on communication research:

- **Capture and maintain the audience's attention.** The more you can engage the audience to think about the message, the more likely that they are to change knowledge attitudes and behaviors. Consider using emotionally involving scenes, vivid visuals, and lively language.

- **Give the strongest points at the beginning of the message.** This is the information most critical for convincing the audience to adopt the recommended behavior.
- **Have a clear message.** People should easily understand the actions you are asking them to take and the incentive for taking them.
- **Specify a reasonably easy action.** Instead of telling people to stop smoking, which many people will ignore as unattainable, you could ask them to sign a pledge card or commit to a smoke-free week, or give them tips for the first day of smoking cessation. It also helps to show role models demonstrating the desired behavior.
- **Use incentives effectively.** Use a variety of incentives, including physical, economic, psychological, social, and moral. Make sure that the audience cares about the incentives and thinks that they are likely to occur if the behavior changes.
- **Provide good evidence for threats and benefits.** People who are already interested tend to respond to expert quotes, documentation, and statistics. People who are not involved are more likely to respond to dramatized case examples and testimonials.
- **Use believable messages.** Do not make extreme claims or use extreme examples.
- **Use an appropriate tone for the audience and topic.** A serious tone is the safest, but do not preach or dictate. Some audiences may respond to a light, humorous, ironic, or dramatic tone.
- **Use an appropriate appeal for the audience.** Consider rational appeals for audiences already interested in the topic and emotional ones for the apathetic.
- **Do not offend.** Do not blame the victim for unhealthy behavior. Help people overcome their environments instead.
- **Display the organization's identity prominently with each message.** Identity elements could include an organization's name, a positioning statement or platform, a logo, a slogan, and, sometimes, an image. Identity elements that help people remember and link the campaign messages over time.
- **Choose messengers who are viewed as credible sources of information by the intended audiences.** Messengers are those who deliver information, demonstrate behavior, or provide a testimonial. Messengers could range from celebrities to public officials to victims to successful role models. Messenger credibility is enhanced by perceived expertise and honesty, as well as being viewed as similar to the target audience.

One approach to communicating health risk communication messages is commonly called single overriding communication health objective (SOCHO). The idea is that the risk should be boiled down into a single most important message and that the message should be repeated until the audience takes some action (stops smoking, joins a community advisory group, evacuates, etc.). Knowing the most important message can help to keep a risk communicator focused, particularly during face-to-face interactions (see Chapter 15) and interactions with the news media (see Chapter 16). It can also help determine the central intended message. Unfortunately, the SOCHO approach has two main drawbacks:

1. It is difficult to distill complex situations into a single message.
2. Flogging the audience with a single message incessantly will not answer questions, resolve issues, or further the goal of supporting continued sharing on all sides

(Bennett and Calman 1999). Indeed, many members of the news media now recognize the practice and will only push harder for another answer.

It is better to understand what your audience wants to know and craft a set of messages that cover all aspects of the risk than to rely on a single message to see you through.

CRISIS COMMUNICATION

In crisis communication, the tone should create a sense of urgency to take action when appropriate and reassure people that answers are being sought, without confusing or alarming them. In some crises, the danger is known in advance (for example, in a flood or hurricane). For example, the UK Environmental Agency (No Date) found that flood warning messages were more effective if they were tailored to local situations, centered on response, and considered people with special needs.

> In the absence of consistent, trustworthy messaging from government authorities, members of the public may act in ways that put them in harm's way.
> —Center for Biosecurity (2012, p. 3).

Dennis Mileti, Professor Emeritus at the University of Colorado at Boulder, and his colleges at the National Hazards Research and Applications Information Center, have conducted research on how people heed warnings of impending danger. Looking across warning research, regardless of hazard, he found that audiences need five things from message content: information on hazard, location, guidance on actions to take, timing, and source (the organization providing the warning). Warning messages were particularly effective when they were:

- Specific
- Consistent
- Certain
- Clear
- Accurate
- Sufficient (contained enough details)
- Repeated frequently (Mileti and Peek 2000)

More information was always better. Audiences crave information in a crisis.

The Centers for Disease Control and Prevention et al. (2003) recommend the following process for message development during a crisis, especially one involving an infectious disease or other public health issue:

- **Describe the audiences you want to reach.** This includes their relationship to the event, their demographics (age, language, education, culture), and their level of outrage, based on risk communication principles.
- **Define the purpose of the message(s).** Purposes could include giving facts and updates, rallying people to action, clarifying event status, addressing rumors, and responding to media requests.
- **Identify the delivery method.** This could include media, the Internet, a spokesperson, recorded phone messages, public meetings, and other methods.

Once you have identified the audience, message purpose, and delivery method, construct messages using the following components:

- Expression of empathy
- Facts and/or call to action
- What is not known and the process needed to get answers
- Statement of commitment from the organization
- Referrals to others for more information
- Next scheduled update

Keep the content easy to understand and free of jargon. Convey what science knows and does not yet know about the risk.

Experiences after the Oklahoma City bombing and other mass casualty events show that people's ability to process information decreases significantly in the face of tragedy. In such cases, messages need to be simple and few in number, and any verbal messages must be reinforced in writing (Blakeney 2002). In addition, use repetition and include relevant and practical actions to help people understand what they should do to protect themselves and their loved ones. What people most want is actionable advice (Center for Biosecurity 2012).

In a media or hotline situation, spokespeople and other organization representatives should practice answering questions that reflect the messages. It is especially important to make sure that everyone is "singing off the same page" when more than one agency or organization is involved because receiving conflicting information from various agencies confuses people and diminishes trust. For more information about responding to questions in crises, see Chapter 21.

CHECKLIST FOR DEVELOPING MESSAGES

When developing and delivering messages:

- ☐ The purpose of the message has been identified.
- ☐ The appropriate approach for care, consensus, or crisis communication has been chosen.
- ☐ Audiences and their knowledge, beliefs, concerns, and barriers to action have been analyzed, with their input.
- ☐ Special attention has been paid to misperceptions and knowledge gaps, as well as beliefs to be reinforced.
- ☐ Message content and delivery are based on an understanding of the audience and are pretested.
- ☐ Evaluation methods are in place to determine the effectiveness of messages and modify them if necessary.

REFERENCES

Bennett, P. and K. Calman. 1999. *Risk Communication and Public Health.* Oxford University Press, New York.

Blakeney, R. L. 2002. "Providing Relief to Families after a Mass Fatality: Roles of the Medical Examiner's Office and the Family Assistance Center." *OVC Bulletin*, November 2002.

Center for Biosecurity. 2012. *After Fukushima: Managing the Consequences of a Radiological Release*. University of Pittsburgh Medical Center, Baltimore, Maryland.

Centers for Disease Control and Prevention, Agency for Toxic Substances and Disease Registry, Oak Ridge Institute for Science and Education, and The Prospect Center of the American Institutes of Research. 2003. Emergency Risk Communication CDCynergy (CD-ROM). http://www.orau.gov/cdcynergy/erc/ (accessed January 23, 2013).

Fischhoff, B. and J. Downs. 1997. "Accentuate the Relevant." *Psychological Science*, 8(3):1–5.

Green, J. A., N. K. Choudhry, E. Kilabuk, and W. H. Shrank. 2010. "Online Social Networking by Patients with Diabetes: A Qualitative Evaluation of Communication with Facebook." *Journal of General Internal Medicine*, 26(3):287–292.

Kahlor, L., S. Dunwoody, R. J. Griffin, K. Neuwirth, and J. Giese. 2003. "Studying Heuristic-Systematic Processing of Risk Communication." *Risk Analysis*, 23(2):355–368.

Leiss, W. and D. Powell. 2004. *Mad Cows and Mother's Milk: The Perils of Poor Risk Communication*, 2nd ed. McGill-Queen's University Press, Montreal, Quebec, Canada.

Lion, R., R. M. Meertens, and I. Bot. 2002. "Priorities in Information Desire about Unknown Risks." *Risk Analysis*, 22(4):765–776.

Mileti, D. S. and L. Peek. 2000. "The Social Psychology of Public Response to Warnings of a Nuclear Plant Accident." *Journal of Hazardous Materials*, 75(2000):181–194.

Morgan, M. G., B. Fischhoff, A. Bostrom, and C. J. Atman. 2002. *Risk Communication: A Mental Models Approach*. Cambridge University Press, Cambridge, United Kingdom.

Petts, J., S. McAlpine, J. Homan, S. Sadhra, H. Pattison, and S. MacRae. 2002. Development of a Methodology to Design and Evaluate Effective Risk Messages; Electroplating Case Study. Prepared by the University of Birmingham for the Health and Safety Executive of Great Britain. Edgbaston, Birmingham, U.K. Contract Research Report No. 400.

Tucker, W. T., S. Ferson, A. M. Finkel, and D. Slavin, eds. 2008. "Strategies for Risk Communication: Evolution, Evidence, Experience." *Annals of the New York Academy of Sciences*, 1128:1–7.

UK Environmental Agency. No Date. "Improving Institutional and Social Responses to Flooding." Science Summary SCHO0509BQBM-E-P. UK Environmental Agency, London.

Witte, K., G. Meyer, and D. Martell. 2001. *Effective Health Risk Messages: A Step-by-Step Guide*. Sage Publications, Thousand Oaks, California.

ADDITIONAL RESOURCE

New York State Health Department. No Date. "How to Craft Effective Health Messages for People with Disabilities." http://www.health.ny.gov/publications/0957/ (accessed January 31, 2013).

10

DETERMINE THE APPROPRIATE METHODS

You know what you are trying to communicate and why and to whom you will be sharing the risk information. Now you need to decide how to communicate. Which methods of communication will best meet both your purpose and objectives and your audience's needs? (Usually, no one method will meet the needs of every segment of your audience.) The basic categories to choose from include information materials, visual representation of risk, face-to-face communication, working with the media, stakeholder participation, social media, and technology-assisted communication.

How each of these relates to the purpose and objectives and audience's needs is discussed later in this chapter, as well as the time involved and how much technical knowledge is needed to effectively share risk information. More information on using each of the methods can be found in Part III.

INFORMATION MATERIALS

Information materials are those that your audience will need to read and are generally printed. These materials may have pictures and other graphical elements, and they range in size from a partial-page advertisement to a multivolume environmental impact statement. Examples include newsletters, fact sheets, brochures, booklets,

> Information materials are those that your audience will need to read and are generally printed. Examples include newsletters, fact sheets, brochures, booklets, pamphlets, displays, advertisements, posters, trade journal articles, popular press articles, and technical reports.

Risk Communication: A Handbook for Communicating Environmental, Safety, and Health Risks,
Fifth Edition. Regina E. Lundgren and Andrea H. McMakin.
© 2013 The Institute of Electrical and Electronics Engineers, Inc., and Regina E. Lundgren and
Andrea H. McMakin. Published 2013 by John Wiley & Sons, Inc.

pamphlets, displays, advertisements, posters, trade journal articles, popular press articles, and technical reports.

Information materials have the advantage of being able to include a wealth of information. They also can be expanded or condensed to meet audience needs (for instance, a technical report can be condensed into a journal article or news release, depending on the audience). Information materials used to be some of the most inexpensive to produce, but rising paper and distribution costs can make the widespread availability expensive, unless you opt for delivering the materials electronically.

However, information materials can be a comfortable form of communication for some members of the audience; they are more familiar than a computer station and allow the reader to carry away something for later reference. So, if your objectives are to disseminate a large amount of risk information and to economically meet the needs of various segments of your audience, then information materials may be a good choice.

On the other hand, some information materials can be difficult for certain members of the audience to comprehend. When writing such messages, it is too easy to resort to jargon and overly technical language ("if they don't understand it, they can look it up"). In addition, the length of information materials (either too short or too long) can deter some readers. Then, too, many people today cannot find time to read or no longer seek information in printed materials. So, if your purpose is to raise awareness of an issue or to communicate with people who have difficulty reading, information materials may not be the best choice.

Depending on their length, the amount of research needed, the approval process, and the method of printing, information materials can usually fit into any schedule. For broad, quick dissemination, newspaper articles are excellent because newspapers have a large readership and generally are published daily; however, you often do not control content or timing (see Chapter 16 for additional information). Fact sheets and pamphlets can also usually be prepared relatively quickly. For a risk that will take some time to resolve and that includes audience involvement throughout, such as the cleanup of a waste site, a newsletter may be a good choice, supplemented by a variety of other methods.

To prepare any form of information material, some technical knowledge is necessary. A technical writer can prepare the information and have experts review it for technical accuracy, or the experts can prepare the message and have a technical communications specialist or risk communicator review it to ensure that it meets the audience's needs. Although there may be a fine line between being overly technical and overly simplistic, it is possible to present technical information in a way that the public can understand. Lay readers may not understand the information in the same way that an expert with 35 years of experience in the field understands it, but they can understand it at a level that allows them to make an informed decision about the risk. See Chapter 20 for advice on how to pretest such materials to ensure that they meet audience needs.

VISUAL REPRESENTATION OF RISK

Risk can also be communicated through the use of graphical elements and relatively little text to carry simple risk messages. Examples include posters, displays, direct advertising, videos, and television; however, virtually all forms of risk communication make some use of pictorial representation of risk.

Visual representations have the advantage of being memorable. Think of traffic safety signs. The use of graphical elements like color, shape, and imagery along with compelling language can bring simple risk messages to life with stunning clarity. Visuals may be culture specific; however, because they contain very little written information, they can usually be more easily translated into another language than other information materials. They can be placed where your audience lives, works, and plays: on television in programs or commercials, on posters in work cafeterias, and on buses and bulletin boards. So, if your purpose is to raise awareness, visual representations may be the best choice.

> Risk can also be communicated through the use of graphical elements and relatively little text to carry simple risk messages. Examples include posters, displays, direct advertising, videos, and television; however, virtually all forms of risk communication make some use of pictorial representation of risk.

Visual representations, by their very nature, can carry only limited information. Therefore, they cannot answer as many of the questions audiences may have about the risk as some other forms of risk communication. Because the style of these pictures is sometimes associated with persuasive communications (like product advertisements), they may fail to attract or may even put off certain elements of the audience ("oh, it is just more hype"). If they are overused, they lose their impact and tend to be ignored. Even television programming and online videos, which can overcome some of these problems, are often viewed as merely entertainment so that the risk messages gets lost among the commercials for crunchy munchies or the latest viral dance video. So, if your purpose is to inform your audience, visual representations cannot be your only choice of method.

> Visual representations, by their very nature, can carry only limited information. Therefore, they cannot answer as many of the questions audiences may have about the risk as some other forms of risk communication.

Visual representations can take some time to produce. Although organizations that regularly produce public service announcements for television can turn them out in a few weeks, those of us with more limited staff and budgets often cannot match this schedule. Coordinating the production of graphics or video can take a significant amount of time, when you are trying for quality. (Certainly, many online videos are shot more simply.) Nonetheless, to meet your audience's needs and your purpose and objectives, you may want to investigate how much time and money is necessary for the production of visual messages. Contact your public affairs or communications departments, or look in the telephone directory under graphic artists, photography, or advertising services, to get cost and time estimates.

Even though the technical message is usually somewhat limited, some technical knowledge is necessary to produce most visual representations. To ensure that the message is technically correct and does not imply something that is incorrect, have people with technical backgrounds review the message and graphics. Knowledge of graphic design, however, is critical to ensure that these messages carry the impact intended. If you do not have a background in graphic design, contact your organization's art department or consult the telephone directory under advertising, graphic art services, or graphic designers. See Chapter 14 for more information on the visual representation of risk.

FACE-TO-FACE COMMUNICATION

Face-to-face communication involves someone speaking directly to the audience or listening while the audience speaks. Usually, the audience and the speaker do not interact, except perhaps to ask questions. (Cases in which the audience can interact are classified in this book as stakeholder participation.) Examples include one-to-one discussions such as between a doctor and a patient or between employees; presentations to clubs, societies, and citizens groups, whether as a single speaker or as part of a speakers bureau; talks in educational settings such as grade school, college classes, or training courses; tours and demonstrations; video; audience interviews; and information fairs. Face-to-face communication has the advantage of having an identifiable human representative of the organization or another credible person presenting the risk information, thus personalizing it. (Of course, if the presenter lacks credibility, the presentation can have a negative effect!) Face-to-face communication generally offers the opportunity for immediate audience feedback, if not through questions then through the audience's visible reaction to certain statements. Some audience analysis information is generally available beforehand to the presenter (who can ask questions of the person arranging or hosting the presentation), allowing each presentation to be tailored. You can also target specific groups to receive an oral message, whereas you may have no way of knowing whether the people to whom you sent a written message ever read it. Face-to-face communication can also be presented in the language of the audience. So, if your objectives are to present information in a forum that allows immediate feedback and to target specific groups, face-to-face communication may be your best choice.

> Face-to-face communication involves someone speaking directly to the audience or listening while the audience speaks. Usually, the audience and the speaker do not interact, except perhaps to ask questions.

On the other hand, face-to-face messages can also be easily misunderstood. Audiences may be too overwhelmed or hostile to ask questions that would clarify misunderstandings. Particularly angry audiences can turn a presentation into a political forum and generally refuse to listen. Oral presentations alone also give the audience nothing to refer to later. So, if you have a particularly angry audience, or one that needs long-term information, face-to-face communication may not be satisfactory or sufficient.

Face-to-face communication can fit into nearly any schedule. For quickly disseminating information, a press conference or radio announcement is useful. For a longer-term effort, a continuing series of presentations to a variety of organizations and societies might serve to reinforce the message and keep the audience up to date. Another way to reinforce face-to-face communication is to encourage health care professionals to disseminate risk information to their clients at risk.

Advanced technical knowledge is usually necessary to present oral risk information convincingly. However, the ability to speak in a manner that entertains as well as informs is also necessary. No matter how well educated, an expert who has trouble speaking before groups is a poor choice for spokesperson. So is a professional speech maker with only superficial knowledge of the risk. The best person to present oral risk information is one whom the audience will find credible and whom the organization responsible for communicating risk finds an acceptable representative. For more information on choosing a spokesperson, see Chapter 15.

WORKING WITH THE NEWS MEDIA

Mass media methods usually involve the use of sources such as television, newspapers, radio, magazines, and the Internet to communicate risk information to broad audiences. Such sources can be powerful because they reach large audiences and can be memorable and credible for many people. Television, radio, and the Internet are particularly useful sources in crisis communication situations in which people need continuously updated information quickly.

A key disadvantage of mass media is that, except in paid advertisements and other limited circumstances, the media source controls the content and timing of the story. Because of time and space constraints and the missions of media organizations, aired or published stories on risk-related issues may not contain the emphasis or depth of information that those who communicate risk would like to see. Thus, the media should not be relied upon as the sole source of information in planning risk communication efforts.

> Mass media approaches usually involve the use of sources such as television, newspapers, radio, magazines, and the Internet to communicate risk information to broad audiences. Such sources can be powerful because they reach large audiences and can be memorable and credible sources for many people.

Because of its wide reach and powerful impact, the use of mass media should be considered carefully as risk communication plans are being prepared. Even a small amount of negative coverage can torpedo the best-planned effort and destroy trust and credibility among the audiences and participants you are trying to reach. On the other hand, productive relationships with media representatives can lead to a more informed, solution-oriented public.

Schedules for planning and implementing media interactions and products can vary widely. On one hand, it may take only a half hour to talk with a reporter about a specific topic and perhaps follow up by sending background materials. In a crisis situation, you are usually working within a short time frame to prevent or mitigate a problem. The time involved in a crisis communication situation may be driven in part by a prescribed series of steps designated in company procedures. The unplanned nature of a crisis, however, makes timing unpredictable. Media follow-up after a crisis event, especially if an organization was perceived as mishandling the situation, may require much more time than it took to communicate during the actual crisis.

A formal event such as a press conference may require several days to plan and coordinate. Working cooperatively with a local newspaper to communicate about an issue that affects the community may span several weeks or months. Creating your own media messages for a public health campaign usually requires weeks or months of coordination and production. See Chapter 23 for more information on public health campaigns.

Developing mutually productive working relationships with reporters is an ongoing effort for many organizations. Though this means vigilantly contacting media representatives to update them on breaking news or to suggest story ideas at appropriate times, the resulting quality of the media coverage often is well worth the time involved.

Many organizations require approvals of internal management when employees are working with external media professionals. When planning media interactions, include appropriate time up front for approvals and review of any information materials you plan to give to the media. Be aware that additional time may be necessary to search for or revise materials to be suitable for a particular medium. For example, newspapers may want a "people" photo rather than a technical illustration. Television reporters may ask if you can provide high-quality video clips or, more likely, ask to shoot their own footage.

Several kinds of costs are involved in media interactions, depending on the type and duration of the activity. First are labor costs. Often, a specialist trained in media relations is needed to plan and coordinate efforts, including creating media plans. There is also the time of those who are talking with media representatives. Production costs are necessary for creating materials such as press releases, press kits, advertisements and public notices, photos, and video footage. Formal events such as press conferences may require room and equipment rental.

Working with the media does not have to be expensive. Even a small organization can communicate important risk information to an audience of thousands through one radio, television, or newspaper interview, for only the few minutes it takes to speak with a reporter. The key in budgeting for media interactions, as in other types of communication activities, is to define your goals and target specific activities to meet them. Media specialists or consultants can help find the greatest value within a budget that you specify. Chapter 16 gives examples of ways to work with media representatives in specific situations.

Those who are communicating directly with media representatives should have at a minimum a basic technical knowledge of the risk situation so that they can answer technical questions with accuracy and credibility. Equally important, however, are an understanding of media organizations and the ability to use language that the media outlet's audience will understand and relate to. When the person speaking is officially representing the company's position, follow the guidance in the section "Choose the Appropriate Spokesperson" in Chapter 15. Chapter 16 provides guidance in understanding media practices and in using appropriately targeted language.

Those who are producing materials for media use should have the professional skills needed to create high-quality products for the target medium. A person with subject-matter knowledge of the issue being portrayed should review the materials for technical accuracy.

STAKEHOLDER PARTICIPATION

Stakeholder participation involves the audience in some way in the discussion, analysis, or management of the risk. Examples include advisory committees, facilitated deliberation, alternative dispute resolution, focus groups, community-operated environmental monitoring, and formal hearings in which the audience is invited to give testimony.

> Stakeholder participation involves the audience in some way in the discussion, analysis, or management of the risk. Examples include advisory committees, facilitated deliberation, alternative dispute resolution, focus groups, community-operated environmental monitoring, and formal hearings in which the audience is invited to give testimony.

The advantage of stakeholder participation is that the audience can see for themselves exactly what is known about the risk, how the risk will be managed, and how decisions are reached. Because they can participate in the risk decision, it is likely to be more acceptable and lasting. Stakeholder participation can be structured to accommodate a variety of audiences, including those that are hostile or have difficulty reading or understanding other forms of communication. So, if one of your objectives is to increase the chances that your risk decision will be one that meets the needs of the audience, stakeholder participation may be the best choice.

However, stakeholder participation can be a frightening proposition to some risk managers. They fear the loss of control over the risk decision, instead of seeing that the audience's input can be invaluable to a lasting, equitable decision. If there is no commitment to stakeholder participation from those who are analyzing the risk, managing the risk, or making the decision, the effort can be devastating to an organization's credibility and hamper any future risk communication or management efforts. Stakeholder participation is generally more costly than simply issuing a technical report or holding a press conference. So, unless your organization is completely committed to letting the audience interact in a way that is meaningful to that audience, stakeholder participation is a very poor choice. See Figure 10-1 to determine whether stakeholder participation can be used effectively in your situation.

Stakeholder participation is usually a long-term proposition. Unless the structure for interaction is already functioning (such as an advisory committee), one cannot be put in place quickly enough to release urgent information. Stakeholder participation can usually only be used effectively when the risk management and risk communication effort will occur over time.

Little technical knowledge about the risk is required to set up stakeholder participation; however, technical staff and management must participate for the interaction to have meaning. In addition, knowledge about stakeholder participation (sometimes called public involvement) is necessary to structure the interactions effectively. Groups such as the International Association for Public Participation evolved to develop and disseminate information concerning stakeholder participation. See Resources at the back of this book for contact information. See Chapter 17 for additional information on involving stakeholders in risk efforts.

TECHNOLOGY-ASSISTED COMMUNICATION

Technology-assisted communication uses technology, often computer based, to discuss or disseminate risk information, or allow a member of the audience to query and receive a variety of information about the risk. For example, one software application allows audience members to evaluate a number of factors to help experts identify the technologies that are more acceptable to the public for cleaning up a waste site.

> Technology-assisted communication uses technology, often computer based, to discuss or disseminate risk information, or allow a member of the audience to query and receive a variety of information about the risk.

Technology-assisted communication has the advantage of being able to disseminate an incredible amount of information, which members of the audience can tailor to their individual needs. It appeals especially to the "technophiles," those among us who always have to have the latest toys and gadgets technology has to offer. Once developed, technology-assisted communication can be updated and revised more easily than materials developed through any of the other risk communication methods so that it is always current, a plus in the area of risk, which can change hourly. If graphic elements are properly built in, these applications can be as eye-catching as full-color ads or displays, yet carry as much information as traditional

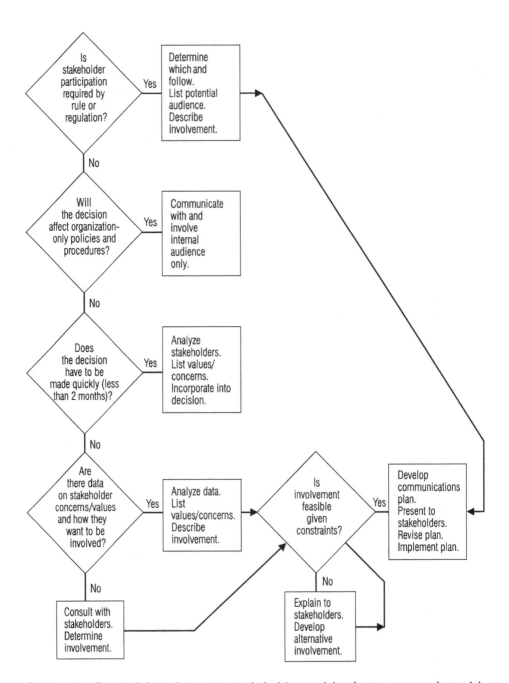

Figure 10-1. Determining when to use stakeholder participation to communicate risk.

information materials or even more. Speeches, video, animations, and other multimedia elements can be incorporated, making the applications incredibly versatile. So, if your objectives are to allow people to see all the data and develop their own perceptions of the risk, to disseminate information quickly, or to cost-effectively involve your audience, technology-assisted communication may be a good choice.

However, technology-assisted communication has several disadvantages. Those applications that must run on a fairly sophisticated computer make mass dissemination impractical. In addition, computer use in the United States is still less than universal, making it difficult to reach all audiences with a computer-based application. So, unless your audience has ready access to the appropriate devices and software, your may not want to rely on technology-assisted communication as your primary method.

Technology-assisted communication is becoming easier to produce every day. However, some applications still cannot be developed in time to meet the goals of short-term schedules. In addition, unlike information materials that are eventually used up, these applications must be kept up to date or audiences will lose interest.

Some applications, such as websites, require relatively little technical knowledge of computers to develop. Depending on the purpose of the site, little risk information is needed as well (for example, setting up a computer bulletin board to log in stakeholder comments on an environmental impact statement). However, a computer application that incorporates risk modeling requires a great deal of technical knowledge, not only about the risk but also about the application. Often, such models are developed by teams, with each member specializing in a particular area such as technical communication, computer programming, and risk assessment. If your organization does not have such experts, contact colleges and universities for help. See Chapter 18 for additional information.

SOCIAL MEDIA

Social media involves using the Internet to share opinions, thoughts, and other information via text, graphics, and video on a risk found relevant to the audience. For example, teens facing diabetes might join a social media network group to share diet and exercise tips and support each other in meeting health goals.

> Social media involves using the Internet to share opinions, thoughts, and other information via text, graphics, and video on a risk found relevant to the audience.

Using social media to communicate risk has several advantages. Because the audience chooses to engage in a conversation about the risk, their level of interest is high, and for care and crisis communication, that interest might translate into a willingness to change behavior that improves health and safety. Social media also has the expectation from the audience's side that it will be immediate. Information can be posted quickly and updated as needed. Feedback is also instant, and changes in opinions, as measured by topics of conversation, are relatively easy to track over time. So, if your objectives are to engage your audience in conversation and to determine how their perceptions shift, then social media could be a good choice.

While many tout social media as the perfect mechanism to share information, the medium has some drawbacks. The audience generally has to find your page or profile, and research has yet to confirm whether those on social media actually seek health and safety information there. That audience also has the growing expectation that a page is always "on." That is, if your organization has a presence on a social media site, someone will always be available to answer questions or other engage in conversation. On the other hand, not everyone can access social media, and certain demographics are less likely to use the sites than others. In addition, the fact that the information is in the audience's control can concern some organizations charged with communicating risk. So, if your audience does not seek information about your risk in social media or your organization is unwilling to make the commitment to the medium, then social media may be a poor choice.

Social media looks deceptively easy. Most accounts can be set up in minutes. But extra time is needed before joining the sites to ensure that organizational policies and procedures are in place to maintain the effort. In addition, social media requires maintenance on a daily basis, and once the engagement starts, there may be an expectation from the audience that it will never end.

Little technical knowledge is needed to interact with the audience on social media; however, being able to tap into a wealth of information from a variety of disciplines can be very helpful to address audience questions and position your organization as an expert on your risk. In addition, not everyone is comfortable representing their organization on social media. Poor social skills will be visible immediately, alienating the audience. You may need to look beyond those traditionally tasked with risk communication to find the right person to lead the social media engagement. High school and college interns who themselves are frequent users can sometimes be encouraged to take part with appropriate training in the organization's expectations and requirements. For more information on using social media to communicate risk, see Chapter 19.

CHECKLIST FOR DETERMINING APPROPRIATE METHODS

If the purpose of the risk communication effort was:

☐ To increase the audience's awareness of the risk, the methods considered included:
  ☐ Visual representation of risk
  ☐ Face-to-face communication
  ☐ Technology-assisted communication
☐ To inform the audience, the methods considered included:
  ☐ Information materials
  ☐ Face-to-face communication
  ☐ Technology-assisted communication
  ☐ Social media
☐ To build consensus between the audience and the organization assessing or managing the risk, the methods considered included:
  ☐ Stakeholder participation
  ☐ Social media
☐ To change behavior for the risk, the methods considered included:
  ☐ Information materials
  ☐ Face-to-face communication
  ☐ Visual representation of risk
  ☐ Social media

ADDITIONAL RESOURCES

Sachsman, D. B., M. R. Greenberg, and P. M. Sandman, eds. 1988. *Environmental Reporter's Handbook.* Rutgers University, Cook College, Environmental Communication Research Program, New Jersey Agricultural Experiment Station, New Brunswick, New Jersey.

Sandman, P. M., D. B. Sachsman, and M. R. Greenberg. 1988. *The Environmental News Source: Providing Environmental Risk Information to the Media.* New Jersey Institute of Technology, Hazardous Substance Management Research Center, Risk Communication Project, Newark, New Jersey.

Santos, S. L. 1990. "Developing a Risk Communication Strategy." *Management and Operations*, November, pp. 45–49.

11

SET A SCHEDULE

Once you know what you hope to accomplish, who you are doing it for, and how you will do it, you need to determine when you will do it. Setting a schedule for risk communication efforts requires that you consider a number of factors, such as legal requirements, organizational requirements, the scientific process, other ongoing activities, and audience needs.

LEGAL REQUIREMENTS

Legal considerations are usually the first item that must be considered if the organization conducting the risk communication effort is to avoid litigation. As discussed in Chapter 3, many risk communication efforts are the result of laws, which usually prescribe schedules as well. One of the better delineated schedules can be found in the Superfund laws and associated advice from the U.S. Environmental Protection Agency (see Figure 3-1 in Chapter 3; EPA 1992).

> Legal considerations are usually the first item that must be considered if the organization conducting the risk communication effort is to avoid litigation.

Federal, state, and local laws and regulations may specify schedules for risk communication milestones. Consult your legal department, a law firm, or a local risk-related government agency to determine which legal requirements apply in your case.

Risk Communication: A Handbook for Communicating Environmental, Safety, and Health Risks,
Fifth Edition. Regina E. Lundgren and Andrea H. McMakin.
© 2013 The Institute of Electrical and Electronics Engineers, Inc., and Regina E. Lundgren and
Andrea H. McMakin. Published 2013 by John Wiley & Sons, Inc.

ORGANIZATIONAL REQUIREMENTS

After legal requirements have been determined, consider the requirements of your organization. What kinds of reviews will be necessary for any risk communication message to be sent? How long will each review take? Make a calendar for yourself so that you can follow your way through these requirements and plan for them. Look for requirements such as the following:

- **Scientific review.** To verify that the information about the risk is correct and the most current available
- **Editorial review.** To make sure that the way the information is presented will be understood by the audience
- **Management review.** To make sure that the information being presented is what the organization wants to present
- **Sensitivity review.** To make sure that the information being presented does not compromise national security or business interests
- **Legal review.** To make sure that the information does not compromise the organization relative to a law with which the organization must comply
- **Patents review.** To make sure that the information being presented does not give away intellectual property and make it difficult for the organization to patent a certain idea or device
- **Public affairs review.** To make sure that the information will not embarrass the organization

The purpose and objectives you have developed for your risk communication efforts also affect the schedule. What frequency or duration is needed to achieve this purpose? If your purpose is to raise awareness, a short-term burst of activity that focuses on several methods would provide the most visibility. If your purpose is to build consensus so that your audience and the organization analyzing or managing the risk can come to a decision, you may want to conduct the activities over a longer time, providing different information as it is needed in the consensus-building process.

> The purpose and objectives you have developed for your risk communication efforts also affect the schedule.

THE SCIENTIFIC PROCESS

Your communication efforts will most likely have to coincide with certain aspects of the scientific process. Once a risk has been identified, you must alert those at risk, then follow up with more information as the assessment continues. While those who are assessing the risk may not be able to give you exact dates when they will know certain information, they should have a schedule showing what steps will take place. Make sure that you are aware of that schedule and any changes to it as you plan (and update) your risk communication schedule.

> Your communication efforts will most likely have to coincide with certain aspects of the scientific process. Once a risk has been identified, you must alert those at risk, then follow up with more information as the assessment continues.

ONGOING ACTIVITIES

The next item to consider is what else is going on within your organization, the community, and the nation, so that you can put your risk communication efforts in context. What other kinds of information will be released from your organization at the same time you are planning to release your risk communication information? For example, if the organization plans to announce the results of a 10-year study on cancer-causing chemicals, you may want to delay releasing your information about one of the chemicals so that you can incorporate the latest data. On the other hand, if the results of the 10-year study confirm your information on risk, you may want to bring your message out first and use the study as reinforcement.

What will be happening in the community during your risk communication efforts? In one case, we were trying to determine when to hold the first public meeting near a Superfund site in central Alaska. We hoped for a certain week in October, only to find that that week was part of moose-hunting season, an extremely popular event for the community. So, even though we did not want to wait to release information, we tried for another time to meet the needs of our audience and to make our efforts more effective.

What will be happening in the nation during your communication efforts? This is much harder to predict. However, if your organization is dependent on federal funding, and an election in which your funding is an issue is imminent, you should wait until after the election to promise funds to a citizens group to act as advisors in your risk assessment, management, and communication activities. Although other national trends are harder to foresee, try to at least determine your organization's agenda for the next few months at the national level.

AUDIENCE NEEDS

To determine audience needs, consider the information from your audience analysis and the timing and severity of the risk. If you are in a crisis situation, obviously, the audience needs as much pertinent information as possible as soon as possible. They need straightforward answers to such questions as:

> To determine audience needs, consider the information from your audience analysis and the timing and severity of the risk.

- What happened?
- How dangerous is it?
- How could it affect me or people or things I care about?
- What can I do?

If the risk is a longer-term one, such as those related to the cleanup of a hazardous waste site, your schedule will be longer and more complex. You will need to answer the same types of questions, but with more detail and in a variety of ways. The answers are likely to change over time as more information becomes available, so you will need to continue to issue messages. (This happens during a crisis as well, but on a much smaller scale and in a compressed time frame.) See Chapter 9 on more on the kinds of information audiences need to make an informed decision about risks in care, consensus, and crisis communications.

What your audience knows about the risk can also affect your schedule. An audience first needs to be aware of a risk and its relevancy to them before being willing to take any action associated with the risk. Many theories have been proposed as to the process by which people at risk take in information about the risk and decide to act (for example, by moving to more healthy behaviors, partnering with others to develop a more effective risk management approach, or taking immediate protective actions for care, consensus, and crisis communications, respectively). Earlier versions of this risk communication handbook described decision processes that have since been overcome by newer models, and those models are likely to be overcome again before the next edition.

For risk communicators, one of the important considerations is that the audience's process of choosing to take some action about a risk may take minutes to years, depending on the individual and the circumstances. For care communication, the process can take weeks to months, unless the health or safety issue becomes more obviously urgent to the audience. For consensus communication, the process can take months to years as audience members learn more about the situation and become comfortable interacting with other stakeholders. For crisis communication, it can take as little as a few minutes to make the decision to evacuate, particularly when audiences have been involved with emergency planning efforts. For less aware and less involved members of the audience, or members of the audience who choose to ignore risk management advice for ethical or religious reasons, actions may be taken too late or not at all.

The other important consideration is that the audience's information needs and the methods of meeting those needs change as audience members move along the continuum from awareness to action. Visual representations, face-to-face communication, information in the news media, and engagement on social media in short, concentrated bursts can help raise awareness. Face-to-face communication, social media, and stakeholder participation can increase trust in the organization and confidence in the risk assessment. Information materials and technology-assisted communication along with additional engagement of the audience through stakeholder participation or social media can share more information as audiences begin to look for choices in managing the risk.

To determine where your audience sits along the decision spectrum, conduct thorough audience analysis, pretest risk messages, or otherwise consult with audience representatives.

REFERENCE

EPA (U.S. Environmental Protection Agency). 1992. *Community Relations in Superfund: A Handbook*. U.S. Environmental Protection Agency, Office of Emergency and Remedial Response, Washington, DC. EPA/540/G-88/002.

CHECKLIST FOR SETTING SCHEDULES

The schedule is based on:

☐ Legal requirements associated with the risk communication effort
☐ Number and timing of organizational reviews
☐ Purpose of the risk communication effort
☐ Objectives of the risk communication effort
☐ Schedule of the risk assessment
☐ Activities within the organization
☐ Activities within the community
☐ Activities within the nation
☐ Audience's point in the decision process
☐ If the purpose of the risk communication effort is to increase awareness, the schedule is based on a concentrated burst of information, factoring in other constraints and audience needs.
☐ If the purpose of the risk communication effort is to inform the audience or change behavior, the schedule allows for the:
    ☐ Introduction of the risk
    ☐ Additional information given over time

If the purpose of the risk communication effort is to build consensus, the schedule allows for the dissemination of risk information in support of the consensus-building process:

☐ Before activities
☐ During activities
☐ After activities

12

DEVELOP A
COMMUNICATION PLAN

Now that you have determined your purpose and objectives, analyzed your audience, developed your messages, chosen your methods, and set your schedule, you need to put them all together into a comprehensive plan. Why not just keep the information in your head, notes on your desk, or a file in your computer? There are several reasons:

- A written plan is less likely to be lost than miscellaneous notes or files.
- At some point in the risk communication process, someone either inside or outside the organization may challenge your methods or approach. Having a comprehensive plan is a good defense.
- Having a formal plan that has been accepted (through signatures if necessary) by management can be handy in setting priorities and in getting timely approvals for specific activities.
- Because formal plans are more organized, easier to review, and imply a more formal effort, the work they describe is more likely to receive the necessary funding and resources than those in nebulous plans.
- It is easier to evaluate the results of formal plans because you can relate results directly back to your purpose and objectives, schedule, and audience.

> Information management and communication should be part of planned design and execution, and be integral to an organization's risk and disaster management plans. Improvised communication can be costly and have unsatisfactory results.
> —S.A. Barrantes et al. (2009, p. 8).

Risk Communication: A Handbook for Communicating Environmental, Safety, and Health Risks, Fifth Edition. Regina E. Lundgren and Andrea H. McMakin.
© 2013 The Institute of Electrical and Electronics Engineers, Inc., and Regina E. Lundgren and Andrea H. McMakin. Published 2013 by John Wiley & Sons, Inc.

In addition, studies are beginning to show that having a risk communication plan well integrated with the risk assessment effort can save time and money as well as increase the success of the risk assessment and management process (Barrantes et al. 2009). The elements to include in a communication plan and how to bring all the elements together are discussed in the remainder of this chapter.

WHAT TO INCLUDE IN A COMMUNICATION PLAN

A variety of information contributes to a comprehensive communication plan, as shown in the outline in Figure 12-1. You may also want to include other elements, depending on your organization's requirements and the type of risk communication (care, consensus, or crisis). For example, Chapter 21 includes specific aspects for communication plans related to emergency response.

The introduction of the plan should discuss why you are writing the plan (purpose), what kinds of activities are covered by the plan (scope), background material on the risk being communicated, the reason your organization is communicating about the risk (authority), and the purpose and objectives of your effort (see Chapter 7 for more

---

Introduction
  Purpose of the plan
  Scope of the plan
  Background on the risk
    What is the risk?
    Who is affected by it?
  Authority
    Under what authority (law or organizational mandate) is the risk being
      communicated?
  Purpose of the risk communication effort
  Specific objectives

Audience Profile
  How audience information was gathered
  Key audience characteristics

Risk Communication Strategies

Evaluation Strategies

Schedule and Resources
  Detailed schedule that identifies tasks and people responsible for completing them
  Estimated budget
  Other resources to be used (equipment, meeting rooms, etc.)

Internal Communication
  How progress will be documented
  Approvals needed/received

Signoff Page
  Names, job titles, and signatures of key staff acknowledging that they have read
    and concur with the plan

Figure 12-1. Outline of a risk communication plan.

information on setting these). In addition, be sure to note any policies that your organization may have on risk communication or information dissemination. This information leads into the next section on audience.

> The introduction of the plan should discuss why you are writing the plan, what kinds of activities are covered by the plan, background material on the risk being communicated, the reason your organization is communicating about the risk, and the purpose and objectives of your effort.

The audience profile section describes what you know about the audience and how you learned it (see Chapter 8). The audience profile section can also discuss how segments of the audience differ. For example, in a care communication effort in which you are informing a community about the potential of fecal coliform bacteria in their well water, the audience might be segmented by level of risk: those who have wells with high readings of the bacteria, those who have wells with minimum readings, those who have wells free of the bacteria, and other interested community members. Each of these groups could differ in the kinds of information they need to make informed decisions regarding their risk. In a consensus communication effort, audiences might be segmented by the amount of involvement they desire in the decision-making process. Each of these groups could differ in the risk communication method used. In other cases, audiences might be segmented by comprehension level (for example, uneducated to highly educated), geographic distribution (for example, those who can come to local public meetings and those outside the local area), or method accessibility (Internet access, comfort dealing with computers). In these cases, the difference between the audience segments is not so much kind of information, but the way in which it is delivered. This section of the plan notes such key differences that will affect the risk communication effort described in the following section.

Risk communication strategies take what you learned about the audience and your purpose and objectives, and lay out the methods you will use to reach each of the segments of your audience (see Chapter 10 for more information on methods). These strategies are closely aligned with the evaluation strategies (see Chapter 20 for additional information).

The schedule and resources section describes what you need to implement the strategies you laid out and how long it will take to fully implement them. Make sure you include resources other than funding. For example, if you are conducting a consensus communication effort, you might need space for the group to meet, audiovisual equipment to make presentations, and clerical support to create meeting minutes, among other resources. For additional information on setting a schedule, see Chapter 11.

The internal communication section describes how your organization will be kept apprised of the risk communication effort. Will you share monthly reports, send electronic mail messages at key points in the effort, or give presentations to interested groups within the organization? Even if your organization does not require you to provide such updates, consider doing so to help ensure that your activities are visible and valued by your colleagues and management. Such visibility can help when additional resources are needed for this effort or future efforts. In addition, factor in what training is needed by staff. Who needs to understand how to interact with the news media? Who has not had risk communication training? Be sure to include this information in the resources section as well.

The last item in the figure, the sign-off page, is particularly important because it confirms organizational support. Get acceptance of the plan (as indicated by the signature) from all those involved in the risk communication effort or having to approve of the risk communication process or materials. Discuss the plan with and get acceptance of it from staff who are conducting the assessment and anyone involved with communicating the risk or approving its communication. Get signatures from managers of technical staff conducting the risk assessment, those in charge of making any decision based on the risk assessment results, managers in the public affairs and/or communications departments, and managers of staff involved in the risk communication effort. You want the acceptance of these managers as well as their staff because the managers will have to approve staff time and resources (which their staff may not be able to commit to alone), the managers need to be aware of what their staff is doing, and the managers are likely to be the ones who have to answer questions from the public or outside organizations regarding your efforts and, therefore, need to be informed about them.

> The last item in the figure, the sign-off page, is particularly important because it confirms organizational support.

Once you have been through the process of gathering information and crafting it into a risk communication plan, it is tempting to cut corners and simply cut and paste from one effort to the next. While organizations that communicate with the same audience about the same risk over time may be able to reuse portions of the plan, always review the information before recycling it into another plan. Audiences, messages, and methods change over time, as does the organization communicating about the risk. Using outdated information can at a minimum hinder effective communication and could significantly decrease audience trust in the organization. For example, the BP Regional Oil Spill Response Plan for its work in the Gulf of Mexico failed to specify communication mechanisms in a crisis and included so many errors (such as the need to protect walruses, sea lions, and seals, none of which is found in the Gulf) that it was embarrassing when it was released to the public following the Deep Water Horizon oil spill (Galloway and Kwansah-Aidoo 2012).

DEVELOPING RISK COMMUNICATION STRATEGIES

A number of methods can be used to plan complex projects. Four of the most useful methods in planning a risk communication effort are using storyboarding, following the guidelines recommended by the U.S. Environmental Protection Agency (EPA) for Community Relations Plans under Superfund, using an audience focus, and using a technique that combines the elements of strategic planning and public involvement.

> Four useful methods in planning a risk communication effort are using storyboarding, following the guidelines recommended by the U.S. Environmental Protection Agency (EPA) for Community Relations Plans under Superfund, using an audience focus, and using a technique that combines the elements of strategic planning and public involvement.

Storyboarding as a Planning Tool

Storyboarding is a technique that has been applied to organizing the content of a variety of communication products. It can also be used to develop the entire communication effort. This technique is most useful in planning care or consensus communication efforts and probably cannot be used in crisis communication because it takes time and requires a team of staff to work together, which may be difficult to accommodate in some schedules.

A storyboard can be any large flat surface to which pieces of paper can be affixed with tape, nonpermanent adhesive, pins, or tacks. The process starts with all those involved (those who are communicating the risk as well as other technical experts, managers, support staff, and even members of the audience, if you can arrange it) brainstorming ideas. Someone not involved in the risk communication effort should facilitate the session. The ideas can relate to any part of the process: audience, particular messages, or how to distribute them. The facilitator should let the ideas flow and not censor them, even if they are impractical or impossible. The facilitator or participants write each idea on a separate piece of paper and affix it to the board.

When the group runs out of ideas, the facilitator helps them begin to organize the ideas into the elements discussed in the outline in Figure 12-1. Use larger or different-colored cards to indicate categories. The facilitator will move the ideas into the categories, encouraging discussion. From the discussion, the group can begin to weed out infeasible or contradictory ideas.

Once the ideas have been categorized, the facilitator works with the group to organize the ideas within each category. The ideas can be put in any order that makes sense for your situation. For example, if one category was audience information, you could organize it by various segments of the audience. You now have a detailed list of strategies to be covered in the plan and to be used in your risk communication effort.

Communication Planning Using the Comprehensive Environmental Response, Compensation, and Liability Act (CERCLA) Approach

The EPA's (1992) publication, *Community Relations in Superfund: A Handbook* (originally published in 1992 and updated since with a number of tools and examples), is an invaluable resource for developing community relations plans (and communication plans) that meet the requirements of the CERCLA or Superfund. This publication lists elements that should be included (these elements look much like the outline in Figure 12-1) and gives an example of a community relations plan. While the method described in the publication is most applicable to planning a risk communication effort related to CERCLA, it can also be used to plan care, consensus, and crisis communication.

> In general, communication planning under CERCLA requires that those who are responsible for cleaning up a Superfund site interview members of the public and organizations who might be interested in or are affected by the cleanup.

In general, communication planning under CERCLA requires that those who are responsible for cleaning up a Superfund site (or conducting the care, consensus, or crisis communication) interview members of the public and organizations who might be interested in or are affected by the cleanup (or the risk being communicated). Much the same information is gathered as during audience analysis, and it is used to plan appropriate strategies for communicating risk information. The number of interviews needed depends on the number of people who may be interested or affected. (For crisis communication, this method is best used in planning a crisis response rather than during the actual crisis.)

A more in-depth approach similar to this one is to develop a "mental model" of how the audience and experts view the risk and then focus on reinforcing appropriate messages and on correcting misperceptions. This approach is particularly useful for care communication. More information on this approach can be found in Chapter 2 and Chapter 9.

Using an Audience Focus

Another technique to identify strategies is to focus on the needs of various segments of your audience. This can be accomplished in several ways. James Creighton, public involvement specialist and consultant responsible for many of the Bonneville Power Administration's successes in risk communication, uses an "onion diagram" (Figure 12-2). The center of the onion represents the risk, and each of the rings represents audience segments. In care and crisis communication, the audiences closest to the center are those most at risk. In consensus communication (as in the example in Figure 12-2), the audiences closest to the center are those who want to be the most heavily involved in making the decision.

For any risk communication effort, a slightly different set of specific audiences will be in each of the rings. For example, in a consensus communication effort in which citizens of the Pacific Northwest would be asked to determine potential ways to mitigate risks to the salmon population:

1. The innermost circle would include the agency charged with making the decision and most likely the Native American tribes who rely on the fish for subsistence,

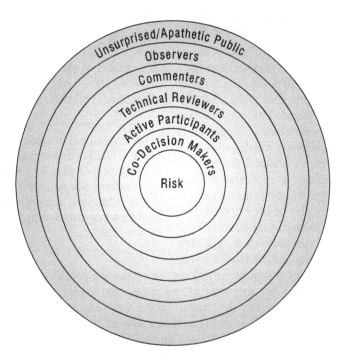

Figure 12-2. Audience-centered technique for communication planning. (Adapted from various works by James Creighton.)

economic development, and ceremonial purposes, but who must be consulted under a government-to-government agreement (co-decision makers).

2. The next circle (active participants) would include others with economic interests (other commercial fishers or those in the tourist industry) and environmental groups.
3. The next circle (technical reviewers) would include other related regulatory agencies.
4. The next circle (commenters) would include interested residents of communities along the rivers.
5. The next circle (observers) would include the rest of the residents of the Pacific Northwest.

Each of these audience segments will want a different type of involvement.

In general, as you move inward through the rings, the more communication effort is expended. Using the example described earlier, the residents of the Pacific Northwest (observers) would probably be satisfied to read information in the news media about the ongoing activities. If something they read increases concerns, those who are most concerned will join the audience in the next ring. The residents of river communities will probably want additional information, such as direct-mail pamphlets and presentations to local organizations, whereas those with economic or environmental interests will want more direct involvement in making the decision, such as participating in workshops or on advisory committees. The Native American tribes will expect to be consulted and to actively help make the decision.

Another way to plan for audience needs is to use a situational assessment. According to researchers David Dozier and James and Larissa Grunig, who studied excellence in public relations and communications organizations (Dozier et al. 1995), audiences can be segmented into the following groups (adapted for use in risk communication) based on their situation in relation to the risk and the organization communicating about the risk:

- **Nonpublic.** Those members of the audience who are not affected by the risk or the organization communicating the risk and vice versa
- **Latent public.** Audiences affected by the risk but do not realize it
- **Aware public.** Audiences both affected by the risk and knowledgeable of the fact
- **Active public.** Audiences organized to do something about the risk

Dozier and his colleagues warn that organizations often focus communication efforts solely on the last group. By the time ignored members become active, their views are often negative, entrenched, and oppositional. Although each segment of the audience may be reached using different methods and schedules, it is better to cultivate all segments early in the risk communication process.

> Ignoring any segment of the audience can have serious consequences.

Another way to view audiences is by what causes them to be interested in the risk. One crisis communication manual developed for the state of California by a team of experts, including risk communication luminary Peter Sandman (OES 2001), divides audiences into the following segments:

- **Residential community.** People living near the risk, who have a personal or familial interest in health, safety, the environment, or quality of life in the area
- **Business/commercial community.** Businesses that may be affected by the risk, who are concerned about loss of revenue, infrastructure availability, and protection of employees, as well as personal safety
- **Industrial community.** Businesses that could be affected and who could affect the risk (for example, chemical tanks being targeted by terrorists), who are concerned with the same things as other businesses, as well as security issues
- **Your organization.** The staff and management who are charged with analyzing, managing, and/or communicating risk, who need to understand the effort as well as the risk
- **Other organizations.** The staff and management of organizations teaming with you in analyzing, managing, and/or communicating about the risk, who need to be able to explain and implement their roles

The types of information and communication mechanisms may differ for each of these groups as well.

Using one of these three audience-focused approaches for your risk communication effort can help ensure that all segments of the audience are considered in planning and implementation.

Strategic Planning for Risk Communication

Social scientists at Battelle developed a method that combines elements of strategic planning with the ideals of public involvement to create plans for a variety of efforts, including risk communication. Strategic planning identifies an organization's strengths and weaknesses, opportunities to reach goals, and threats to those opportunities. Public involvement is incorporated by interviewing people who are interested in or affected by an organization to learn whether they view the organization in the same way as the organization views itself. This method is most useful in planning care communication efforts but can also be used in planning for a response to a crisis.

To use this method, start by identifying the values of your organization. What does your organization believe in, what is its mission, what is its purpose for being? Then state what the organization hopes to accomplish and how you will know when it has been accomplished. Next, list the purposes and objectives of your risk communication effort. Looking at these and the information you developed in the previous steps, you can begin a situational analysis. What is the current situation? What assumptions did you make when planning for it? What are the strengths and weaknesses of your organization? What opportunities are available to you as you try to communicate risk? Is there anything that could prevent you from taking those opportunities? Who is your audience? What are the key issues?

From this analysis, your strategies for risk communication should start to emerge. Discuss your strategies and the information you have developed in the previous steps with representative members of your audience. Do they see the organization in the same way you did? Do they agree with your strategies for communicating risk? What suggestions can they offer to make your program stronger? Incorporate these suggestions into a final plan, using the outline in Figure 12-1.

CHECKLIST FOR DEVELOPING A COMMUNICATION PLAN

☐ The plan includes the elements outlined in Figure 12-1.
☐ All segments of the audience have been considered in planning.
☐ The plan has been agreed to by:
    ☐ Those who are communicating the risk
    ☐ Those who are assessing the risk
    ☐ Those who are managing the risk
☐ The plan has received the signatures of the:
    ☐ Managers of those who are communicating the risk
    ☐ Managers of those who are assessing the risk
    ☐ Managers of those who are managing the risk

REFERENCES

Barrantes, S. A., M. Rodriguez, and R. Pérez. 2009. *Information Management and Communication in Emergencies and Disasters*. Pan American Health Organization, Regional Office of the World Health Organization, Washington, DC.

Dozier, D. M., L. A. Grunig, and J. E. Grunig. 1995. *Managers Guide to Excellence in Public Relations and Communication Management*. Lawrence Erlbaum Associates, Mahwah, New Jersey.

EPA (U.S. Environmental Protection Agency). 1992. *Community Relations in Superfund: A Handbook*. U.S. Environmental Protection Agency, Office of Emergency and Remedial Response, Washington, DC. EPAJ540/G-88/002.

Galloway, C. and K. Kwansah-Aidoo. 2012. "Victoria Burning: Confronting the 2009 Catastrophic Bushfires in Australia." In A. M. George and C. B. Pratt, eds., *Case Studies in Crisis Communication: International Perspectives on Hits and Misses*. Routlege, New York, pp. 279–282.

OES (Governor's Office of Emergency Services). 2001. *Risk Communication Guide for State and Local Agencies*. Office of Emergency Services, Sacramento, California.

ADDITIONAL RESOURCES

McCallum, D. B. 1995. "Risk Communication: A Tool for Behavior Change." *NIDA Research Monograph*, 155:65–89.

Shyette, B. and S. Pastor. 1989. "Community Assessment: A Planned Approach to Addressing Health and Environmental Concerns." In *Superfund '89: Proceedings of the 10th National Conference*, pp. 635–641. Hazardous Materials Control Research Institute, Washington, DC.

III

PUTTING RISK COMMUNICATION INTO ACTION

Each method (information materials, visual representation, face-to-face communication, working with the media, stakeholder participation, technology-assisted communication, or social media) has its own idiosyncrasies when used as a method for risk communication. Knowing these differences can help make the risk communication process more effective.

> The farther away we get from individual contact, the more room there is for confusion and misunderstanding.
> —Thomas Wilson (1989, p. 78).

13

INFORMATION MATERIALS

The written message in the form of information materials has long been a staple of the communication world. The advantages and disadvantages of using information materials are discussed in Chapter 10. There are a number of ways to communicate risk via information materials, including newsletters; pamphlets, booklets, and fact sheets; posters, advertisements, and displays; articles in professional or trade journals, popular press magazines, blogs, and newspapers; and technical reports. Specific advice on each of these types of information materials is provided later in this chapter, focusing on how these materials differ from those used in other types of communication efforts. First, however, there are some principles for constructing information materials.

CONSTRUCTING INFORMATION MATERIALS

Whatever the form of the information material, those who are communicating risk must consider what information to include, how to organize messages, appropriate language, and the use of the narrative style.

Information to Be Included

For information materials, a number of techniques can be used to present content in a way that your audience will understand. Remember, however, that the primary rule is to know your audience. Take or leave the other rules as they fit your audience's needs and situation. Figure 13-1 lists the kinds of information that might be included in information

Risk Communication: A Handbook for Communicating Environmental, Safety, and Health Risks,
Fifth Edition. Regina E. Lundgren and Andrea H. McMakin.
© 2013 The Institute of Electrical and Electronics Engineers, Inc., and Regina E. Lundgren and
Andrea H. McMakin. Published 2013 by John Wiley & Sons, Inc.

- Goals and content of the information material
- Nature of the risk
- Alternatives to the action that is causing the risk and any risks associated with these alternatives
- Uncertainties in the risk assessment
- How the risk will be managed
- Benefits of the risk
- Actions that the audience can take to mitigate or manage exposure to the risk
- Contact point
- Glossary
- Conversion table
- "Helpful hints"
- Index
- List of related information

<u>Figure 13-1.</u> Information to include in written risk communication messages.

materials. The following list describes this information in more detail. Depending on the length and scope of the information material (for example, an advertisement vs. a technical report), these elements may be a single word or symbol or represent entire sections or chapters:

- For materials over two pages long (one page on a web screen), summarize the **goals and content** of the information material. What is the purpose of this communication? What does it contain? This information is particularly important when you are using a variety of methods because it allows the audience to choose the risk communication message that best meets their needs. For example, if a member of the audience wants information on methods for testing a home for radon, and your pamphlet states that it describes what radon is and why the audience should test for it, the audience member will know to look elsewhere for the information on testing. Information on goals and content also helps guide readers through the risk information.
- Describe the **nature of the risk and what the risk entails**. What is the risk? Who is at risk? Put the risk in context. For example, is the risk similar to other risks with which the audience is more familiar?
- Discuss **alternatives to the action that is causing the risk** and any risks associated with these alternatives. For care communication, what alternatives does the audience have to living with the risk? Are there any risks associated with those alternatives? For consensus communication, what alternatives are being evaluated in making a decision? What criteria are being used to evaluate the alternatives? For crisis communication, alternatives focus on different actions that can be taken to mitigate the risk.
- Discuss **uncertainties in the risk assessment**. How were the data collected? How were they analyzed? See Chapter 6 for more information on uncertainties. See Chapter 14 for ways to portray uncertainties visually.

- Explain **how the risk will be managed**. Risks are managed differently in care, consensus, and crisis communication. For care communication, it is usually the audience who must manage the risk, for example, by following safety procedures (although government agencies may be charged with monitoring a public health risk). For consensus communication, the audience and the organization charged with making a decision about the risk work to build consensus on how the risk will be managed, for example, by agreeing to operate a network of measurement devices around a hazardous waste site to monitor releases. For crisis communication, local, state, and federal agencies as well as individuals and organizations may be involved in managing the crisis situation.

- Describe any **benefits of the risk**. Will anything good result from people being exposed to this risk? For example, hormone therapy in postmenopausal woman helps maintain bone density and regulates hormonal systems but may contribute to greater chances of breast and uterine cancer. Be very leery of how you use this information. If the benefits are all to one group or the nebulous "science" or "humankind," and the risks are all to another group, the group at risk may not care how much their risks benefit others. Let your audience's desire to know this information be your guide in deciding whether to include it in an information material. See Chapter 9 for more information on the use of benefits in risk communication.

- Present **actions your audience can take to mitigate or manage exposure to the risk**. Can they change their behavior? Can they write to Congress? Can they provide comments on the material? Can they be involved in the risk assessment or risk management process? Knowing what they can do empowers your audience. The less they feel like victims, the less hostility you will have to combat.

> Always include the name, phone number or e-mail address, and physical address of someone to contact directly. If you cannot provide a personal contact, a hotline or website will sometimes suffice.

- Always include the **name, phone number or e-mail address, and physical address** of someone to contact directly. If you cannot provide a personal contact, a hotline or website will sometimes suffice. List the same contact in all your risk communication materials because the audience will feel more comfortable knowing that there is one reliable source available to answer their questions. If you include a phone number, make sure that number is answered 24 hours every day, either by a person or by an answering system. Answering only during business hours is not enough because many people do not have access to a phone on which they can make personal calls during that time. In addition, especially for care and crisis communication, it can be potentially life threatening for someone to wait for an answer to a question.

- For anything more than 20 pages long in hard copy, and even fewer pages online, include a **glossary** (with definitions of all necessary abbreviations, acronyms, and technical terms). (For those fewer than 20 pages, avoid acronyms and abbreviations and define technical terms in the text.) In general, for any risk communication message, avoid acronyms and abbreviations. The exceptions are terms that are commonly used more frequently as acronyms than as the spelled-out versions (for example, many people readily identify NASA, but they have to think about National Aeronautics and Space Administration) and phrases used several times per page. When in doubt, spell it out.

- Include a **method to convert metric units**. Although many schools are teaching the metric standard, many adults not directly involved in scientific pursuits are still uncomfortable with metric units. However, some federal agencies require that the primary units used in their information materials be metric. If you have only a few measurements, use analogies or metric units with other units in parentheses. If doing this conversion repeatedly will severely interrupt your text, use metric units and include a conversion table in your glossary.
- In information materials that cannot avoid being technical, include a section of **"helpful hints"** that explains things like scientific notation, uncertainties in tables and graphs, use of "less than" and "greater than" symbols, and other technical conventions. Readers who have a high level of technical knowledge can ignore the section while those with a lower level can use it to decipher the information.
- In documents longer than 40 pages, include an **index** so that the readers can find specific information. Most members of your audience have limited time to devote to reading, so anything you can do to help them find the information they want quickly will encourage them to read your information. In general, indexes should contain key words, phrases, and ideas from the text.
- End with a **list of related information**. Telling your audience where to find other sources will help them obtain the exact information they want and will encourage them to learn more about the risk. Point them to both your organization's publications and those prepared by other organizations.

Organizing Material for Information Materials

The way you organize your material will in large part depend on the form of the information material and your audience's needs. Fact sheets and technical reports are necessarily organized differently. However, one point is important for all forms: discuss how the risk was determined before you present extensive data on the risk itself. Provide a summary of results by all means, especially for those readers who want only that information. However, do not jump into information on exposure calculations before you have discussed the process by which that information was developed or you will confuse your readers. Explaining the risk assessment process before the data gives your audience a context for the data and allows them to make a more informed decision regarding the risk. For consensus communication, particularly in cases in which the audience has a high level of skepticism or hostility toward the risk or the organization charged with managing the risk, it is equally important to discuss the process by which the decision will be made. Include information on how the audience has been and will be involved in the decision process.

> The way you organize your material will in large part depend on the form of the information material and your audience's needs.

Language for Information Materials

The language you use depends primarily on your audience. You may have to translate materials to reach all segments of your audience. Studies have shown that non-English speakers in America are often more at risk both at work and in leisure activities because of an inability to understand health and safety warnings in English. For example, a study on fish advisories for Lake Michigan found that whereas 81% of the English-speaking fishers knew about potential chemical contamination in the lake's fish, only 30% of the non-English-speaking fishers had the same knowledge. Joseph McFadden, a consultant specializing in the effects of language and culture on occupational safety and health, warns that translators must know the culture as well as the language. Strict translation of safety terms may miss nuances of particular cultures. For example, the word for safety in Spanish can sometimes be taken as security instead (TrainingOnline 2006).

Reading level, education, feelings concerning the risk, and experience with risks and science in general and with this risk specifically affect the audience's ability to process the message. Some general rules apply to most situations:

- Avoid any kind of language that might give your audience the feeling that they have no control. "Victims" process information less effectively and react with greater hostility. If you are trying to build consensus about the location of a hazardous waste incinerator, do not say, "The facility will be sited in . . ." when you mean, "The facility would or could be sited in . . . ." If the decision has not yet been made, keep verbs conditional.

- Do not present estimates as facts. Many experts seem to confuse the results of their computer modeling with real life, making statements like "The resulting harm to the affected population was 10 additional deaths from cancer per year." This makes it sound as if 10 people have already died, when, more likely, the information being communicated is an estimate produced by a computer modeling effort, given certain assumptions and a range of uncertainties. Give your audience the information to judge what the model results mean.

- Avoid scientific notation, mathematic formulas, and exponents. Although you can explain some of these to a certain extent, just the fact that they are used at all will scare some readers into avoiding your message. It is true that tables of very small and very large numbers quickly become unwieldy if scientific notation (1×10^{-6}) or engineering notation (1E-6) is not used. However, depending on audience perceptions, a table with endless rows of zeroes can make the risk numbers much more obvious and frequently less threatening than a table filled with exponents. Also avoid the convention used in many technical journals of using a superscript -1 to mean "per" ($1 \ d^{-1}$ for 1/d or 1 per day). Many members of your audience will find this notation incomprehensible. Instead, use the phrase "one per day."

> Avoid scientific notation, mathematic formulas, and exponents. Although you can explain some of these to a certain extent, just the fact that they are used at all will scare some readers into avoiding your message.

- Define how you are using the word "conservative" (or avoid using it) in relation to exposure calculations. Although the scientific community usually uses this term to indicate that the risk was overestimated, common usage, for example, in the finance industry, is to indicate an underestimate ("He's worth, conservatively, $1.5 million"). Therefore, many members of your audience will interpret the term as the opposite of what is intended. Either define the term or use terms like "cautious" or "overestimated" to explain the process.

- Use culture-appropriate terms. Know your audience and what they consider acceptable words. For example, members of the Confederated Tribes and Bands of the Yakama Indian Nation are Native Americans, not Indians. And many tribes do not want to be called stakeholders because they have a government-to-government relationship with state and federal agencies. As in any form of communication, avoid sexist and racist terms.

Some organizations encourage the use of reading levels to gauge the appropriateness of information materials to present a risk message to an audience. Various formulas, such as Dale–Chall, Fry, Flesch Grade Level, Flesch Reading Ease, Fog Index, SMOG, FORCAST, Powers–Somner–Kearl, Spache, and Fleisch–Kincaid, allow the writer to calculate the comprehensibility of text at certain grade levels. Some common software programs enable users to test the readability of their text. Most formulas look at number of syllables or letters per word, number of words per sentence, and number of sentences per paragraph. Others factor in phrases and clauses. As mentioned in Chapter 8, such formulas are not the best indicators of whether material is comprehensible. Pretesting risk messages is usually the better choice.

> Over 250 studies show that health-related print materials are far exceeding the reading ability of the average adult.
> —Green et al. (2007, p. 106).

Narrative Style in Information Materials

One approach to developing information materials is to use the "narrative" style. This style consists of presenting the risk information in the form of a personal story instead of or in addition to presenting exposure calculations or other data. The story structure helps the audience understand the risk by simplifying it and by focusing on cause and effect. International health researcher Dawn Hillier advocates stories as a powerful way to relay health risk information across cultures (Hillier 2006).

Golding et al. (1992) evaluated the narrative style versus what they call "the technical style" in each's ability to (1) encourage the audience to continue reading, (2) enhance knowledge, and (3) motivate the reader to action. They found that both forms enhanced knowledge equally well but that more readers kept reading the narrative. Unfortunately, neither style was likely to motivate the reader to action. In the case of the narrative style, one possible reason for this lack of motivation is that the audience may not have identified themselves closely enough with the person in the story, pointing to the need to understand your audience before using this technique. George Cvetkovich of Western Washington University, a researcher in the area of public perceptions of risk and public policy, offers some suggestions for optimizing the power of the narrative in risk communication. He advises checking narratives for the following:

- **Involvement.** Is the message interesting to the audience?
- **Relevance.** Does the audience think that the message applies to them?
- **Ability.** Can the audience understand the information being presented and can they act in the way being modeled in the story?

To check narratives for these factors, use audience analysis information to guide you in preparing a prototype message and to take the prototype to representative members of your audience for review. Ask them the questions related to involvement, relevance, and ability, and redesign the narrative as necessary to meet their needs.

GUIDELINES FOR SPECIFIC TYPES OF INFORMATION MATERIALS

Information materials can range from the short two-page fact sheet or quarter-page ad in a magazine to the multivolume environmental impact statement. What follows are guidelines for some of the more commonly used types of information materials for communicating risk. For information on websites, see Chapter 18. For information on blogs, see Chapter 19.

Newsletters

Newsletters are especially good for long-term projects with a relatively stable audience (stable meaning that the audience consists of mostly the same people throughout the communication process) who are interested in the project/risk being described. They can be mailed or e-mailed to audience members or handed out at other events.

Each issue of a newsletter consists of a series of articles related to a specific risk or type of risk. For example, many Superfund sites send a newsletter to interested members of the communities near the site to keep them informed of the progress of cleanup and the risks entailed. Many health care centers also have newsletters describing healthy lifestyles for their patients. Although the exact content of a newsletter will depend on the audience and the risk, some general rules apply:

> Newsletters are especially good for long-term projects with a relatively stable audience (stable meaning that the audience consists of mostly the same people throughout the communication process) who are interested in the project/risk being described.

- When first developing a newsletter, **allow time in your schedule for approvals** from all agencies involved. Because a newsletter often serves as a reflection of the organization over a long period of time, this type of risk communication often requires a number of approvals before the first issue is published, and sometimes for subsequent issues.
- **Develop and maintain mailing lists.** Be sure to include as many members of your audience as possible in your distribution. Include in each issue a method for requesting to be added to or removed from the newsletter distribution, or, for online newsletters, to subscribe and unsubscribe. Maintain an accurate mailing list by updating names and addresses at least quarterly. Some e-mailed newsletters tell their readers that if their e-mail address is "returned undeliverable" for a certain number of times, it will be dropped. Consequently, subscribers are encouraged to notify the newsletter themselves if their e-mail addresses change. If your audience is already hostile, spelling names wrong, sending information to the wrong address, or forgetting some members entirely will not help.
- **Avoid the use of acronyms and abbreviations.** Like newspapers, newsletters are seldom read straight through from beginning to end. Audience members pick stories and headlines that interest them. They will not know where to look for a definition if they first come across an acronym on page 6 and the acronym was spelled out in some other story on page 3.

- **Encourage people to read by using compelling headlines and graphic elements.** Use design elements appropriately to encourage your audience to keep reading.
- **Be consistent.** One of the advantages of newsletters is that subsequent issues will be sent to the same audience (although one that grows with your mailing list) over time. Use the same words to describe the same place or situation (for example, do not call your environmental cleanup areas source areas in one issue and sites in another). For consensus communication efforts involving a decision process, show readers the process each time with the current stage highlighted. Also, watch the content. If in the first issue you run a story about the new sewer plant opening next April, do not forget to follow up with the grand opening. Lack of consistency can lead to lack of credibility for your entire effort.

Pamphlets, Booklets, and Fact Sheets

> Pamphlets, booklets, and fact sheets are good for short-term, one-message communication efforts or for covering one aspect of a complex risk.

Pamphlets, booklets, and fact sheets are good for short-term, one-message communication efforts or for covering one aspect of a complex risk. Because they are short, they attract those who are put off by longer information materials. The following are a few points to remember for risk communication messages:

- Focus these short forms to meet specific needs. By their very nature, pamphlets, booklets, and fact sheets have a limited amount of space. Cover only one limited subject in each. Your audience's needs will determine the subjects.
- Make these forms self-contained. Pamphlets, booklets, and fact sheets are meant to be picked up, carried away, and read quickly, or reviewed online. Although information on contact points and ways to get additional information should be part of the message, your audience should need nothing more than what they see to understand the point you are trying to make about the risk.
- If the message is part of a series, strive for visual consistency. Try for a "family look" to the publications (similar use of type styles and design) so that your audience will begin to recognize them and, with any luck, become comfortable with them. The higher the comfort level, the more likely your audience will be to read them.
- Distribute these materials where your audience lives. Use a direct mail approach, but do not overlook the power of placing packets in locations where your audience is likely to pick them up and read them. Medical offices, libraries, local businesses, community centers, church vestibules, and even local chambers of commerce are places where your audience may see and pick up such information. Some audiences want to read or download these materials online.

Posters, Advertisements, and Displays

American society is increasingly visually oriented. Posters, advertisements, and displays are a form of information material geared to appeal to this visual orientation. Although the message they can carry will be limited to the space available and the creativity of the designer, these information materials can strongly reinforce key concepts if certain guidelines are followed:

> Posters, advertisements, and displays are a form of information material geared to appeal to the increasingly visual orientation of American society.

- Any written message in posters, advertisements, or displays should be in the language of the audience. Text should be at the audience's reading level and address audience concerns. The message should also be written in a language other than English if the audience speaks English only as a second language. If different members of the audience speak different native languages, similar displays should present the information in each language.
- The message should be simple and clear. This seems obvious, especially given the limited space in display visuals. However, it is frequently violated, sometimes with disastrous results. For example, to advertise its safety campaign, an industrial plant posted a sign outside its gates along the main route for commuters. The sign was covered in text, all in capital letters, and several different colors, with slogans and statistics about lost-workday accidents. Although its placement suggested that it was meant to be read quickly as workers drove home, workers had to slow down to see what it had to say so that the safety sign actually caused accidents. In general, display visuals are meant to be read quickly, so keep them simple.
- All graphic elements should reinforce the risk message. Too many display visuals offer mixed messages. A billboard outside a chemical manufacturing firm warned workers to remember safety as they went to work. The bold, black lettering and sober message spoke of a serious concern; however, the pastel painting of two happy children picking wildflowers that appeared beneath the lettering contradicted the seriousness of the message. A simple, easy-to-remember slogan would have carried more weight.
- Put display visuals where the audience will see them. Where does your audience work and play? Where are they most likely to be exposed to the risk? That is where your messages belong. Putting a message about the dangers of unsafely operating a forklift in the administrative offices may reach a number of people, but not likely the workers who are at risk. Better to put the poster in a place where your audience will be thinking about the risk, such as where the machines are parked.
- Always include sources of additional information. Although display visuals are particularly good at raising audience awareness of a risk, many members of the audience will find the limited message inadequate and want more information. Always be sure to point them in a direction that will reinforce your message. This could be an address, website, or phone number for your organization or a sheet that they can tear off with various contact points. Including this information helps empower your audience to take action.

Articles

If written by someone knowledgeable about the risk who can communicate well, articles in professional or trade journals, popular press magazines, blogs, and newspapers can be very effective in communicating risk to a variety of audiences. However, in some cases, the actual content is often out of the control of those who are communicating the risk. (See Chapter 16 for information on working with the media.) If you are writing the article:

> If written by someone knowledgeable about the risk who can communicate well, articles in professional or trade journals, popular press magazines, blogs, and newspapers can be very effective in communicating risk to a variety of audiences.

- Make sure that the publication's readers match your intended audience. If you do not reach the people you were trying to reach, it does little for your risk communication efforts. Most journal and newspaper publishers will tell you who reads their publications.
- Consider professional journals. For certain occupational risks, for example, carpal tunnel syndrome, your audience may well be members of professional societies or unions. A well-written article in their publication may reach your intended audience faster than some other forms of communication.

Technical Reports

Although intimidating for many readers, technical reports are needed by those who want to see more detailed data to form their own opinions. Technical reports can meet the needs of several segments of your audience: the expert who has extensive technical information or wants it, the reader with some background in the risk who wants additional in-depth information, and the neophyte who is interested in this type of information. To meet the needs of these segments of your audience, organize the document from the back to the front. That is, place the technical detail—computer runs, tables of supporting data, lists of standards, and quality-assurance data—in appendixes or supporting documents. (In online reports, you can include hyperlinks that directly link summary information and references to the supporting detail.) This information will serve the expert. Then, use the information as a basis for a report that can be read at the tenth-grade level; this will generally serve the reader with some background in the risk. Use the report information as a basis for a summary with a minimum of technical terminology; this summary should be able to be read at the sixth-grade level. This will serve the needs of the neophyte. (We use reading levels here as a guide to content and style, not to imply that all interested readers read at the tenth-grade level or that all neophytes read at the sixth-grade level.)

> Although intimidating for many readers, technical reports are needed by those who want to see more detailed data to form their own opinions.

As mentioned in the section on what to include in written risk communication messages, be sure to include helpful hints, a glossary, and an index to help all readers. Provide additional information-identifying devices, such as introductions that summarize key points for each major section; transitions between sections, paragraphs, and sentences; and paragraphs with topic sentences. Craft sentences and paragraphs so that familiar information comes first, with more difficult and newer information coming later.

CHECKLIST FOR INFORMATION MATERIALS

☐ The information is tailored for the intended audience.
☐ The information material includes information on any of the following that will meet audience needs:
  ☐ Goals and content
  ☐ Nature of the risk
  ☐ Alternatives
  ☐ Uncertainties
  ☐ Risk management
  ☐ Risk benefits
  ☐ Audience actions
  ☐ Contact information
  ☐ Glossary
  ☐ Metric conversion
  ☐ "Helpful hints"
  ☐ Index
  ☐ List of related information
☐ The message discusses how the data were developed before the data themselves.

The wording in the message:

☐ Does not present the audience as victims
☐ Distinguishes between estimates and facts
☐ Avoids scientific notation
☐ Avoids mathematic formulas
☐ Avoids exponents
☐ Avoids or defines the term "conservative"
☐ Does not use racist or sexist terms or other terms the audience might find offensive

If the message uses narrative style, it will:

☐ Involve the audience
☐ Be relevant to the audience
☐ Be within the audience's ability to understand and act upon

For newsletters:

☐ Time has been allowed for approvals.
☐ The mailing list has been developed.
☐ There is a mechanism for updating the mailing list.

The text of the newsletter:

☐ Avoids acronyms and abbreviations
☐ Uses compelling headlines and graphic elements
☐ Is consistent from issue to issue

(continued)

CHECKLIST FOR INFORMATION MATERIALS (*continued*)

For pamphlets, booklets, and fact sheets:

☐ Each has been focused to meet specific audience needs.
☐ Each is self-contained.
☐ All strive for consistency.
☐ Each has been distributed where the audience lives.

For posters, advertisements, and displays

☐ The text portions are written in a language that the audience will understand.
☐ Text messages are clear and simple.
☐ All graphics reinforce the message.
☐ Visuals will be displayed in locations where the audience will see and heed them.
☐ Information is included about where the audience can get additional information.

For articles:

☐ The article is written for publications that will reach the audience.
☐ The article will be in professional journals where appropriate.

For technical reports:

☐ The report is structured to meet audience needs.
☐ The report uses language and organization that lead the reader through the report.

REFERENCES

Golding, D., S. Krimsky, and A. Plough. 1992. "The Narrative versus Technical Style in Risk Communication." *Risk Analysis*, 12(2):27–35.

Green, M., J. Zenilman, D. Cohen, I. Wiser, and R. Balicer. 2007. *Risk Assessment and Risk Communication Strategies in Bioterrorism Preparedness*, NATO Security Through Science Series-A: Chemistry and Biology. Springer, Dordrecht, Netherlands.

Hillier, D. 2006. *Communicating Health Risks to the Public: A Global Perspective*. Gower Publishing, Aldershot, Hampshire, England.

TrainingOnline. 2006. "The Impact of Language and Culture on Job Safety." *Reliable Plant Magazine*, January.

Wilson, T. 1989. "Interactions between Community/Local Government and Federal Programs." In V.T. Covello D. B. McCallum, and M. T. Pavlova, eds., *Effective Risk Communication*. Plenum Press, New York, pp. 77–81.

ADDITIONAL RESOURCES

Kolin, J. L. and P. C. Kolin. 1985. "Instructions." In *Models for Technical Writing*. St. Martins Press, New York.

No author. 1982. "Indexes." In *The Chicago Manual of Style*. The University of Chicago Press, Chicago, Illinois, pp. 511–560.

14

VISUAL REPRESENTATIONS
OF RISKS

From hieroglyphics on cave walls to movies on portable electronic devices, humans have always gravitated toward visual ways to communicate. It is not surprising, then, that when people are trying to understand and make decisions about risks, they often want to see various aspects of the risk in visual or graphic formats. Those who communicate risk should be aware of the power of well-chosen visuals to help people understand and think about risks. The human brain has a remarkable capacity to assimilate visual information. People have been shown to "take in" more than 600 pictures without any particular effort and then, with more than 98% accuracy, distinguish them from different pictures that are added to the original 600 (Shepard 1967).

> What is important is the ability to intuit and communicate the human meaning of data.
> —Lawrence Wallack (1993, p. viii), Director, Berkeley Media Studies Group.

Visuals have been shown to help people understand and remember content (Graber 1990; Lang 1995; Shepard 1967). For example, consumer comprehension of nutrition information on product labels was improved when bar graphs and pie charts, rather than words only, were used (Geiger et al. 1991). Carefully chosen pictures make information transmission more rapid, realistic, and accurate than is possible in purely verbal messages (Graber 1990).

Visuals help clarify abstract concepts, which often are inherent in risk-related information. One study found that people making mental comparisons involving abstract concepts increased their response times when pictures, rather than words, were used (Paivio 1978). Good visuals help audiences construct mental models of abstract or complex concepts

Risk Communication: A Handbook for Communicating Environmental, Safety, and Health Risks,
Fifth Edition. Regina E. Lundgren and Andrea H. McMakin.
© 2013 The Institute of Electrical and Electronics Engineers, Inc., and Regina E. Lundgren and Andrea H. McMakin. Published 2013 by John Wiley & Sons, Inc.

(Graber 1990). A typical study tested people's comprehension and problem solving, with and without graphics, about how lightning is created (Mayer et al. 1996). People who were given captioned figures along with explanatory text showed increased comprehension and problem solving than those who received the text only.

Beyond improving comprehension and recall, visuals can help people put facts into context. Numerical information in pictorial formats such as charts can make it easier to get a more holistic (bigger picture) view than with numbers alone, helping users gain more insights into the information (Lacerda 1986). Graphics also reveal data patterns that may go undetected otherwise (Tufte 1990).

This chapter describes ways to represent risk-related information visually, whether in photos, pictures, illustrations, graphs, charts, tables, labels, or other forms. Those who are communicating risk often use visual representations of risks in explanatory materials such as displays, posters, fliers, fact sheets, flip charts, presentations, newsletters, booklets, product labels, videos, websites, and other multimedia sources. There are almost an unlimited number of options for portraying risk information visually. The key is to tailor the design and use of the format to the needs of the interested individuals or groups. No one presentation format fits all people and situations. Our intent is to give those who are communicating risk some ideas, tools, and guidelines for communicating risk information pictorially.

One important caveat is that the way risks are presented is only one aspect of the way people perceive and act on risks. Some risk experts believe that if they can just find the right way to portray a risk, the public will draw the same conclusions about the risk as do the technical experts (or policy experts, risk managers, plant managers, or government and public health officials). As described in Chapter 4, many other factors affect the way people respond to risks, such as the nature of the risk and the trustworthiness of those communicators explaining it (Bord and O'Connor 1992; Johnson et al. 1992; Slovic 1987). The way quantitative information is presented, though important, is only one contributing factor.

Regardless of the role of pictorial representations, they can serve as powerful tools to help people understand various aspects of risks and their alternatives. In portraying risks visually, those who are communicating risk should aim for approaches that are clear, comprehensible, nonmanipulative, and useful for making decisions. This chapter draws on research and practice to recommend practical approaches for communicating risk. For simplicity, the words *graphic*, *visual*, and *pictorial representation* are used interchangeably unless specified.

DESIGN VISUALS FOR SPECIFIC AUDIENCES AND USES

In deciding which aspects of a risk to portray and how to present them, you will need to identify three things: what people want to know, what they need to know to make an informed decision, and how the visual information will be used.

The first step is to analyze your audience's information needs. Different people may want different kinds of information about a risk. In one study of local river contamination, state regulators wanted to know whether the contamination levels would rise over the legal limits, and under what conditions. Farmers wanted to know the

> Good design should take into account how, when, and where the information is used.
> —Edward R. Tufte (1997, p. 115).

potential effects of irrigating their crops with the water. Native American tribes were concerned about how the environmental problems could affect future generations. They also wanted to know whether tribal members whose diets were heavily dependent on locally caught fish were more at risk. Those involved with the river contamination study worked closely with the stakeholders throughout the study to make sure that the results would portray all of these aspects of the risk. Thus, the study contained graphics for different conditions throughout time, showing contaminant levels for various kinds of river uses.

Follow the advice for audience analysis in Chapter 8. The results will help you determine what you may want to show visually, and in what format.

In many communication efforts, especially in care communication efforts, it is necessary to go beyond what people want to know. Include what people need to know, but may not think to ask, about the risk to knowledgeably evaluate it. The mental models research approach described in Part I is one way to identify important factual misperceptions or information gaps about a given risk that must be addressed in communication materials.

Carnegie-Mellon University used the mental models approach in developing a brochure about electric and magnetic fields. The brochure covers the typical issues about possible health effects and how to avoid them. But the brochure also visually depicts specific information to help people understand risks from electric and magnetic fields, including the strength of fields from various sources, how the strength is affected by proximity to the source, how fields are measured, the stages of an electric power system, ways science gets evidence about health effects, and the concept of dose and exposure (Carnegie-Mellon University 1995). Figure 14-1 shows one such depiction from the brochure. This particular graphic and accompanying explanation were added when it was discovered through testing that laypeople typically underestimated the rate at which the strength of fields decreased with increasing distance from their sources (Morgan et al. 1990).

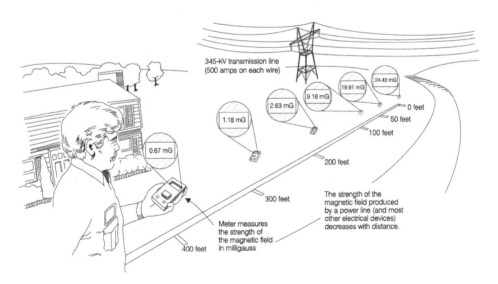

Figure 14-1. Illustration from a brochure on health risks of electric and magnetic fields. The figure shows how the strength of a 60-Hz magnetic field decreases with distance from a 345-kV transmission line. (Source: Carnegie-Mellon University 1995; used with permission of the author.)

In addition to identifying what people want and need to know, another factor for designing visuals is to determine where and how the visual will be used. Table 14-1 shows some considerations for various presentation options.

MATCH THE VISUAL PORTRAYAL TO THE INFORMATION TO BE CONVEYED

Not every type of risk or aspect of it lends itself to pictorial representation. How do you determine when a visual representation of risk is needed, and how do you decide the best way to show it? Some common kinds of risk information that lend themselves to pictorial representations include the following:

- The nature of the risk and its effects
- How large or significant the risk is
- How likely the risk is to occur and the chances that it will affect people

Table 14-1. Considerations for showing visuals in various media

Where graphics will appear	Considerations
Printed information materials (newsletters, fliers, fact sheets, brochures, booklets)	Such materials usually provide the most detailed explanation of risks; people can read on their own "turf" at a convenient time. Can typically use more detailed visuals, and a wider variety, than in other media, because there is room for explanation. Often the only media where number-intensive graphs, charts, flowcharts, and tables are appropriate.
Posters and displays	Typically designed to get attention and convey a few key messages quickly. If displayed at a public forum to discuss the risk, can contain more explanation. Graphics must be clearly legible from at least a couple of feet away. Message of graphic should be quickly apparent. Pictures and simple graphics are most effective.
Presentations	Make sure the entire audience can see all aspects of the visuals being shown, or use handouts. Tailor visual content to background, knowledge, and interests of majority of audience members. Use supplementary print materials for those who want more information.
News media	Usually aimed at a broad, general audience. Graphics are typically designed to attract attention and convey a single key point. Keep visuals simple, uncluttered. For TV, consider showing people dealing with the risk. For newspapers, consider picture-oriented visuals (such as icons or symbols) that represent how the risk is carried or mitigated. See Chapter 16 for more information.
Technology-assisted communication, multimedia	Often interactive and tailored to specific interests. Good opportunity to present technology-related visuals at several levels, such as a summary graphic, more detailed visuals, communication, and explanations for those who want to "drill down" to learn more. Good for multimedia showing movement, including animations, virtual reality, video clips, and streaming video. See Chapter 18 for more information.

- How much the risk has increased or decreased over time
- Alternatives to the risk and the benefits and dangers associated with alternatives

Table 14-2 shows potential ways to address each of these factors visually. As described in the next sections, select the approach(es) based on an audience analysis and on pretesting.

One way to plan visuals is to pose a series of questions that address the key issues, then determine how to answer them visually. An example of how to use that process is shown here. Key questions were used to guide the selection of visuals for an annual school safety report (St. Clair 1956):

- How do the accident statistics of our locality compare with those on a national basis or with those in similar communities? A line graph was suggested for comparing the local accident rate with the national rate.

Table 14-2. Options for portraying various aspects of risk visually

Risk information	Options for visual format
The risk and its effects	If the effects of the risk can be seen (such as a visible health effect, effect on plants and foods, etc.), depict them in a photo or illustration to help people identify the risk. Also, consider showing conditions leading to or indicating a risk, such as blocked fire doors in an industrial plant, high-power lines for electromagnetic fields (Figure 14-1), skull and crossbones indicating poison on warning labels, or people demonstrating unhealthy or unsafe behaviors and their consequences.
Size and significance of the risk	Show the risk in the affected population, using numbers or charts. Show the risk over time, increasing or decreasing, such as in a line graph or bar chart. Compare judiciously with other similar risks to show relative magnitudes (see Table 14-3 and Chapter 6). Consider including a recommended "action" level—a point at which people may want to take action to mitigate the risk (see Table 14-3).
Likelihood of risk for specific people	Show probabilities and uncertainties for various conditions. Consider phrasing as "X in Y chances of occurring" under certain conditions. Tables, charts, and graphs can show various risk levels for various situations (see Figure 14-6). "If–then" flowcharts can help people walk themselves through the risk probabilities (see Figure 14-8 and Figure 14-9).
Change over time	Use graphs, charts, or pictograms (small pictures representing the risk) to indicate trends over time. Consider several different representations if many variables are involved, such as conditions that change the risk over time.
Alternatives to the risk, with corresponding benefits and dangers	Compare alternatives and list pros and cons of each. Consider using tables if there are shared variables among the alternatives and the alternatives are being compared in a similar way (costs, environmental effects, health effects, etc.). If the alternatives are not easily comparable, use formats that do not invite comparison on the same scales.

- Has progress in accident prevention over the last year been made? A bar graph was suggested for comparing relative magnitudes: reduction in accidents in the current year compared with past years.
- What shortages are noticeable in our safety education? A line graph to show the norm and a bar chart to show deviations were suggested for showing accident trends that may need attention in the upcoming year.
- What special emphasis should be made during the next school year? A pictograph was suggested for symbolizing the curriculum areas requiring more safety emphasis.

As the example suggests, sometimes the best approach is to show as many aspects of the risk as possible, giving people more information and choices. It may be helpful to use several graphics, each highlighting a certain aspect of the risk. One study found that a presentation which included the most information scored as well as or better than other formats on almost all measures of communication success (Weinstein et al. 1989). There was no evidence that respondents were confused by the amount of information.

PRETEST GRAPHICS WITH THOSE WHO WILL USE THEM

After you have researched your audience, identified the uses of your graphics, and prepared an initial set of graphics, the next step is to pretest them with people who represent your target audiences. Pretesting graphics is usually done as part of a broader evaluation of messages and/or materials. For example, it is common to pretest an entire brochure or oral presentation containing graphic elements. Pretesting typically provides feedback on the graphic elements as well as other aspects of the message and presentation. Here, we focus on how to get the most out of pretesting for graphics.

> Poorly designed or produced visuals are worse than no visual at all.
> —Peter J. Hager and H. J. Scheiber (1997, p. 171).

Use interviews, discussion groups, and other techniques to get maximum information. (See "Whenever Possible, Pretest Your Message" in Chapter 6.) Ask people what they think each pictorial graphic means. The answer will help you determine whether your graphics are meeting their objectives. Ask whether anything in the pictures, tables, or charts is unclear, misleading, confusing, incomplete, or inaccurate. Ask what feelings are evoked and what words come to mind when viewing specific graphics. In addition, ask people about their overall reactions to the graphics. You may discover that some graphics come across as patronizing, scary, or overly technical, or that they carry an unintended message. For example, the U.S. Department of Homeland Security replaced the visual alerts of its Homeland Security Advisory System in 2011 after public and agency comments. Those who commented wanted alerts that were specific and time dependent, clearly delineated the affected areas, and provided guidance on actions to take.

After hearing from your audience, make decisions about how to modify your graphics, add more, or eliminate them to meet your audience's needs and your communication objectives.

What does pretesting reveal about graphic representations? The following is an example of some comments and changes resulting from pretesting of an information booklet about environmental risks (Pacific Northwest National Laboratory 1994, 1995). The booklet was designed for the general public. Feedback came from students, teachers, agricultural representatives, state regulators, health department officials and practitioners, environmental advocates, and other community members:

- People wanted to compare the existing contaminant amounts with an existing standard. Wherever applicable, regulatory safety limits for contaminants in drinking water, air, and food were included in the applicable illustrations and tables. Figure 14-2 shows an example of a redesigned map.

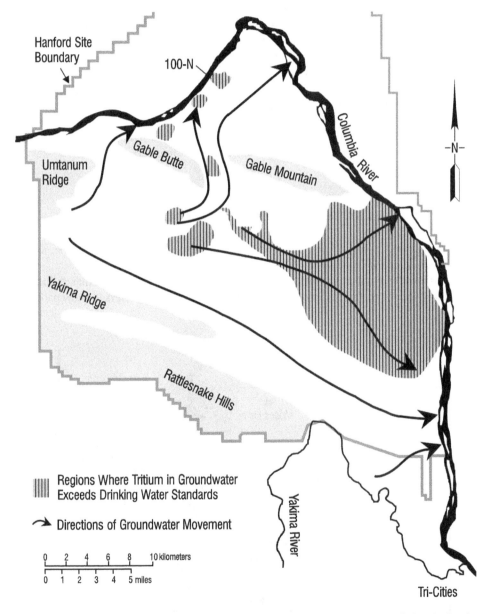

Figure 14-2. Redesigned map of groundwater contamination on a federal site in 1994. The map shows areas where drinking water standards were exceeded in the past years. (Source: Pacific Northwest National Laboratory 1995.)

- People wanted more orientation and clarification of maps. Maps showing contamination locations were clarified to show the direction of contaminant movement (see Figure 14-2). A small state map was included to help the reader locate the smaller geographic area shown on the map (Figure 14-3).
- Some people thought that photos of families enjoying the outdoors were not representative of actual conditions and that they appeared to put an overly "happy face" on a serious issue. Such photos were replaced with more informative ones such as workers gathering river water samples.
- People had trouble understanding illustrations that included numbers such as contaminant levels and quantitative exposure effects over time, regardless of which graphic format was used to convey this information. These illustrations were eliminated and discussed narratively in the text instead.
- People wanted to put events in a historical context and see at a glance when certain cleanup actions would be completed. Visual time lines (Figure 14-4) were added to show key dates and events.
- A table of numerical environmental monitoring results was seen as reference-type information that would be useful only for certain readers. The table was moved to the back so that its more detailed nature would not interrupt the narrative "story" before it.

USING VISUALS TO PERSONALIZE RISK INFORMATION

People often want to know what risks mean to themselves or to their family members. Visual displays can help personalize risk information.

Photos are effective for realistically depicting visible risk characteristics that can affect individuals. Health professionals often use photos in brochures or posters to show patients

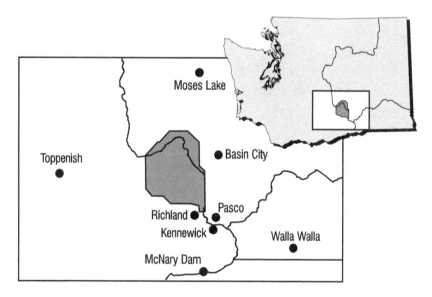

Figure 14-3. "Locator" state map used in conjunction with a more detailed map. (Source: Pacific Northwest National Laboratory 1995.)

Hanford timeline

Hanford Site established for
plutonium production for
World War II nuclear
weapons.

Monitoring of the
Columbia River
and sagebrush begins.

President Truman declares the start
of the Cold War. Hanford production
expands to keep pace.

| pre-1943 | 1943 | 1944 | 1945 | 1945 (to 1947) | 1947 | late 1940s |

Native American tribes,
then miners, farmers,
and others live in the
Columbia Basin.

First nuclear facilities
begin operating.

Largest amount of radioactive
iodine released to the air from
plutonium extraction.

Monitoring of animals
and groundwater begins.

Hanford nuclear production reaches
its peak. Operations release large
amounts of radioactive and chemical
waste materials to soil, groundwater,
and the Columbia River.

Hanford's nuclear
production facilities
close down.

A scientific study estimates radiation
doses the public may have received
as a result of past Hanford releases.

| 1955 (to 1965) | 1960 (to 1975) | 1964 (to 1989) | 1965 | 1987 (to 1990) | 1989 (to present) |

After leaks are discovered from
single-shell tanks holding highly
radioactive waste, the waste is
evaporated or pumped into safer,
new double-shell tanks.

President Johnson announces
a decreased national need for
weapons materials.

Health study begins to determine whether
there is more thyroid disease than usual
in people who lived downwind of Hanford
in the 1940s and 1950s. The study is
scheduled for completion in 1998.
Hanford's new mission is cleanup and
development of new technologies to
support national needs.

Figure 14-4. Time line of dates and events. The time line provided a historical context for explaining risks associated with a contaminated site. (Source: Pacific Northwest National Laboratory 1995.)

warning signs of certain health conditions that they should watch for. For example, photos can be used to accurately show early gum disease and potential skin cancer signs. County extension agents and pesticide control officials use photos to indicate damage done by certain pests, as well as to help people recognize the effects of improperly applied pesticides on plants.

Another way to personalize risk information is to show people various conditions associated with the risk and how those might apply to specific individuals. One information booklet used the chart in Figure 14-5 to help people determine their own exposure to radiation from various sources. During pretesting, people consistently rated this chart as one of the most useful pieces in the booklet.

The University of Rochester used a similar personalized approach in its self-test for nicotine addiction, which appears in a smoking cessation guide (Figure 14-6). A score of five or higher indicates possible nicotine addiction.

A regional study funded by the U.S. Centers for Disease Control and Prevention used the "road map" in Figure 14-7 that people could follow to find their own radiation dose from past radiation releases, based on demographic and lifestyle factors. Once people identified "their" category, they could find their own radiation dose in the ranges given in a corresponding chart (Figure 14-8). People who viewed these graphics indicated that they were easy to follow and provided helpful information.

COMPARING RISKS IN VISUAL FORMATS

Laypersons and experts often want to know how risks, especially unfamiliar ones, compare with a more common risk or with alternatives. Visuals can be used to compare magnitudes, effects, and alternatives on a common numeric scale, for example.

What's My Radiation Dose?

Use this simple chart to see how much radiation exposure you receive each year. Fill in the numbers in the right-hand column. The total gives an estimate of your average annual radiation dose.

		Your Average Annual Dose (millirem)
Where You Live	• Cosmic radiation at sea level (from outer space)	26
	• For your elevation (in feet) add this number of millirems:	____

0-1000 ft = $2^{(a)}$ 5-6000 ft = 29
1-2000 ft = $5^{(b)}$ 6-7000 ft = 40
2-3000 ft = $9^{(c)}$ 7-8000 ft = 53
3-4000 ft =15   8-9000 ft = 70
4-5000 ft = $2^{(d)}$

(a) Includes the Tri-Cities, Walla Walla, Seattle, and Portland
(b) Includes Las Vegas
(c) Includes Spokane
(d) Includes Salt Lake City

• Terrestrial (from the ground):

▨	Add 23	____
☐	Add 46	____
▰	Add 90	____

	• House construction: — If you live in a stone, brick or concrete building	Add 7 ____
	• If you live within 50 miles of Hanford, dose from Hanford operations	Add 0.03 ____
What You Eat, Drink, and Breathe	• Internal Radiation (in your body): — From food and water — U.S. average — From breathing in air (radon) — U.S. average	40 200
How You Live	• Jet plane travel: — For each 1000 miles you travel in an average year	Add 1 ____
	• If you watch TV	Add 1 ____
	• If you have annual medical diagnostic exposures — (for example, dental, chest x-rays) — U.S. average	Add 50 ____
	• If you have had nuclear medical procedures (radiation therapy) — average per procedure	Add 430 ____
	• If you have an annual mammogram	Add 75 ____

My total estimated annual radiation dose ____

Note: 360 mrem is the average for a U.S. resident in a year.

Figure 14-5. Chart used to help people determine their own radiation dose from various sources. (Source: Pacific Northwest National Laboratory 1994.)

Nicotine Addiction Self-Test

Circle one answer for each question.

Do you usually smoke your first cigarette of the day within 30 minutes of waking up?	No Yes
Do you find it hard not to smoke in places where it's not allowed, such as at the library, theatre or doctor's office?	No Yes
Do you smoke 10 or more cigarettes per day?	No Yes
Do you smoke 25 cigarettes per day?	No Yes
Do you smoke more during the morning than during the rest of the day?	No Yes
Do you smoke even when you are so ill that you are in bed most of the day?	No Yes

Give yourself one point for each question answered "Yes." What was your total score? _____ points.

The higher your score, the higher your addiction level. If you scored five or higher, you may be highly addicted to the nicotine in cigarettes. Nicotine replacement therapy or Zyban® may be especially helpful for you.

No matter how addicted you are, you *can* stop smoking!

Figure 14-6. Addiction self-test. (Source: University of Rochester School of Medicine and Dentistry 2001; used with permission.)

In one study, people appreciated seeing the health risks from geologic radon and asbestos compared with health risks from smoking (Weinstein et al. 1989). The subjects stated that the comparison helped them understand the data and reduced ambiguity about the risk. The authors noted that the comparison was appropriate because smoking increases the risks from both radon and asbestos.

Table 14-3 shows another way to depict radon risk for smokers by comparing it with other common hazards. The U.S. Environmental Protection Agency (EPA) uses this table to help citizens understand their risk of radon in the home and to recommend action. (An accompanying table, not shown here, depicts risks for the nonsmoker.)

It is important to note that using comparisons to clarify risks can lead to confusion and outrage rather than illumination, as researchers and practitioners have discovered.

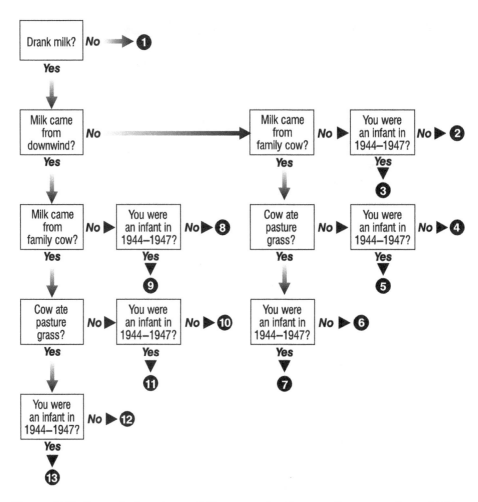

Figure 14-7. Example "road map." By answering the questions, people could identify the category into which they best fit, as indicated by numbered circles. (Source: Technical Steering Panel 1990.)

Follow the advice in "Principles for Comparing Risks," in Chapter 6. And, as we have emphasized, the best foundation for making a decision about comparisons is to analyze your audience and pretest the information before disseminating it.

STATIC VERSUS INTERACTIVE VISUALS

Another dimension of visual formats is presenting the risk information in an interactive or static format. It seems intuitive that asking the user to actively process a graphic, such as by creating, changing, or completing it, would improve comprehension and decision making more than just passively viewing the graphic.

However, this is not necessarily the case. In one study, people were asked either to look at a static copy of, or fill in a graph of, the risks from two thyroid cancer treatments

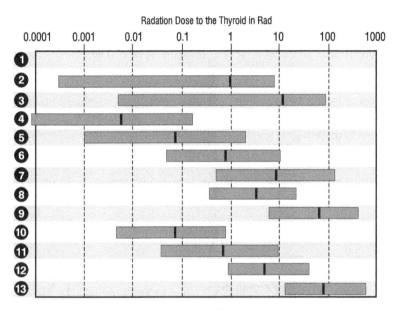

Radiation Dose to the Thyroid in Rad

Vertical lines in the bars are the medians. The median is the dividing point showing where half of the people in that category received a larger dose than the median dose and half the people received a smaller dose.

Figure 14-8. Range of possible radiation doses by category. The numbers in the circles correspond to those in Figure 14-7. (Source: Technical Steering Panel 1990.)

Table 14-3. Home radon risk for smokers and corresponding recommendations

Radon level	If 1000 people who smoked were exposed to this level over a lifetime. . .	The risk of cancer from radon exposure compares to. . .	What to do: Stop smoking and. . .
20 pCi/L	About 135 people could get lung cancer	100 times the risk of drowning	Fix your home
10 pCi/L	About 71 people could get lung cancer	100 times the risk of dying in a home fire	Fix your home
8 pCi/L	About 57 people could get lung cancer		Fix your home
4 pCi/L	About 29 people could get lung cancer	100 times the risk of dying in an airplane crash	Fix your home
2 pCi/L	About 15 people could get lung cancer	2 times the risk of dying in a car crash	Consider fixing between 2 and 4 pCi/L
1.3 pCi/L	About 9 people could get lung cancer	(Average indoor radon level)	(Reducing radon levels below 2 pCi/L is difficult)
0.4 pCi/L	About 3 people could get lung cancer	(Average indoor radon level)	

Source: U.S. Environmental Protection Agency et al. (1992).

and then identify the better treatment option. However, the interactive risk graphic backfired. Fewer people completed it, and they were also less likely to recognize the best treatment option than those who viewed the static version. The authors concluded that interactivity, though visually appealing, can present a cognitive burden and distract people from understanding relevant statistical information (Zikmund-Fisher et al. 2011).

It is important to note that this research does not rule out the use of online collaborative or other technology-based visual formats. In fact, virtual worlds and other immersive environments have been shown to be useful for learning and training, especially in health care professions (Hansen 2008). Nevertheless, the static versus interactive study shows that greater technology sophistication is not always better. As always, we recommend tailoring the format to the audience and purpose, and pretesting the format in advance.

DEPICTING PROBABILITY AND UNCERTAINTY

> Probability is how likely the event is to occur. Uncertainty represents the factors about a hazard that are not completely known.

Risk communicators often struggle with how to present two particular characteristics of risks: probability and uncertainty. Deaths, disease, or injuries can happen at various levels or under various conditions.

Probability is how likely the event is to occur. An example of explaining probability is telling someone that they have a 1 in 10 chance of getting a certain kind of cancer in their lifetime. Identifying the probability that something will happen is often based on known and unknown factors. The known factors for predicting cancer occurrence could include age, gender, and smoking habits.

Uncertainty represents the factors about a hazard that are not completely known. For example, it is often impossible to say what caused a specific person to contract cancer because the disease may have been triggered by numerous factors, many of which medical science does not yet fully understand.

Research in how people respond to probabilities and uncertainties varies widely in its results, and some of the results conflict with each other. This conundrum is both frustrating and intriguing for those who must communicate such characteristics! Our approach here is to present some of the salient research results and suggest guidance that may be considered for various risk communication situations.

Presenting Probability

In explaining probability, an odds ratio often is used, meaning a fraction with the numerator depicting the chance of something happening and the denominator depicting the total number of possibilities. For example, the odds ratio of 1/10 indicates that there is a 1 in 10 chance that a certain thing will happen.

Even if people understand the magnitude of the probability, there is no way to guarantee how they will respond. In genetic counseling, for example, some clients focus on the denominator of the odds ratio—the large number of people who do not get a negative genetic trait—and are reassured. But other clients focus on the numerator and are frightened by the image of that one person among the many who does suffer harm (Weinstein et al. 1994). In addition, a growing body of research in medicine, public health, and even climate change suggests that people's perception of probability varies with the severity of the outcome. That is, when viewing risk portrayed in the format of "1 chance in X of

Y occurring," they will consider X larger as Y gets worse. For example, a 1 in 20 chance of catching the flu will be seen as a lesser probability than a 1 in 20 chance of getting colon cancer (Pighin et al. 2011).

Studies have shown that people prefer probability estimates using human figures (Figure 14-9) rather than bar graphs (Figure 14-10) (Goodyear-Smith et al. 2008; Schapira et al. 2001, 2006). People said that they could identify with human figures and that the information was more understandable and carried more impact than the bar graph. However, when comparing more than one kind of risk at the same time (cancer, heart disease, stroke), people preferred a vertical bar graph such as that shown in Figure 14-11.

One caution when using human figures such as in Figure 14-9 is not to cross out victims with the letter "X." In a study by the Duke Risk Communication Laboratory, women did not like the idea of having the stick figure women who were affected by breast cancer "x'd" out

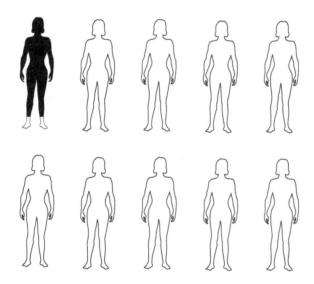

Figure 14-9. Human figures used in risk estimates. The highlighted female figure among a total of 10 represents the 9% lifetime risk of breast cancer for a 50-year-old woman. This depiction was preferred over the bar chart in Figure 14-10. (Source: Schapira et al. 2001; used with permission.)

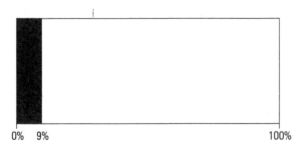

Figure 14-10. Risk estimate in bar graph format. This figure represents the same information as in Figure 14-9, but in a different format. (Source: Schapira et al. 2001; used with permission.)

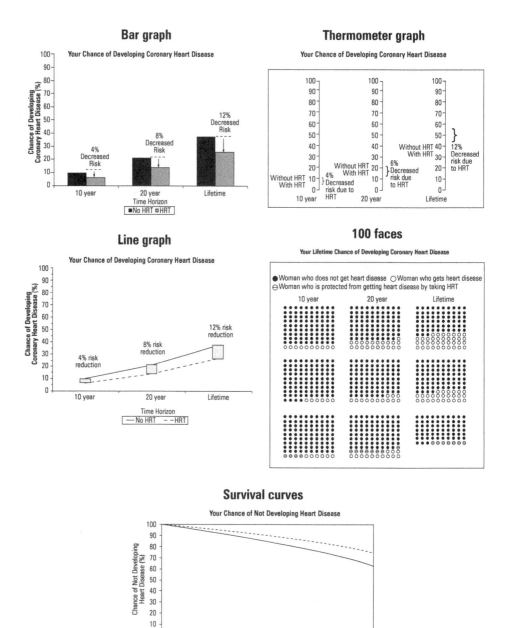

Figure 14-11. Graphical displays of heart disease risk with and without hormone replacement therapy (HRT). Women preferred the bar graph (first display) over the other formats. (Source: Fortin et al. 2001; used with permission.)

(Lipkus and Hollands 1999). It is better to color them in, as shown in Figure 14-9.

In another study, women were asked to choose among several graphical formats for communicating the risk of heart disease with and without hormone replacement therapy (Fortin et al. 2001). Women were shown the same risk information portrayed in bar graph, line graph, thermometer graph, 100 faces, and survival curve formats (Figure 14-11). Respondents overwhelmingly choose the bar graph, saying that it was basic, simple, and clearer than the other formats. They also preferred lifetime risk estimates over 10- or 20-year horizons and absolute over relative risks. And they wanted to see a narrative explanation with the graphic displays.

> Human figures may be best for single estimates of risk probabilities, whereas bar charts are more effective for risk probability comparisons.

These studies suggest that for single estimates of risk probabilities, human figures may be best, whereas bar charts are more effective for risk probability comparisons.

For explaining medical risks associated with the likelihood of treatment effects, researchers have found that it is not sufficient to say something like "You have a low chance of experiencing a side effect." Patients interpret "low" and "high" differently. Instead, it is helpful to show patients the difference between baseline and treatment risk visually. Figure 14-12 shows the number of women who would get cataracts without tamoxifen treatment, with a separate color showing the additional number of women who would experience the side effect of cataracts with tamoxifen treatment. Researchers found that in a study of more than 600 women considering taking tamoxifen as chemoprevention, this visual reduced worry about medication side effects and reduced perceived likelihoods of experiencing a side effect (Zikmund-Fisher et al. 2008).

The Presidential/Congressional Commission on Risk Assessment and Risk Management (1997) suggests several approaches to explaining low-risk probabilities that have been found helpful in practice. One is to use analogies: one in a million is equivalent to 30 seconds in a year, 1 inch in 16 miles, or 1 drop in 16 gallons. Another approach is to

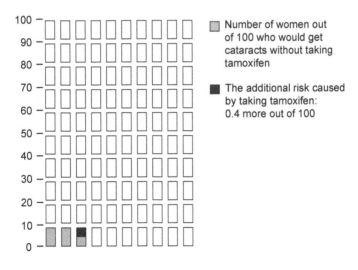

Figure 14-12. Added risk of side effects from medical treatment among 100 women. This visual reduced patients' worry about side effects (cataracts) and the likelihood of experiencing them. (Source: Zikmund-Fisher et al. 2008; used with permission.)

express risk in terms of the number of persons who might be affected per year or per hypothetical 70-year lifetime. In explaining and portraying these kinds of probabilities, it is important to clarify that the one-in-a-million probability is not an estimate of actual risk but a statistical upper boundary.

Another approach for conveying probabilities is to convert units of population to periods per event, such as 1 death expected in 3500 people. The city of Columbus, Ohio, did an analysis estimating that one death would occur in Columbus in 204 years from an additional cancer risk at the theoretical one-in-a-million level. The analysis compared that risk with frequencies of several deaths per day or every few days for measurable risks, such as ordinary rates of heart disease, cancer, homicide, and automobile collisions. The mayor of Columbus stated that the analogy helps citizens understand the magnitude of the effects on the community caused by federal or state regulations concerning the environment, transportation, labor, or education (Presidential/Congressional Commission 1997). This approach is similar to the one portrayed in Table 14-3.

A single denominator should be chosen for comparisons (for example, 1 in 10,000 and 337 in 10,000). It is easier for people to understand whole numbers (for example, 1 in 10,0000) rather than fractions or decimals (0.01 in 100); thus, if risks are very small, larger denominators will be necessary (Fischhoff et al. 2011).

We have two cautions in showing probabilities. Despite these and other examples showing that portraying risk is effective in practice, conflicting studies show that people respond differently to portrayal of risk probabilities than they do to other kinds of risk information (Schapira et al. 2001; Weinstein et al. 1994). Pretesting your information, as described in the section "Pretest Graphics with Those Who Will Use Them," will provide insights for determining the best approach. The second caution is to avoid attempting to influence people by downplaying or highlighting the magnitude of the risk. This is especially true in consensus communication efforts, where the goal is to present the risk as objectively as possible so that people can evaluate it for themselves.

> Avoid attempting to influence people by downplaying or highlighting the magnitude of the risk.

Presenting Uncertainty

Risk assessment is not an exact science and as such carries with it many uncertainties. Uncertainty is often shown as a range of risk estimates or potential consequences, depending on various factors such as the demographics of the population at risk.

Unfortunately, people are unfamiliar with uncertainty in risk assessment and in science in general, making the job of the risk communicator all the more challenging. In one study, up to 20% of respondents reading news stories about risks had difficulty recognizing the presentation of uncertainty in the form of a range of risk estimates (Johnson and Slovic 1995), as opposed to a single number that represented the risk level. Pretesting various visual formats (and narratives) that reflect uncertainty should help reveal confusion and suggest potential visual solutions.

As with probabilities, the way the information is presented affects how people perceive it. Research has shown that people see risks with uncertainties as greater (1) if the risks are more ambiguous, (2) when the unfavorable risk evidence is presented last, (3) when the most unfavorable risk studies were performed most recently, and (4) when some aspect of the risk is substantially negative (Viscusi et al. 1991). Risk communicators should be aware of these factors when presenting uncertainty. The goal is not to present the uncertainty in the most favorable format to persuade people, but to present it as objectively as possible.

In dealing with uncertainty, risk managers and communicators also grapple with the issue of credibility. Acknowledging uncertainty has been shown to increase the perceived trustworthiness of the information sources (by admitting that they do not know the exact number) but less competent (they are not smart enough to figure it out) (Johnson and Slovic 1995). One respondent, upon viewing uncertainty presentations hypothetically published by a U.S. government organization, labeled the agency "honest imbeciles"—a dubious distinction! We agree with practitioners who advocate dealing with uncertainty head-on. It is best to acknowledge uncertainty; explain why it exists; describe what, if anything, can be done to get a better handle on it; and explain how the risk can be reduced in the meantime.

Probability plus Uncertainty

Some risks contain characteristics of both probability and uncertainty, and both must be shown. In one situation, for example, the National Hurricane Center started with a hurricane warning graphic that focused on uncertainty, then, based on public feedback, added another to show probability. The initial graphic (Figure 14-13) is known as the "cone of

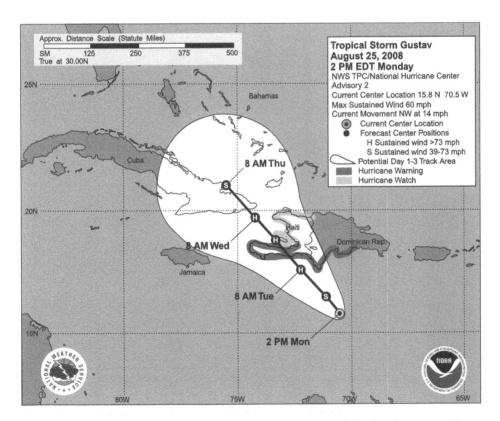

Figure 14-13. A typical "cone-of-uncertainty" graphic used for hurricane warnings. Many people mistakenly assumed that if they were not directly in the path of the black line, they were safe. (Source: National Hurricane Center.)

uncertainty." Interviews, surveys, and testing showed that many people mistakenly assumed that if they were not directly in the path of the black line, they were safe (Broad et al. 2007). In reality, even if the line predicted perfectly the hurricane eye's path—something it rarely does—many people forget that a storm's devastating winds can spread out dozens of miles in all directions. So residents in the entire white area of the cone could be at risk, depending on the severity and impact of the storm.

To overcome the public's overemphasis on the black line, the National Hurricane Center added a graphic in 2006 that shows the probability that tropical storm or hurricane-force winds will threaten a specific area over a period of time, such as a 24-hour period (Figure 14-14). Tested for 2 years, this additional graphic gives residents a clearer picture of whether and when they could experience storm-force winds and helps them make important decisions, such as whether to evacuate.

TV weather forecasters, emergency managers, and the public now have access to the full-color graphic, a series of ovals spreading outward from the storm's center. The ovals closest to the center show the highest probability for storm-force winds within a certain time period, and the probabilities get smaller with each bigger oval that surrounds the center.

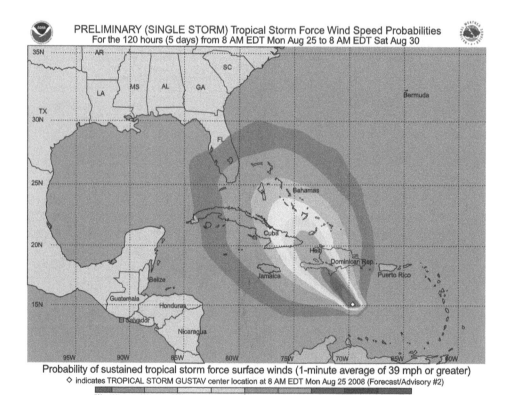

Figure 14-14. Additional graphic for hurricane warnings, showing probability by geographic area. The ovals closest to the center of the storm have the highest probability for dangerous winds. (Source: National Hurricane Center.)

Some formats show probability and uncertainty together, such as the chance of a certain disease showing up over time in a given community near a toxic waste site. The chances that someone in the community will get cancer from exposure to the waste site may vary from small to large, with the most likely point on the range shown with a certain degree of confidence.

One statistical format that does this is called a cumulative distribution function, as shown in Figure 14-15. The horizontal axis depicts a range—it could be a range of exposures, potential health effects, or other type of data. The vertical axis depicts the percent of elements in the range that corresponds to a particular number in the range. For example, this figure shows that 50% of people received an exposure between 1 and 10 (units are not specified in this generic example).

Risk communicators should carefully consider and pretest such graphics to determine whether they add value or increase confusion. Ibrekk and Morgan (1987) found that cumulative distribution functions with no explanation were dramatically misinterpreted by laypeople. In cumulative distribution functions that used the statistical concept of a mean, people thought that the mean was much higher than indicated on the graph and often misidentified the maximum as the mean.

The difficulty in communicating with the cumulative distribution function was corroborated in a multiyear environmental risk study funded by the U.S. Centers for Disease Control and Prevention (Technical Steering Panel 1990). Citizens typically did not use or refer to the data shown in cumulative distribution functions. Not one media outlet printed or aired that form of graphic, though it was provided to dozens internationally. Instead, many people focused on the highest (worst-case) number for the risk, regardless of the repeated emphasis of how unlikely it was and how few people were affected by it at that level. A more used portrayal was a series of ranges with the median risk levels within each range, as shown in Figure 14-8.

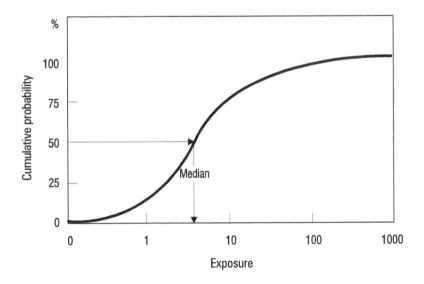

Figure 14-15. Example of a cumulative distribution function.

WARNING LABELS

Warning labels that include visual elements deserve special consideration. A robust body of research exists on the design of warning labels such as those for prescription drugs, herbal remedies, and household chemicals.

Educate yourself about mandates regarding the use of graphics to convey warnings. For example, a number of countries now mandate large, photo- or picture-based warnings on cigarette packages, based on research showing that they cut down on smoking (Hammond et al. 2007). Many are purposely revolting, to gain attention and show the seriousness of the risk. Figure 14-16 shows some examples.

A worldwide voluntary guideline exists for labeling chemical hazards: The Globally Harmonized System of Classification and Labeling of Chemicals (see Chapter 3). This is a voluntary hazard classification system developed by the United Nations Economic Commission for Europe and its partners. At least 67 countries had adopted the system at the time this book was published. The U.S. Occupational Safety and Health Administration is in the process of aligning its requirements to this system. The system requires specific symbols, or pictograms, on labels to indicate certain hazards. For example, the symbol for carcinogens and other health hazards is a white jagged "star" inside a dark silhouette of a human frame (Figure 14-17). It also requires words such as "danger" or "warning" to appear, along with standard phrases assigned to each hazard class. At the time this book was published, the most recent labeling requirements were shown on the Globally Harmonized System web page (http://www.unece.org/trans/danger/publi/ghs/pictograms.html). Be aware that even when such a system is adopted, education may still be necessary to ensure that people understand what the symbols mean. In Zambia, people appreciated nonabstract warning symbols they could relate to, such as skulls and crossbones, flames, or ghost-like images. But other symbols sometimes used to indicate hazards, such as an exclamation point and the St. Andrews cross (a tilted X), were not well understood (Banda and Sichilongo 2006).

Colors have meaning as well. In the agricultural sector, red is associated with high toxicity, while other colors, such as yellow and blue, may not be. Industry terminology for hazards varies as well. The industry and transportation sectors tend to use the word "danger" rather than "toxic" (Banda and Sichilongo 2006).

In the United States, the American National Standards Institute (ANSI) has set a standard, known as Z535.3, for developing and evaluating symbols and warnings, most often used for workplace and commercial hazard signage (ANSI 2002). The standard specifies a minimum 85% correct interpretation and a maximum 5% critical confusion, meaning interpretations that may lead to the direct opposite of the sought behavior.

For example, using the ANSI standard, researchers pretested four warning labels for pharmaceutical packages intended to keep women from taking certain medication that could harm a fetus (Goldsworthy et al. 2008). Four graphic symbols, shown in Figure 14-18, were tested with one of the target audiences, adolescent girls who could become pregnant.

Pretesting showed mixed results, demonstrating the dilemmas risk communicators sometimes face. Participants judged symbol C as being the most effective because of the universally recognized symbol, the skull and crossbones. But the symbol with the highest correct interpretation was symbol B. Because all the symbols rated fairly high by ANSI standards but differed in preference among the target audience, the researchers declined to choose the "best" one.

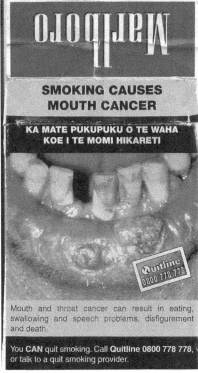

Figure 14-16. Examples of tobacco warning labels shown to be effective. These cigarette packages from New Zealand are labeled in English and Maori.

However, a key finding was that when text was added to the symbols, as in Figure 14-19, the correct interpretation increased. Specifically, adding the phrase "may become pregnant" made that concept salient to the adolescents, whereas it was not with just the symbol. One way to approach this would be to include the skull and crossbones icon in the best-interpreted symbol (B), add the text warning, and pretest again against the other choices.

Figure 14-17. Warning symbol for respiratory sensitization, cancer, and other human health hazards. This symbol, with the diamond shape, is part of the Globally Harmonized System of Classification and Labeling of Chemicals.

(A)   **(B)**   **(C)**   **(D)**

Figure 14-18. Four versions of medication warning labels that were pretested with adolescents. (Source: Goldsworthy et al. 2008; used with permission.)

Figure 14-19. Medication warning label with text added to increase understanding. (Source: Goldsworthy et al. 2008; used with permission.)

CONSIDER USING ACTION LEVELS

For crisis and care communication efforts, where it is prudent to take a specific action even when uncertainty exists, some specialists advise using an action standard. Table 14-3 shows an example of action standards, where recommendations corresponding to various radon levels are given. If the risk gets to that level, the reader is advised to take action.

A graphic called a risk ladder has been shown to be effective in explaining risks and in recommending associated action levels. Risk ladders help people "anchor" a risk to upper- and lower-bound reference points. Including an action standard with the risks increases the likelihood that people will follow recommendations (Weinstein et al. 1989). Figure 14-20 shows a risk ladder conveying a range of radon risks and associated action levels.

> Risk ladders help people "anchor" a risk to upper- and lower-bound reference points.

This risk ladder communicates both risk magnitude, relative risk, an action standard (4 picocuries per liter), and advice about how to interpret the risk and what action to take, if any. Studies showed that the risk ladder in Figure 14-20 helped people distinguish among risk levels, identify appropriate mitigation intentions in accordance with their level of risk, and feel confident that they understand the risk (Weinstein et al. 1989).

One interesting finding was that people's perceptions of threat are influenced by the location of the risk on the ladder, more so than the numbers themselves (Sandman et al. 1994). Thus, if the communication goal is to get people to pay attention to a risk that they may be apathetic about, placing a risk closer to the top of the ladder will increase perceived risk. A downside of risk ladders is that they may suggest a dichotomy, whereby people may feel that everything up to the action level is safe, and everything beyond it is dangerous. The actual situation is more often a continuum, and those who communicate risk should convey this. For example, the risk ladder in Figure 14-20, despite including an action level at 4 picocuries per liter, also contains explanations with the advice given at each stage.

ETHICAL PORTRAYAL OF RISK INFORMATION

Many risk communication experts feel that persuasive messages such as fear appeals are manipulative and that people should simply be given the facts and allowed to make their own decisions. Yet beyond being blatantly persuasive, risk information can be portrayed in ways that are arguably deceptive. Here, we focus on several ethical factors that risk communicators should consider when portraying the visual aspects of risks.

> The format used to present statistical information influences people's perception of the likelihood of events.

Researchers have found that the format used to present statistical information influences people's perception of the likelihood of events (for example, Britton 1991; Halpern et al. 1989). In his classic books describing the visual display of quantitative information, Yale University professor Edward Tufte describes how various design "tricks" are used to circumvent what he calls graphical integrity (Tufte 1983). Two of the most common faults that risk communicators should be aware of are

1. Using pictorial representations that are out of proportion to the actual numerical quantities represented, especially when depicting increases or decreases

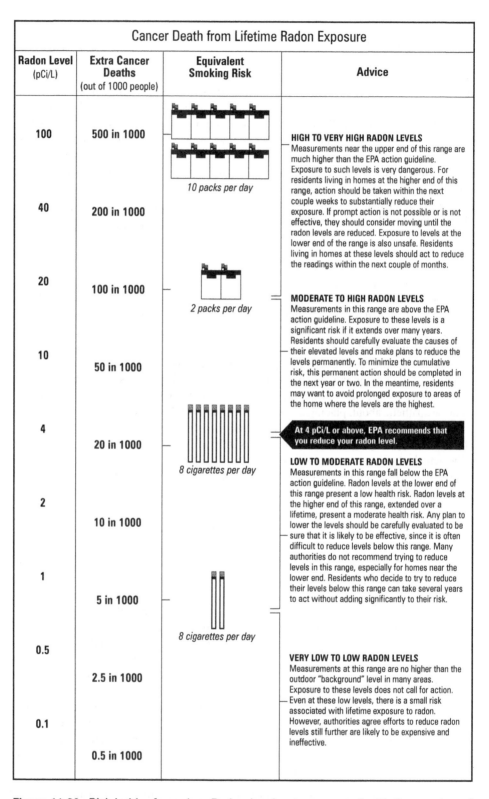

Figure 14-20. Risk ladder for radon. Radon levels are compared with the number of extra cancer deaths and, for perspective, an equivalent number of cigarettes smoked. The "advice" column recommends associated action (or no action) levels. (Source: Lipkus and Hollands 1999, p. 152; used with permission.)

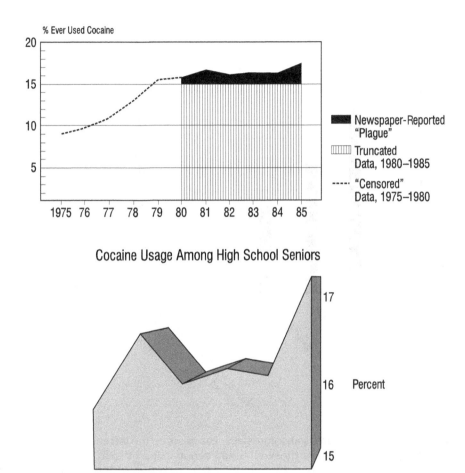

Figure 14-21. Deceptive use of data. The figure on the top shows the original data depicting a trend over time. The figure on the bottom shows how these data were improperly used to depict a different trend. Only certain data from the figure on top were used, and the scale was compressed to accentuate the peaks and thus "reveal" a cocaine "epidemic." (Source: Orcutt and Turner 1993; used with permission.)

2. Using purely decorative design elements (which Tufte calls "chartjunk") that obscure the meaning of data.

Orcutt and Turner (1993) provided an interesting example of data manipulation in a health-related situation. They showed how major news magazines selectively used and displayed government survey data to manufacture a youth "cocaine epidemic" in the mid-1980s. Figure 14-21 shows an adapted version of the data and the treatment imposed by the news magazines.

Another problem that can lead to data manipulation, especially when describing the effect of various medical treatments, is using different reference classes or, more specifically, using relative risks versus absolute risks. Figure 14-22 shows these comparisons.

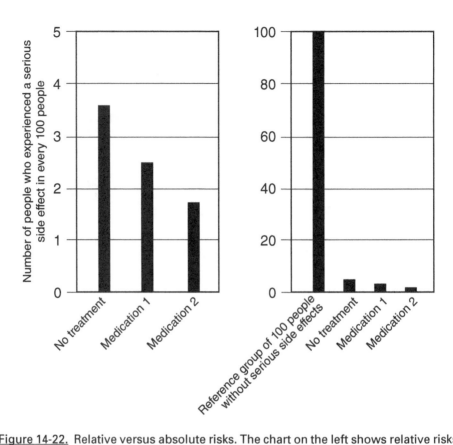

<u>Figure 14-22.</u> Relative versus absolute risks. The chart on the left shows relative risks only within a small group of those who had serious side effects. The chart on the right shows absolute data by including the reference group of 100 people who had no side effects. Because the left-hand chart appears to show a much greater risk reduction, it could be used deceptively to persuade people to choose medication over nontreatment.

The two charts both show the effect of two kinds of medications to treat a group of patients. The chart on the left shows only those who experienced a serious side effect as a result. It appears that both medications decreased the chance of having serious side effects by up to 50%, making it more likely that patients would choose to be treated with medication. In fact, pharmaceutical claims are all too often guilty of this kind of comparison because the risk reduction appears so significant.

The chart on the right, however, which includes the reference group of 100 people, gives a different perspective. It shows that, compared with the reference group of 100, only about 2% reduced their side effects by taking either of the two medications. One could thus conclude that taking medication does not make much of a difference because most people in the reference group suffered no ill effects, whether they received treatment or not. In fact, many risk communication professionals advocate using absolute versus relative risks because absolute risk is less misleading and allows people to make a more informed decision (Gigerenzer and Edwards 2003).

Beyond the deliberate misuse of graphics to manipulate, graphics can be used carelessly. This can lead to the obscuring of important information and, consequently, faulty and even tragic decisions. Edward Tufte makes a convincing argument that improperly designed pictorial displays linking O-ring damage and temperature failed to clearly convey the risk that led to the 1986 space shuttle *Challenger* explosion (Tufte 1997). As evidence, he shows a series of O-ring-related figures exemplifying the lack of clarity in depicting cause and effect, improper ordering of data, and deceptive "chartjunk."

Tufte also blasts the misuse of PowerPoint® slides as a potential contributor to the *Columbia* space shuttle disaster in 2003. The National Aeronautics and Space Administration (NASA) used the slides, which summarized engineering studies about possible tile damage on the *Columbia*, as a rationale not to investigate the tiles further during flight. Calling the slide presentations a "PowerPoint festival of bureaucratic hyperrationalism," Tufte says that the overwhelming levels of bulleted hierarchies, "grid prisons" surrounding spreadsheet entries, and the fact that the reasoning is broken up into "stupefying fragments" among many slides helped obscure what the data really showed—that the tile problem could indeed cause significant heat damage (Tufte 2003).

More humorously, a somewhat notorious graphic made the rounds in 2010 as an example of a military tool that spun out of control. It showed a PowerPoint slide that was meant to portray the complexity of American military strategy, showing more than 60 color-coded "nodes" with various connections to each other. When General Stanley McChrystal, the leader of American and North Atlantic Treaty Organization (NATO) forces in Afghanistan, was shown the slide, he was reported to have quipped, "When we understand that slide, we'll have won the war" (Bumiller 2010).

The lesson here is to realize that design can affect people's perception of the information being conveyed. If you feel that a certain graphic may be perceived differently as a result of various formats, pretest different variations of the graphic to determine to what extent the design affects audience perceptions. The goal is to design a graphic that presents risk-related information as clearly and objectively as possible. At the same time, the graphic form should portray factual information, such as measured quantities of contaminants, in ways that are consistently understood across your target audiences.

USING VISUAL INFORMATION IN GROUP DECISION MAKING

Showing risk information and alternatives visually can be a very powerful tool for consensus communication that involves group decision making. For example, community members may be considering installing a dam on a local river. They may wish to visualize the potential effects of the dam on the local economy, recreational use, and fish spawning habits. Viewing these effects visually can provide a common foundation for members to discuss alternatives and trade-offs.

Computer-assisted decision software systems also can be effective. With interactive computer graphics, such systems can show, sometimes in real time, the effects when certain factors are varied. For instance, using the example described earlier, with proper input data, the system could create a graph estimating the decrease or increase in the fish population with a new dam. It could chart the amount of electricity produced over time. It could show estimated costs and savings with and without the dam. (Chapter 18 gives more information on using computers for consensus communication.)

One company specializing in environmental work has had some success with a user-friendly graphic interface that allows nonspecialists to see the workings of a risk assessment computer model. The interface shows the process by which the input (factors that influence the risk, such as lifestyle habits) affects the resulting risk assessment, and by how much.

CHECKLIST FOR VISUAL REPRESENTATION OF RISK

In depicting risk-related information in visual formats:

☐ The target audiences and uses of graphics formats have been identified.
☐ Visuals have been designed to portray specific aspects of a risk.
☐ Visuals have been personalized to the extent possible.
☐ Visuals with quantitative elements remain true to the original data.
☐ Probability and uncertainty have been depicted appropriately.
☐ Graphics have been pretested and modified in response to comments.

REFERENCES

American National Standards Institute (ANSI). 2002. *Criteria for Safety Symbols, Z535-3-Revised.* National Electrical Manufacturers Association, Washington, DC.

Banda, S. F. and K. Sichilongo. 2006. "Analysis of the Level of Comprehension of Chemical Hazard Labels: A Case for Zambia." *Science of the Total Environment*, 363:22–27.

Bord, R. J. and R. E. O'Connor. 1992. "Determinants of Risk Perceptions of a Hazardous Waste Site." *Risk Analysis*, 12:411–416.

Britton, R. L. 1991. *The Influence of Presentation Format on Interpretation of Paternity Test Results: Due Process or Deception?* Unpublished master's dissertation, University of California, Irvine, California.

Broad, K., A. Leiserowitz, J. Weinkle, and M. Steketee. 2007. "Misinterpretations of the 'Cone of Uncertainty' in Florida during the 2004 Hurricane Season." *Bulletin of the American Meteorological Society*, 88(5):651–667.

Bumiller, E. 2010. "We Have Met the Enemy and He Is PowerPoint." *The New York Times*, April 26, 2010. http://www.nytimes.com/2010/04/27/world/27powerpoint.html?_r=1&ref=technology (accessed January 25, 2013).

Carnegie-Mellon University. 1995. *Fields from Electric Power.* Department of Engineering and Public Policy, Pittsburgh, Pennsylvania.

Fischhoff, B., N. T. Brewer, and J. S. Downs, eds. 2011. *Communicating Risks and Benefits: An Evidence-Based User's Guide.* Food and Drug Administration, U.S. Department of Health and Human Services, Silver Spring, Maryland. http://www.fda.gov/downloads/AboutFDA/ReportsManualsForms/Reports/UCM268069.pdf (accessed January 25, 2013).

Fortin, J. M., L. K. Hirota, B. E. Bond, A. M. O'Connor, and N. F. Col. 2001. "Identifying Patient Preferences for Communicating Risk Estimates: A Descriptive Pilot Study." *BMC Medical Informatics and Decision Making*, 1:2. http://www.biomedcentral.com/1472-6947/1/2 (accessed January 25, 2013).

Geiger, C. J., B. W. Wise, C. R. M. Parent, and R. G. Hanson. 1991. "Review of Nutrition Labeling Formats." *Journal of the American Dietetic Association*, 91(7):808–815.

Gigerenzer, G. and A. Edwards. 2003. "Simple Tools for Understanding Risks: From Innumeracy to Insight." *British Medical Journal,* 327:741–744. http://bmj.bmjjournals.com/cgi/content/full/327/7417/741#REF21 (accessed January 25, 2013).

Goldsworthy, R. C., N. C. Schwartz, and C. B. Mayhorn. 2008. "Interpretation of Pharmaceutical Warnings among Adolescents." *Journal of Adolescent Health*, 42(6):617–625.

Goodyear-Smith, F., B. Arroll, L. Chan, R. Jackson, S. Wells, and T. Kenealy. 2008. "Patients Prefer Pictures to Numbers to Express Cardiovascular Benefit from Treatment." *Annals of Family Medicine*, 6:213–217.

Graber, D. 1990. "Seeing Is Remembering: How Visuals Contribute to Learning from Television News." *Journal of Communication*, 40:134–155.

Hager, P. J. and H. J. Scheiber. 1997. *Designing and Delivering Scientific, Technical, and Managerial Presentations.* Wiley, New York.

Halpern, D. F., S. Blackman, and B. Salzman. 1989. "Using Statistical Risk Information to Assess Oral Contraceptive Safety." *Applied Cognitive Psychology*, 3:251–260.

Hammond, D., G. T. Fong, R. Borland, K. M. Cummings, A. McNeill, and P. Driezen. 2007. "Text and Graphic Warnings on Cigarette Packages: Findings from the International Tobacco Control Four Country Study." *American Journal of Preventive Medicine*, 32(3):202–209.

Hansen, M. M. 2008. "Versatile, Immersive, Creative, and Dynamic Virtual 3-D Healthcare Learning Environments: A Review of the Literature." *Journal of Medical Internet Research*, 10(3):e26. http://www.jmir.org/2008/3/e26 (accessed January 25, 2013).

Ibrekk, H. and M. G. Morgan. 1987. "Graphical Communication of Uncertain Quantities to Nontechnical People." *Risk Analysis*, 7:519–529.

Johnson, B. B. and P. Slovic. 1995. "Presenting Uncertainty in Health Risk Assessment: Initial Studies of Its Effects on Risk Perception and Trust." *Risk Analysis*, 15:485–494.

Johnson, B. B., P. M. Sandman, and P. M. Miller. 1992. "Testing the Role of Technical Information in Public Risk Perception." *Risk: Issues in Health and Safety*, 3:341–364.

Lacerda, F. W. 1986. *Comparative Advantages of Graphic versus Numeric Representation of Quantitative Data.* Unpublished doctoral dissertation, University of Virginia Polytechnic Institute and State University, Blacksburg, Virginia.

Lang, A. 1995. "Defining Audio/Video Redundancy from a Limited-Capacity Information Processing Perspective." *Communication Research*, 22:86–115.

Lipkus, I. M. and J. G. Hollands. 1999. "The Visual Communication of Risk." *Journal of the National Cancer Institute Monographs*, 25:149–163.

Mayer, R. E., W. Bove, A. Bryman, R. Mars, and L. Tapangco. 1996. "When Less Is More: Meaningful Learning from Visual and Verbal Summaries of Science Textbook Lessons." *Journal of Educational Psychology*, 88:64–73.

Morgan, M. G., H. K. Florig, I. Nair, C. Cortes, K. Marsh, and K. Pavlosky. 1990. "Lay Understanding of Low-Frequency Electric and Magnetic Fields." *Bioelectromagnetics*, 11:313–335.

Orcutt, J. D. and J. B. Turner. 1993. "Shocking Numbers and Graphic Accounts: Quantified Images of Drug Problems in the Print Media." *Social Problems*, 40:190–206.

Pacific Northwest National Laboratory. 1994. *In Summary: Environmental Report 1993.* Prepared for the U.S. Department of Energy, Richland, Washington.

Pacific Northwest National Laboratory. 1995. *Hanford: Your Environment and Your Health.* Prepared for the U.S. Department of Energy, Richland, Washington.

Paivio, A. 1978. "Mental Comparisons Involving Abstract Attributes." *Memory and Cognition*, 6(3):199–208.

Pighin, S., J.-F. Bonnefon, and L. Savadori. 2011. "Overcoming Number Numbness in Prenatal Risk Communication." *Prenatal Diagnosis*, 31:809–813.

Presidential/Congressional Commission on Risk Assessment and Risk Management. 1997. *Risk Assessment and Risk Management in Regulatory Decision-Making.* Final Report, Volume 2, Washington, DC.

Sandman, P. M., N. D. Weinstein, and P. Miller. 1994. "High Risk or Low: How Location on a 'Risk Ladder' Affects Perceived Risk." *Risk Analysis*, 14(1):35–45.

Schapira, M. M., A. B. Nattinger, and C. A. McHorney. 2001. "Frequency or Probability? A Qualitative Study of Risk Communication Formats Used in Health Care." *Medical Decision Making*, 21:459–467.

Schapira, M. M., A. B. Nattinger, and T. L. McAuliffe. 2006. "The Influence of Graphic Format on Breast Cancer Risk Communication." *Journal of Health Communication*, 11(6):569–582.

Shepard, R. N. 1967. "Recognition Memory for Words, Sentences, and Pictures." *Journal of Verbal Learning and Verbal Behavior*, 6:156–163.

Slovic, P. 1987. "Perception of Risk." *Science*, 236:280–285.

St. Clair, R. A. 1956. *Presenting School Safety Facts: A Format for Graphic Presentation of Accident Data in the Annual Safety Report*. Unpublished thesis, Stout State College, Menomonie, Wisconsin.

Technical Steering Panel. 1990. *Initial Hanford Radiation Dose Estimates*. Washington Department of Ecology, Office of Nuclear and Mixed Waste, Olympia, Washington.

Tufte, E. R. 1983. *The Visual Display of Quantitative Information*. Graphics Press, Cheshire, Connecticut.

Tufte, E. R. 1990. *Envisioning Information*. Graphics Press, Cheshire, Connecticut.

Tufte, E. R. 1997. *Visual Explanations: Images and Quantities, Evidence and Narrative*. Graphics Press, Cheshire, Connecticut.

Tufte, E. R. 2003. *The Cognitive Style of PowerPoint*. Graphics Press, Cheshire, Connecticut.

University of Rochester School of Medicine and Dentistry. 2001. *Clear Horizons: A Quit Smoking Guide Especially for Those 50 and Over*. The Smoking Research Program, James P. Wilmot Cancer Center and Department of Community and Preventive Medicine, Rochester, New York.

U.S. Environmental Protection Agency, U.S. Department of Health and Human Services, and U.S. Public Health Service. 1992. *A Citizen's Guide to Radon (Second Edition): The Guide to Protecting Yourself and Your Family from Radon*. U.S. Government Printing Office, Washington, DC.

Viscusi, W. K., W. A. Magat, and J. Huber. 1991. "Communication of Ambiguous Risk Information." *Theory and Decision*, 31:159–173.

Wallack, L., L. Dorfman, D. Jernigan, and M. Themba. 1993. *Media Advocacy and Public Health: Power for Prevention*. Sage Publications, Newbury Park, California.

Weinstein, N. D., P. M. Sandman, and N. E. Roberts. 1989. *Communicating Effectively about Risk Magnitudes*. U.S. Environmental Protection Agency, Office of Policy Planning and Evaluation, Washington, DC. EPA 230/08-89-064.

Weinstein, N. D., P. M. Sandman, and W. K. Hallman. 1994. "Testing a Visual Display to Explain Small Probabilities." *Risk Analysis*, 14:895–896.

Zikmund-Fisher, B. J., P. A. Ubel, D. M. Smith, H. A. Derry, J. B. McClure, A. Start, R. K. Pitsch, and A. Fagerlin. 2008. "Communicating Side Effect Risks in a Tamoxifen Prophylaxis Decision Aid: The Debiasing Influence of Pictographs." *Patient Education and Counseling*, 73(2): 209–214.

Zikmund-Fisher, B. J., M. Dickson, and H. O. Witteman. 2011. "Cool but Counterproductive: Interactive, Web-Based Risk Communications Can Backfire." *Journal of Medical Internet Research*, 13(3):e60. http://www.jmir.org/2011/3/e60 (accessed January 25, 2013).

ADDITIONAL RESOURCES

Covello, V. T., P. M. Sandman, and P. Slovic. 1988. *Risk Communication, Risk Statistics, and Risk Comparisons: A Manual for Plant Managers*. Chemical Manufacturers Association, Washington, DC.

Gray, J. G. Jr. 1986. *Strategies and Skills of Technical Presentations: A Guide for Professionals in Business and Industry*. Greenwood Press, Westport, Connecticut.

Maibach, E. and R. Parrot, eds. 1995. *Designing Health Messages: Approaches from Communication Theory and Public Health Practice*. Sage Publications, Newbury Park, California.

Raines, C. 1989. *Visual Aids in Business*. Crisp Publications, Waterloo, Ontario, Canada.

Sandman, P. M. and N. D. Weinstein. 1994. *Communicating Effectively about Risk Magnitudes: Bottom Line Conclusions and Recommendations for Practitioners*. U.S. Environmental Protection Agency, Washington, DC. EPA-230-R-94-902.

United Nations Economic Commission for Europe. "Globally Harmonized System of Classification and Labeling of Chemicals." http://www.unece.org/trans/danger/publi/ghs/pictograms.html (accessed January 25, 2013).

Wogalter, M. S., D. M. DeJoy, and K. R. Laughery. 1999. *Warnings and Risk Communication*. Taylor & Francis, London.

15

FACE-TO-FACE COMMUNICATION

Another way to communicate risk is face to face through some form of oral presentation. Face-to-face communication includes a wide range of activities such as the following:

- One-to-one interactions (health care professional to patient, employee to employee, peer to peer, neighbor to neighbor)
- Small group settings (speaking before clubs, societies, organizations)
- Speakers bureaus
- Facility tours
- Demonstrations of activities related to preventing, analyzing, or monitoring risk
- Video presentations
- Audience interviews to elicit concerns or perceptions
- Information fairs
- Large formal learning situations (grade school to college courses, continuing education courses, training seminars)

In this book, we differentiate face-to-face communication from stakeholder participation in that usually only one of the groups involved (either those who are communicating risk or the audience at risk) does

In this book, we differentiate face-to-face communication from stakeholder participation in that usually only one of the groups involved (either those who are communicating risk or the audience at risk) does most if not all of the talking (one-way communication).

Risk Communication: A Handbook for Communicating Environmental, Safety, and Health Risks, Fifth Edition. Regina E. Lundgren and Andrea H. McMakin.
© 2013 The Institute of Electrical and Electronics Engineers, Inc., and Regina E. Lundgren and Andrea H. McMakin. Published 2013 by John Wiley & Sons, Inc.

most if not all of the talking (one-way communication). Formal hearings and other kinds of group interactions involving two-way communication are described in Chapter 17.

The advantages and disadvantages of using face-to-face communication are discussed in Chapter 10. This chapter discusses specific aspects of constructing face-to-face messages, focusing on issues important to risk communication, and provides guidelines for specific types of face-to-face risk communication activities.

CONSTRUCTING FACE-TO-FACE MESSAGES

Many people have learned about effective ways to speak or listen to their audiences in face-to-face interactions, perhaps through such organizations as the Toastmasters International. However, for risk communication, a few points should be emphasized. Key to these points is the choice of who will lead the face-to-face interaction.

Choose the Appropriate Spokesperson

Whether the audience will be doing most of the talking (as in an audience interview) or the organization's speaker will be doing most of the talking, those who are communicating risk have several choices as to who will lead the effort. Sometimes those in charge of the risk communication program will be the spokespeople for their organization. In other cases, other managers or experts will represent the organization. In still other cases, those outside the organization will speak to the issue. How do you know when to choose a spokesperson and which kind to choose? The two key criteria in choosing a spokesperson are audience acceptability and organizational acceptability.

> The two key criteria in choosing a spokesperson are audience acceptability and organizational acceptability.

Audience Acceptability

A number of factors affect whether your audience will find the spokesperson acceptable. Will they find the person credible, that is, will they believe what the person has to say? Credibility has to do with credentials related to the risk (Does the person have an advanced degree in the subject matter or many years of experience in the field?), the audience's past experience with the person or organization doing the communicating (Do they trust anyone from that organization?), and the speaker's ability to demonstrate a caring attitude. For example, in studies of public perceptions of trust and credibility in the United Kingdom, government agencies as well as government and industry scientists ranked low as sources of trusted risk information, whereas friends and family ranked high (Bennett and Calman 1999). Credibility is important regardless of whether the audience or the spokesperson will be doing most of the talking. (See Chapter 5 for additional information on choosing a spokesperson.)

Another factor is authority. Is the person able to respond to their concerns? This issue is particularly important in settings where the audience will be able to ask questions. If their concerns are largely technical, a scientist or an engineer is best. If they have management concerns, a manager with accountability for decision making is best. Using multiple experts can be effective if the room is arranged to allow them to interact with each other, and the audience (for example, by sitting at curved or circular tables) and the experts are responsive to each other. However, this approach can fail if audiences see it as the organization's attempt to "gang up on them" or perceive that the experts are arguing with each other. For example, researchers reviewing literature associated with the public's response to warnings of a nuclear power plant accident found that people liked hearing from more than one person to validate the information and to increase confidence in its accuracy. Consistency among the messages, however, was the key to the audience's ability to find the information credible (Mileti and Peek 2000).

Another factor to consider is whether the person can speak in a way that the audience will find acceptable. As noted later in this chapter in the guidelines section, the most appropriate leader of the interaction (regardless of who will be doing most of the speaking) will be one who can speak in the language of the audience. This ability includes the ability to speak in languages other than English if that is preferred by the audience as well as the ability to find innovative ways to describe highly technical information.

In addition, the spokesperson should be cognizant of nonverbal communication—for example, his or her stance, hand movements, and facial expressions. This body language can be just as important to acceptability as content and delivery of the message. For example, Vince Covello, renowned risk communication researcher and consultant, notes that in high-concern situations, audiences will take up to 75% of the message from body language, not the words. He recommends that in risk communication situations in which there is low trust in the communicating organization and high concern over the risk that speakers refrain from nodding their heads while listening to audience concerns. Instead of being perceived as active listening, this behavior is often perceived by concerned audiences as agreeing with an accusation. Researchers in Great Britain found that the audience tended to synchronize facial expressions, voices, and postures with a spokesperson during disasters, even when that person is viewed on a television or over the Internet (Bennett and Calman 1999). The spokesperson must be aware of these subtle clues to be effective in communicating risk information.

From an audience's perspective, then, the best spokesperson is one who is credible, responsive to concerns, and a believable speaker.

> From an audience's perspective, the best spokesperson is one who is credible, responsive to concerns, and a believable speaker.

Organizational Acceptability

A number of factors also affect whether your organization finds the person an acceptable representative. For cases in which the audience will be doing most of the talking, is the person a good listener? Can that person sit still and note concerns even when these concerns seem to contradict scientific precepts or organizational requirements? For cases in which the spokesperson will be doing most of the talking, is the person able to make speeches? Has the person been trained in public speaking, specifically in media relations and in answering tough questions? Can the person speak earnestly? Most importantly from the organizational perspective, does the person understand the organization's rules and philosophy as well as the work that is being done in connection with the risk? Depending on your organization, a host of other factors may also affect the decision. Check with management and the public affairs office of your organization before making a choice.

Finding the Right Person

Once you know what criteria your spokesperson must meet, you can find your spokesperson in a number of places. You can choose health care professionals; recognized experts in the field, either inside or outside your organization; risk managers; line managers; public affairs staff; or celebrities. Each group has its own benefits and liabilities.

Health care professionals can be extremely credible to the audience and can usually be responsive to technical concerns related to health and some environmental risks, although some see environmental risks in much the same way as do their patients. For example, risk researchers found in a large-scale study involving several major cities across the United States that physicians were the most trusted source of chemical risk information (McCallum et al. 1991). In addition, unless health care professionals are concerned with the risk themselves, either by being associated with the organization or by seeing the effects of the risk among their patients, these professionals may find themselves too busy to represent your organization. Their busy schedules especially conflict in the case of activities that require frequent interactions (such as audience interviews) or a long-term commitment (such as being the lead speaker for a speakers bureau).

Experts inside the organization will understand the risk and the organization, but they may not be credible to the audience. Experts outside the organization will probably be credible and will understand the risk, but, like the health care professionals, they may be too busy or too costly to represent your organization. Also, the very act of your employing them may make them less credible in your audience's eyes. In addition, they may not understand organizational concerns.

Risk and line managers understand the risk and the organization. They will be able to address at least some of the audience's concerns. However, they may not be credible to the audience. For situations in which the audience will be doing most of the talking (such as audience interviews), they may be unable to separate themselves from the risk assessment process or organizational needs to listen without trying to correct misperceptions.

Public affairs staff will understand organizational concerns and, depending on the person and the level of technical understanding, may be able to discuss the scientific aspects of the risk. However, in many cases, they will have no credibility in the eyes of a hostile audience because of the unfortunate stereotype of the public affairs person as the manipulative Madison Avenue type. The expertise of public affairs staff in making speeches is often best used to facilitate meetings and presentations, and to coach speakers.

Celebrity spokespersons have been used in care communication situations as well as to lobby for a certain constituency in consensus communication. Celebrities generally neither understand the organization nor the risk, and they may not be particularly credible to the audience, unless they can make some claim to have experienced the risk first hand. If they do not have such a motivation as having experienced the risk (and hence are willing to donate their time to promote its prevention or mitigation), they can be very expensive to hire. However, their high-visibility position can serve to create awareness of a particular risk and even perhaps to motivate an audience to action. A prime example of this power is the highly successful campaign to stop smoking that featured the late Yul Brenner, star of stage and screen, in a television public service announcement in which he discussed how smoking had in fact shortened his illustrious career and his life.

Table 5-3 in Chapter 5 provides additional information on choosing the appropriate spokesperson.

Give the Audience Something to Take Away

One of the drawbacks of face-to-face communication is that, unless the members of your audience are good at taking notes, they will have nothing to take away from the presentation to help them remember key points. Even though some people learn by listening, most need visual reinforcement. Unless you reinforce the presentation with written materials such as fact sheets (one-page handouts that emphasize key points), the audience may not take away, or retain, the information you intended.

> Unless you reinforce the presentation with written materials, the audience may not take away, or retain, the information you intended.

Reinforce Your Message with Visual Aids

Whenever possible, include visual aids as part of the face-to-face interaction. These visuals must be readable from the back of the room in group settings and readable at arm's length for more intimate settings. Regardless of expert design, creative use of color, and clever wording, your visuals are useless if they must be prefaced by the apology, "Now, I know you probably can't see this, but what it shows is. . . ." Use words or short phrases to emphasize key points and to show the audience how concepts fit together. Use photographs, drawings, and graphs to further illustrate key ideas. See Chapter 14 for additional information on using visual representations of risk.

Speak in the Language of the Audience

The speaker should use words and phrases that the audience will understand. At a public meeting concerning the siting of a proposed wind energy farm, the spokesperson for the organization proposing the farm had obviously gone to great trouble to develop slides and an oral presentation that would present the facts to his audience. Unfortunately, when it came to describing the design of the farm, a factor that concerned audience members, he described heights and distances in "rotor diameters." In one of his opening remarks, he had stated that a rotor diameter was so many feet; however, it was rather unrealistic to expect that his audience (1) would have taken note of that fact, which concerned a foreign concept to begin with; (2) would remember the fact until he began using it again; and (3) would be able to do the math in their heads. It would have been far simpler for everyone if he had given the distances in feet. As it was, his audience became more and more hostile with each mention of the term, until it became obvious to even the spokesperson that many of them were no longer listening.

Whenever possible, the speaker should speak in the language of the audience. If the audience speaks a nonstandard English, however, by all means have the speaker use standard English. Your message will not be credible if they feel you are trying to speak down to them. If English is not the audience's main language, as is the case in many Hispanic neighborhoods, for example, find a spokesperson (yourself or someone else who is credible) who can speak in the primary language.

Do Not Promise What You Cannot Deliver

When you as a speaker get a question from your audience that you cannot answer, offer to get back to them with an answer only if you can realistically find that answer. "I don't know, but I'll find out" is a good response to unfamiliar questions. However, too often, speakers promise information that they will not be able to access or that they forget is classified or proprietary. As you try to respond to your audience, it is easy to promise to give information, knowing that you personally probably will not be held to task later by the often unknown audience member. However, that audience member will remember the lapse and let others know that your organization failed to live up to a promise, eroding your credibility. If you do make a promise for information, find out how to contact the person who wanted the information and make it a point to get back to them within 1 or 2 days.

Be careful, too, that you do not promise something that your organization is unable to give. For example, if the request from the audience is to hold more public meetings, and you gladly agree without checking whether your organization has the time or resources needed, you may not be able to keep your promise. Breaking such promises, as you can imagine, is a sure way to increase audience hostility toward the organization and to erode credibility.

GUIDELINES FOR SPECIFIC TYPES OF FACE-TO-FACE COMMUNICATION

The following information is provided for specific types of face-to-face risk communication activities.

Speaking Engagements

Speaking engagements are one of the most common forms of face-to-face risk communication. To ensure that the audience understands the message being delivered, consider the following guidelines:

- **Coach staff with little speaking experience.** Sometimes the spokesperson who will best meet both audience and organizational needs happens to have little or no speaking experience. Such people need coaching to be effective. Simply writing a script for them generally does not work. Those words came from someone else, and the neophyte speaker will probably not be comfortable with them. In addition, repeating someone else's ideas will not help them when they get unanticipated questions from the audience. It is, therefore, best to train the spokesperson in how to be an effective speaker. Courses like those offered by the Dale Carnegie Foundation or Toastmasters can be extremely helpful in getting novice speakers over the fear of facing an audience. In-house courses offered by knowledgeable communicators are also helpful because they can be tailored to specific situations and organizational needs. The speaker should also be prepared with information on how to deal with hostile audiences, working with the news media if that is expected, and the basic principles of risk communication.
- **Practice the presentation.** Even an experienced speaker should practice the presentation if at all possible. Use similar lighting, acoustics, and room size. Include in the practice audience people who will listen for technical details, audience concerns, and legal issues. Have the practice audience listen for content, and watch body language. One speaker had a tendency to push up his glasses with his middle

finger while speaking. When he was told to practice in front of a mirror, he saw the gesture and easily remembered not to use it, for fear of offending his audience.

- **Practice answering questions as well.** Using the information you gleaned from the audience analysis, anticipate what questions might be raised. Answer as many questions as possible in the presentation itself. If the audience might want additional details, develop appropriate answers and have them ready if the audience asks for them. Have the practice audience ask questions. Also practice dealing with hostility and the news media (see Chapter 16).

- **Strive for an accessible, comfortable setting.** These points are particularly important to the credibility of a risk communication presentation. The setting or place of the face-to-face interaction should be accessible to the audience and comfortable for them. By comfortable, we do not mean that the seats should be soft and have plenty of leg room, although this is always nice. Rather, we mean that the setting should be in a neutral location, one that does not evoke negative feelings. For example, a federal agency near our home holds all its public meetings in its own auditorium, which is a nice facility with good lighting and fairly comfortable seats. However, the facility can be used in only one configuration: a raised stage at one end and theater-style seating that encourages an "us versus them" perception. This design, and the fact that the agency often comes under fire for its decisions at these meetings, make other facilities more appropriate.

- **Dress to suit the setting.** For a formal presentation before the city's business leaders, a business suit would be appropriate. For a speech at a lodge picnic, more casual attire is warranted. Speakers should not try to dress exactly like the audience unless it is the way he or she normally dresses. Audiences generally react more favorably to speakers who are being genuine than to ones who are attempting to be someone they are not.

- **Consider acoustics and lighting.** Make sure that the acoustics and lighting are adequate for the kind of presentation planned. Make sure that the speaker can be heard to the back of the audience. If the speaker will need a microphone, make sure that he or she knows how to use it and has the controls within reach. If the speaker is using a projected presentation as a visual aid, make sure that the lights can be dimmed so that the audience can see it and still see each other and the speaker. A completely darkened room can be a safety hazard and can also encourage some members of the audience to fall asleep. Also make sure that the speaker or an associate can reach the dimmer switch in mid-presentation if need be. That way, the lighting can be easily increased to see who just asked a question.

- **Know your equipment.** For a presentation on a projector, a flip chart, or any other kind of visual aid, make sure that the speaker knows the location of any associated equipment and how to make it work. Bring spare parts such as light bulbs, thumb drives, and extension cords. That way, you can help ensure that the presentation will go on despite possible problems with equipment. Also make sure that someone is available who knows how to work the heating, ventilation, and air-conditioning system. Tempers in even the most moderate audience will flare if the room is too hot or too cold.

Speakers Bureaus

Instead of waiting to be asked to make presentations, many organizations in charge of risk communication efforts have developed a speakers bureau, a group of speakers known for their expertise in a given subject who can be contacted to give free or low-cost speeches for local communities and organizations. The speakers can all belong to a given organization, or they may be experts that the organization has contracted to provide the service. When developing a speakers bureau for the purpose of communicating about risks, consider the following:

- **Choose speakers who will be credible with a wide range of potential audiences.** The experts in your bureau must be able to speak before virtually any stakeholder group who might request a presentation. Therefore, it is necessary to select speakers who will be credible to the widest possible audience. Depending on your audience's needs, such speakers might include health care professionals, scientists, engineers, regulators, risk managers, or college or university staff.
- **Ensure that the speakers have been appropriately trained.** The speakers who make up the bureau will by necessity be experts in their fields. They will, however, most likely need additional training in appropriate risk communication principles and techniques as well as the risk communication effort they are supporting.
- **If possible, develop consistent materials to support their presentations.** As with the speaking engagement described earlier in this chapter, ensure that your speakers go out armed with appropriate information materials to leave with their audiences and that their visual aids are appropriate to the situation. If, however, your intended audience is highly hostile to the organization in charge of communicating the risk, avoid providing materials that publicize the organization. Your speakers may be more credible, and the risk message is more likely to be received, if there is some perceived distance between the speaker and the organization.

Tours and Demonstrations

Tours and demonstrations are another way to communicate risk in face-to-face interactions. In a tour, some segment of the audience is invited to view a facility or site that is perceived as contributing to a risk. In a demonstration, the audience views or participates in an activity designed to assess, prevent, monitor, or mitigate a risk. Regardless of the form, several guidelines can be applied:

- **Make sure that tours and demonstrations are open.** Do not exclude some segment of your audience. Hold the tours and demonstrations at times when it is possible for the audience to attend. For example, if many members of your audience work full time, plan the tour or demonstration before or after their work hours or on weekends.
- **Make sure that tours and demonstrations are easily accessible to all members of your audience.** For example, have wheelchairs and make sure that elevators are available to take them between floors. Have someone fluent in sign language available to assist the hearing impaired. Make sure that people know in advance what kind of footwear and/or clothing to wear if that is important. For example, walking over rough terrain can be difficult in high heels that some women might wear to a tour unless prewarned. Tell your audience how much walking is involved and in what kind of situations so they can be prepared.

- **Be as careful in choosing a tour or demonstration leader as you would someone who is giving an oral presentation.** Make sure that the person is acceptable to both your audience and your organization, can answer audience concerns and questions, and is able to speak at your audience's level.
- **Determine what you hope to gain by the tour.** Are you trying to persuade your audience that what you are doing is perfectly safe? (See Chapter 5 for the dangers in using persuasion.) Are you trying to raise awareness of an issue? Are you trying to give your audience information on which to base some decision? Make sure that the tour or demonstration reinforces the purpose and does not detract from it.
- **Consider organizational and audience needs.** Make sure that what you are going to show your audience does not compromise proprietary or classified information and that the information you are presenting meets your audience's needs. For example, if what your audience really wants to know is how your organization trains its workers, showing them the beautifully landscaped grounds will not meet their needs and may make them angry. A manager at a hazardous waste incinerator took the local garden club on a tour of the plant. One of the women asked him just how dangerous the smoke she saw coming out of a stack was to breathe. The manager, acting in an unfortunately patronizing fashion, told her not to worry about it, that it was no more dangerous than eating a peanut butter sandwich. She hit him with her purse. Consider your audience carefully, listen to their questions, probe for underlying concerns, and answer questions honestly and courteously.
- **Practice the tour and demonstration.** Practice what the tour leader will say and how the tour or demonstration will be given. Have the practice audience listen as outsiders would. Sometimes, concepts that seem clear from within the organization can seem foreign or can be totally misconstrued by your audience. A chemical manufacturing plant had been criticized for failing to take safety issues seriously, despite an excellent safety record, well-documented and applied procedures, and a well-trained workforce. Managers opened the plant to a particularly vocal activist group and carefully explained all the precautions that had been taken. After the tour, the leader of the activist group reported to the press waiting outside the gates, "Well, now we know there's something really unsafe. Look at all the attention they're paying to safety!" Analyze your audience carefully in developing your tour or demonstration.

Video Presentations

Video combines many of the features of information materials and face-to-face communication. It can be used in a variety of situations such as formal learning environments or in-home study. A video can carry more information than some other more visual forms of risk communication such as posters or displays, but the narration of the video adds a dimension of human contact like an oral presentation. Therefore, the content of a video should follow the general guidelines for constructing information materials. The delivery, as in the spokesperson and visual aids, should follow the general guidelines for constructing face-to-face messages.

Professional video can be expensive to produce—up to several thousand dollars per minute of finished footage for high-end productions. However, many handheld video cameras are available that can be used to develop simple videos relatively cost-effectively, depending on your purpose and objectives and your audience's needs. Either way, you should have a very specific purpose for the video that cannot be accomplished otherwise (for example, depicting a hazardous situation that would be unsafe to show people directly). This reason should include movement; if you do not need to show movement, you can get a similar affect through a presentation with still photography.

Give your video as long a shelf life as possible by writing the script in ways that do not make your content out of date next week. Focus on longer-term, enduring issues related to your risk so that the video can be used in a variety of situations.

Use experienced script writers and videographers. Remember that videos are like television; audiences focus more on (and remember) what is shown than what is said. Make the script and visuals flow together for greater impact.

For successful risk communication videos, always consider the needs of your audience:

- **How will the video be distributed?** Other types of information materials like pamphlets and newsletters can be easily mailed or handed out at information fairs. The relatively higher costs of producing videos (often up to $10 per copy) generally make mass distribution infeasible unless you plan to distribute it as digital media or on the Internet. Make sure that you have a ready distribution network, such as through health care professionals, the school system, a professional organization, or an active website before developing a video. Remember too that video takes up considerable size digitally. If you plan to distribute it via your own website, make sure that you have the necessary infrastructure in place to allow it.
- **Where will the video be shown?** Just as important as distribution is the setting in which the video will be shown. If the video will be used as part of a formal learning environment (public school system, college, or organization training course), it will need to complement other materials and information such as learning objectives, student workbooks, and tests. If the video will be shown as part of an oral presentation, the speech and other materials should likewise be complementary. If the video will be accessed in a home environment, on the other hand, it will need to be more self-contained and more comprehensive.
- **Will there be a spokesperson present to answer questions?** The chief disadvantage of a video as a form of face-to-face communication is that, unless a spokesperson is present at the showing, there is no opportunity for the audience to ask questions. If the video will always be used with a spokesperson present, it may not have to be as comprehensive as a video that will be used without a spokesperson present.

One other factor to consider in video production particularly is the quality of the video. Again, consider your audience. Although high-caliber video production companies are generally available to produce videos that rival some of Hollywood's most impressive films in the areas of special effects and acting ability, such a high-gloss production may in fact alienate some audiences, particularly those who might be hostile to the organization producing the video. Such audiences will see a high-quality video production as a Madison Avenue cover-up. Also, some audiences still equate videos with high price tags, although, in fact, the price has fallen considerably with the advances in technology. Such audiences will also be alienated by high-end videos, especially when produced by a government organization with taxpayer funding. In these cases, it may be better to opt for a more simple production or to use another method to communicate about risk.

Audience Interviews

Interviewing the audience to understand their concerns and perceptions can be an effective form of face-to-face interaction in that it provides those who are communicating risk the opportunity to better understand their audience, and it provides the audience with the opportunity to share their concerns. Audience interviews can be a particularly effective way to begin consensus communication efforts.

An audience interview is similar to conducting a survey, but the questions should be more open ended. For example, a survey question concerning the audience's perception of home pesticide use might ask the person to rank pest control methods against others on a quantitative scale. In an audience interview, the spokesperson for the organization communicating risk might simply request, "Tell me how you control pests." Like surveys, however, the order of the questions, and the manner in which they are asked, will have a profound effect on the answers given. In general, move from the general to the specific, from topics that are positive or neutral in perception to those that are more likely to be received with hostility or some other negative reaction.

Other guidelines on audience interviews include the following:

- **Be inclusive.** Try to ensure that you are interviewing the full range of your audience. One way to do this is to start with a known list of possible interviewees and, as the final question in each interview, ask if there is anyone else who should be interviewed. The first few interviews will provide a wealth of new names, but gradually you will begin to see the same names coming up again and again. This is a sign that you have reached a good majority of your intended audience. Another approach is to use focus groups (see Chapter 17 for additional information).
- **Make your audience comfortable.** Make sure that the place for the interview is one that is comfortable for each interviewee. It is usually best to conduct each interview separately, in the interviewee's home or location of their choice. Dress casually, but professionally. Let them do most of the talking.
- **Explain the process before starting the interview.** Help them understand why you are conducting the interview, how the results will be used, and how they will know that their concerns have been heard. Ask permission before recording the interview, taking notes, or otherwise visibly capturing the information from the interview. If possible, give them the opportunity to see and correct anything you record regarding their concerns or perceptions.
- **Consider carefully before correcting apparent misconceptions or misperceptions.** It can be very tempting to jump in and correct a misconception or argue over a perception, particularly in cases in which the interviewee is clearly upset about the issue. Correcting the person you are talking with after they make a statement may anger or embarrass them. Either way, they may not want to speak candidly. They may even want to stop the interview. In addition, such correction can be a form of persuasion. Review the information in Chapter 5 on the use of persuasion before interviews and decide how you will handle such situations should they arise.

Information Fairs

An information fair is a grouping of tables or booths staffed by organizations charged with communicating a particular risk or set of risks. Those interested in the risk can circulate through the fair and choose to talk to or select information materials from various organizations of interest to them. These organizations may also put on demonstrations of how risks can be prevented, analyzed, or monitored. The fair can be for employees of a given organization (for safety and health care communication) or for a local community (for environmental, safety, or health care communication or the planning stages of crisis communication). The fair can also be used to provide information to start a consensus communication effort.

Information fairs are another hybrid of information material and face-to-face interaction. Use the guidelines on information materials in Chapter 13 and the guidelines in this chapter on selecting the appropriate spokesperson and tours and demonstrations to ensure successful fairs.

Training

As mentioned in Chapter 3, the Occupational Safety and Health Act requires that employees be trained in the use of hazardous materials associated with their jobs. Training can also be used to teach concepts related to other forms of risk and to build a crisis response unit.

Developing effective training materials is a science unto itself. The primary considerations are the audience, the purpose of the training, and the resources available, including time. For example, the type of presentation and amount of material you can cover will differ greatly between the training of a group of clerical staff on ergonomic issues in the office at a 1-hour lunch meeting and the training of a group of firefighters on hazardous materials in the community at a weeklong retreat. In general, consider these guidelines:

- **Visual is always better than oral, and hands-on is always better than visual.** There is an old teaching adage, "I hear and I forget, I see and I remember, I do and I understand." Given audience, purpose, and resource constraints, try to provide opportunities for those being trained to experience the risk in question. For example, one instructor of basic radiation safety at a nuclear facility brings in a variety of items such as a smoke detector, gas-lantern mantle, and pottery as well as a radiation detector to allow his students the chance to determine for themselves what is radioactive in their world.
- **Focus your training.** Articulate clearly what you want the students to gain from attending the training and how you will provide that information. Too often, training related to preventing, managing, analyzing, or monitoring a risk fails because the content of the course is too broad. Better to ensure that your students learn a few key concepts than to try to cram the equivalent of an advanced degree into a weeklong course.
- **Do not attempt training for a hostile audience.** If your intended audience is highly hostile toward the organization in charge of communicating the risk or the risk in general, training is not your best option to communicate risk. Start by understanding and dealing with the source of the hostility.

Chapter 18 has additional information on computer-based training.

CHECKLIST FOR FACE-TO-FACE COMMUNICATION

The spokesperson communicating the risk or leading the interaction:

☐ Is acceptable to both the audience and the organization
☐ Has written material that the audience can take away to supplement the oral presentation
☐ Has visuals that are easy for the audience to read and understand
☐ Speaks in the language of the audience
☐ Will not promise information or changes in policy unless these can be delivered
☐ Has been coached if inexperienced
☐ Has practiced:
   ☐ The presentation
   ☐ Answering questions
   ☐ Dealing with the media
   ☐ Dealing with hostility in the audience
☐ Will be presenting in a neutral setting
☐ Is dressed appropriately for the setting
☐ Can be heard to the back of the audience
☐ Knows how to work the microphone
☐ Can reach the dimmer switch if needed
☐ Knows how to work the visual aid equipment and has spare parts in case they are needed
☐ Knows who to contact to run the heating, ventilation, and air-conditioning system

The organization's speakers bureau:

☐ Uses speakers who are credible with the audience
☐ Has trained speakers in risk communication
☐ Is armed with consistent messages

Tours and demonstrations:

☐ Are scheduled at convenient times for the audience
☐ Are accessible to the audience
☐ Have appropriate spokespersons
☐ Have clear goals
☐ Consider audience and organization needs
☐ Have been practiced

Video presentations:

☐ Follow general guidelines for information materials as well as selecting spokespersons
☐ Have known distribution networks to reach the audience
☐ Have supplemental materials to support the setting in which they will be shown

(continued)

CHECKLIST FOR FACE-TO-FACE COMMUNICATION (*continued*)

☐ Are comprehensive enough to stand on their own should a spokesperson be absent at a showing
☐ Have quality appropriate to meet audience expectations

Audience interviews:

☐ Cover a wide range of audience perspectives
☐ Are held in settings comfortable for the audience
☐ Begin with an explanation of process
☐ Were preplanned to include how to deal with misperceptions or inaccuracies

Information fairs:

☐ Follow the general guidelines for information materials and choice of spokesperson
☐ Follow the general guidelines for tours and demonstrations

Training:

☐ Emphasizes hands-on or visual communication methods
☐ Is focused
☐ Will be given to a receptive audience

REFERENCES

Bennett, P. and K. Calman. 1999. *Risk Communication and Public Health*. Oxford University Press, New York.

McCallum, D. B., S. L. Hammond, and V. T. Covello. 1991. "Communicating about Environmental Risks: How the Public Uses and Perceives Information Sources." *Health Education Quarterly*, 18(3):349–361.

Mileti, D. S. and L. Peek. 2000. "The Social Psychology of Public Response to Warnings of a Nuclear Power Plant Accident." *Journal of Hazardous Materials*, 75(2000):181–194.

ADDITIONAL RESOURCE

Peters, R. G., V. T. Covello, and D. B. McCallum. 1997. "The Determinants of Trust and Credibility in Environmental Risk Communication: An Empirical Study." *Risk Analysis*, 17(1):43–54.

16

WORKING WITH THE NEWS MEDIA

News media channels that are available to large segments of the population—including television, newspapers, radio, magazines, and the Internet—are arguably the largest source of information in today's society. Many people form their opinions about health, environmental, and safety risks by what they read or hear in traditional or online news outlets.

In communicating risk-related information, administrators, technical and health professionals, and communication specialists often deal with the news media as a key provider, interpreter, gatekeeper, or channel of risk-related information. We have devoted a chapter to working with media representatives—reporters, journalists, editors, and producers—because of the distinct and significant role they play in communicating risk information to the public.

The use of the Internet for risk communication is described in Chapter 18. The concept of engaging journalists via social media is discussed here, but using social media channels to communicate risks directly is discussed in Chapter 19. Media considerations during emergencies are described in Chapter 21. The role of media for public health campaigns is covered in Chapter 23.

THE ROLES OF THE NEWS MEDIA IN RISK COMMUNICATION

Media organizations, such as television producers and newspapers, can choose among several roles, or levels of participation, to address a given risk-related issue. Participation can span a wide range from least to most involvement: (1) reporting existing information, (2) influencing the way an issue is portrayed, (3) independently bringing an issue to the

Risk Communication: A Handbook for Communicating Environmental, Safety, and Health Risks,
Fifth Edition. Regina E. Lundgren and Andrea H. McMakin.
© 2013 The Institute of Electrical and Electronics Engineers, Inc., and Regina E. Lundgren and
Andrea H. McMakin. Published 2013 by John Wiley & Sons, Inc.

public's attention or restricting its coverage, and (4) proposing solutions to a risk-related decision, including taking a stand on an issue. This section describes the levels of participation a media organization can take and corresponding strategies that those who communicate risk may wish to consider. At all levels, it is important to develop and maintain productive relationships with media representatives.

> Media organizations can choose among several roles or levels of participation to address a given risk-related issue.

Many factors affect which role, or combinations of roles, media organizations take. One factor is the type of communication situation: care, consensus, or crisis. When a crisis presents imminent danger, reporters are likely to start with reporting existing information, when the public must be alerted quickly to protect themselves. Later, media organizations may turn to a more investigative role to attempt to uncover the factors that led to the crisis. This role may involve working with official investigating organizations, citizens' groups, policy makers, and others to portray a more complete picture of the risk, its causes, and its potential solutions.

At the lowest-participation end of the scale, those responsible for communicating risks may be interviewed by reporters or otherwise asked to provide information for a story. In addition, those who are communicating risk may need to seek out media representatives to provide information on breaking news or current events. This outreach can be done through press kits, media events, press releases, and other avenues. For providing information to reporters, see the tips later in this chapter under the heading "Guidelines for Interacting with the News Media."

In care communications, media organizations may choose to take an active role in illuminating and reducing a given risk such as gang-related crime, a low child-immunization rate, or a lack of adequate nutrition for seniors. In this role, media organizations often describe the negative consequences of the risk to the community and suggest ways that individuals, groups, and entire communities can act to reduce risks. This approach casts media organizations in a stakeholder role, in which they participate with others in characterizing the problem and its potential solutions. At this level of involvement, policy and technical professionals may work directly with media and other community representatives to characterize risks and their alternatives. If you are in this situation, use the appropriate guidelines later in this chapter for working with media representatives.

At the highest-involvement end of the scale, individual editors or producers occasionally feel that a particular issue is significant enough that their organizations must get more involved, sometimes to the point of going on record with a stated position. This often involves an issue that affects the community, affects many stakeholders with different opinions about the risk, and requires a consensus decision informed by many views. Examples of such consensus-required issues are whether a new federal prison should be sited in the community, whether field burning should be made more or less restrictive, whether a new dam should be built or an existing one removed, or whether two hospitals should merge services to cut costs.

Media organizations may choose to participate more strongly in such issues by taking an advocacy role. For example, editors or producers may participate in discussions with opinion leaders about the nature of the risk, its benefits and consequences, alternatives, various points of view, and potential solutions. They may even establish an editorial position and publish or air stories supporting that position. This approach generally involves editorials or similar commentary rather than news coverage. At this level of participation, media representatives such as editorial boards may work actively and regularly with other stakeholders to fully describe the nature and consequences of a particular risk and propose solutions.

At all levels of media participation, it is important to establish productive relationships with media representatives, as described later in this chapter. Good working relationships between media representatives and those who are communicating risk increase the chances for accurate, balanced coverage. For example, research at McMaster University in Hamilton, Ontario, found that media coverage of infectious diseases such as avian influenza can allow readers to gain a more accurate view of risks. However, the same study found that reporting could just as easily increase feelings of risk in the readers (Young et al. 2008). But when media organizations publish or air something you disagree with regarding a particular risk, they are more likely to listen to your concerns if you have established yourself as a credible, reasonable source.

News Media Contrasted with Other Stakeholders

Two broad characteristics distinguish media representatives from other stakeholders and limit involvement. The first is mission. A media organization's primary purpose is to provide the public with current information, often a combination of news and entertainment. This mission takes priority over involvement in a given risk-related issue. Other participants such as a citizens' group may be held responsible for crafting workable solutions, hammering out a myriad of details in agreements, and even implementing the agreements and evaluating their outcomes. Media organizations are there primarily to report on and illuminate issues. Though media involvement can be a very powerful voice in the ultimate outcome of risk issues, media organizations' responsibility in a decision-making process usually does not extend past proposing solutions.

The second characteristic that distinguishes media representatives from other stakeholders is one of position and representation. Nonmedia stakeholders participating in a risk-related group decision process often represent the views of a specific "constituency" such as health professionals, homeowners, or recreational enthusiasts. In group discussions, these members reflect the values and judgments of those whose views they represent.

Reporters and journalists, in contrast, generally aim for objectivity and balance in their stories. Thus media representatives may be hesitant to become directly involved in an issue, viewing such involvement as a professional conflict that could subject them to allegations of bias. In fact, the U.S. Supreme Court ruled in 1997 that a Washington State newspaper could rightly take a reporter off news assignments because her outside-of-work activities, including activism on various social causes, could be perceived as biasing the new stories she wrote, thus risking the newspaper's reputation of objectivity (Nelson v. McClatchy Newspapers, Inc. et al. 1997).

Productive Interaction, Not Polarization

Regardless of the media's level of involvement, agreeing on how risks should be portrayed can be challenging. Science and policy experts sometimes view news media coverage of health and environmental risks as oversimplified, inaccurate, and sensationalized. Journalists and reporters, for their part, are occasionally frustrated by technical and policy professionals who appear unhelpful, arrogant, or controlling.

> Working productively with reporters and journalists can lead to a more informed, empowered, solution-oriented public.

Nevertheless, experts who deal with issues involving health and environmental risks cannot afford to ignore media representatives or criticize them from afar. Entire industries have discovered the power of the news media to influence their financial bottom lines. In 2012, Beef Products Inc. sued ABC News for defamation, claiming that inaccurate stories about the finely textured beef called "pink slime" forced the company to close three of its four U.S. plants and lay off more than 650 workers. The Washington state apple industry lost about $130 million in sales in the season following a *60 Minutes* television program in 1989 about the dangers of the agricultural chemical Alar (O'Rourke 1990).

On the other hand, some evidence indicates that the more publicized a disaster, the faster donations and rescue help arrives, and the bigger it will be. For example, the Haitian earthquake of 2010 had three times as many stories in print and television news within 10 days of the event than devastating floods the same year in Pakistan, and financial contributions per person for Haiti outpaced those for Pakistan by 10 to 1 (Ferris and Petz 2011).

It is equally true, and this is the position we advocate, that working productively with reporters and journalists can lead to a more informed, empowered, solution-oriented public. A good example of the benefits of this approach is broad public awareness of AIDS and *Escherichia coli* bacteria risks through media coverage.

The first step in working effectively with media representatives is to understand their goals and constraints. Only then can risk communicators apply the guiding principles for care, consensus, or crisis communication. The rest of the chapter deals with news media sources most familiar to the general public: television, newspapers (print or online), and radio. Social media is covered in Chapter 19.

UNDERSTANDING "CULTURAL" DIFFERENCES

A researcher at a national laboratory was interviewed for the first time by a reporter about a technology destined for eventual use by the public. The researcher, flattered by the reporter's interest and questions, carefully explained the detailed technical workings of his invention and its uses. When asked, he freely discussed some of the technology's potentially controversial characteristics, downplaying the potential for public rejection. When the resulting news article was published, the researcher felt betrayed by its tone, viewing it as a misrepresentation of the technology that emphasized its controversial aspects rather than its benefits. "I thought the reporter was my friend," the researcher lamented.

The problem lay in the fact that the researcher had approached the interview as he would an informal discussion with one of his peers. Lacking an understanding of the reporter's mission and without his own goals for the conversation, the researcher inadvertently contributed to an article that disparaged his own technology.

This incident illustrates some of the differences between the ways subject matter experts and news media representatives traditionally approach risk communication. We call them cultural differences because the two groups have their own, sometimes competing, values, traditions, and practices. Understanding the following differences can help risk communicators more effectively work with reporters and journalists.

The News Media Are Event Focused

Reporters, especially when covering the news, are largely reactive, reporting on the facts surrounding an event or an ongoing risk. The Environmental Risk Reporting, for example, found that scientific risk had little to do with the environmental coverage presented on the nightly news. Instead, the coverage appeared driven by the traditional journalistic news values of timeliness, geographic proximity, prominence, consequence, and human interest, along with the television criterion of visual impact (Greenberg et al. 1989).

> News coverage is driven by timeliness, geographic proximity, prominence, consequence, human interest, and visual impact.

Because of limited staff expertise in technical issues and constraints on space and time, the media may give little emphasis to explaining the likelihood of the hazard occurring under various conditions, broader societal or policy issues surrounding the risk, or other contextual information. Technical and policy experts, in contrast, are highly concerned about fostering rational decision making by the public, meaning providing information about immediate and long-term consequences, costs and benefits of a hazard and its alternatives, and the moral and economic issues that are inherent in hazardous processes and events. Thus, risk communicators, especially in dealing with complex decisions, should not rely on the news media alone to fully provide the contextual and background information necessary for well-informed decision making.

Certain Kinds of Risks Get More Coverage

Research has shown that the U.S. news media disproportionately focus on hazards that are catastrophic and violent in nature, new, and associated with the United States (for example, Adams 1986; Combs and Slovic 1979; Singer and Endreny 1993). Drama, symbolism, and identifiable victims, particularly children or celebrities, make risks more memorable. Controversy ensures greater coverage.

Risks are not covered in the news media commensurate with their probability of occurrence. For example, airline crashes with fatalities are covered far more extensively than heart disease, though diseases take 16 times as many lives as accidents. The result, adding to the risk communicator's challenge, is that people consistently misjudge the frequency of certain lethal events, according to studies (for example, Combs and Slovic 1979). Thus, those communicating risks in a crisis situation may find it very easy to get media attention, whereas those in care or consensus communication situations may have more difficulty reaching their audiences through media forums.

> Risks are not covered in the news media commensurate with their probability of occurrence.

Journalistic Independence and Deadlines Affect Content

Many technical professionals view the news media as a conduit or pipeline, responsible for transporting technical information to the public (Nelkin 1994). Regarding the media as a technique to further their own goals, they expect to control the flow of information to the public just as they do within their own domains, and feel betrayed when their views are challenged.

In industry, government, and academia, extensive peer review systems ensure that information is approved by the appropriate authorities before release. This process is designed to eliminate technical errors and ensure that all parties have a chance to agree on how the material is presented. In contrast, journalists are charged with being independent watchdogs of society. Allowing a source to review an article is seen as opening the door to media censorship. Reporters pride themselves in giving their audiences an independent view of a situation, untainted by corporate or academic "propaganda." Though most reporters want to get the facts right, they do not appreciate being told how to say something.

Deadlines are another reason for the lack of source review. Most reporters simply do not have time to check back with each source. Many find that when they do take the time, their sources want to add material, change their quotes, nitpick about wording, argue about the interpretation or theme of the article, and ignore word limitations. Both of these factors—independence and deadlines—mean that the risk communicator often has little control over the published or broadcast story.

The Need for Balance Invites Opposing Views

The American view of fair and impartial reporting is to present divergent points of view on a given issue. Ironically, this attempt at balance sometimes results in an unwitting imbalance—the view of a vocal, self-proclaimed expert or media-savvy special interest group can be made to seem just as valid as that of a peer-reviewed group of scholars or even a worldwide scientific consensus. Many reporters are reluctant to characterize the nonscientific or special interest viewpoint as such for fear of appearing biased.

> Many accusations of inaccuracy follow less from actual errors than from efforts to present complex material about risk in a readable and appealing style.
> —Dorothy Nelkin (1994, p. 233).

Adding to the reporter's challenge is the fact that members of the scientific community sometimes disagree among themselves on the same topic. For example, during the H1N1 outbreak in 2009 in the United States, federal and local entities held competing press conferences that confounded the news media. Such "dueling experts" stories may leave audiences both confused and concerned that the risk is unknowable if not even the experts cannot agree on it.

Information Is Condensed, Simplified, and Personalized

In today's information-glutted world, media stories must grab and keep the audience's attention. Stories for television, radio, web-delivered content, and other news media do not generally lend themselves to long discourses. In addition, many reporters are generalists with little or no background in science and technology.

When describing risks, journalists see their responsibility as informing people of potential dangers and identifying ways people can respond. To do this, they want to give people specific warning signs to alert them to danger and to tell them where to go for help and how to alleviate the risk. Concepts important to technical

> When describing risks, journalists see their responsibility as informing people of potential dangers and identifying ways people can respond.

professionals, such as probabilities, uncertainties, risk ranges, acute versus chronic risks, and risk trade-offs, do not translate well in many news media formats.

To humanize and personalize the risk story, news organizations often use the plight of an individual affected by a hazard, regardless of how representative the person's situation is. National media coverage of breast cancer from 2003 and 2004, for example, focused on personal narratives of cancer patients rather than data and statistics by a two-to-one margin (Atkin et al. 2008).

These approaches of condensing, simplifying, and personalizing information may make information more accessible to the public, but they may result in incomplete and sometimes unbalanced information for making personal risk decisions.

GUIDELINES FOR INTERACTING WITH THE NEWS MEDIA

Understanding the differences between risk experts and media professionals is the first step in effective interaction. We do not mean to suggest that the two groups are mutually exclusive. It is possible to find places where the interests of the two "cultures" converge, fostering a jointly beneficial working relationship. For successful media interactions, we suggest the following broad guidelines. They apply equally to care, consensus, and crisis communications unless specified.

Develop Relationships with Local and Regional News Media Representatives

Journalists need information and ideas for stories that have importance for the local community. Risk communicators should think of themselves as resources who can make it easier for journalists to do a good job. Lawrence Wallack, professor at the University of California-Berkeley's School of Public Health and Director of the Berkeley Media Studies Group, recommends providing journalists with timely, accurate information; examples of local activities; summaries of key issues; and names of potential sources (Wallack et al. 1993).

> Demonstrating that you want to serve as a source of information, rather than merely satisfying your self-interests, is the best way to build a long-term relationship with reporters.

An ongoing working relationship with media representatives makes it much more likely that you will get a fair hearing when a reporter is doing a story about a particular event. Demonstrating that you want to serve as a source of information, rather than merely satisfying your self-interest, is the best way to build a long-term relationship. Identify and meet with reporters and editors who cover your organization and exchange contact information. Hold roundtable discussions and briefing sessions to receive feedback from reporters and to share information about your organizations. Invite reporters to participate in emergency preparedness drills and training exercises. The goal is to build a reputation as a trustworthy, articulate source on one or more topics, so that you will be sought out in the future. Other goals could be determining which news media to include in your outreach efforts, getting feedback from reporters on working together with them, and improving the accuracy of reporting (Hyer and Covello 2005).

Go beyond individual reporters. Joann Rodgers, Director of Media Relations at Johns Hopkins Medical Institute in Baltimore, recommends expanding to include other media gatekeepers such as newspaper editors, editorial boards, and television producers who influence what becomes news and how news is reported (Lebow and Arkin 1993).

Know When to Approach Media Representatives or When They May Approach You

When there is a potential for immediate public health or environmental risk, it is necessary to contact media representatives without delay so that people can be informed about how to avoid or reduce the risk. We recommend that organizations that may face such crisis communication situations preestablish media protocols, including the use of trained spokespeople. The organization's top leaders, including presidents, operations managers, and agency heads, should be ready to speak clearly and candidly to the public through news media channels.

The intent of such preparations is not to downplay the risk situation or deflect blame but to respond as quickly and effectively as possible to alert people to danger. (See Chapter 2 for more information about crisis communications and Chapter 15 for more information on choosing and coaching a spokesperson. See Chapter 21 for advice on working with the news media in crisis situations.)

> The intent of establishing media protocols for potential crises is not to downplay the risk situation or deflect blame but to respond as quickly as effectively as possible to alert people to danger.

Be aware of work you are involved with that may prompt media attention. This work may be a topic that has been featured recently in the media (such as the dangers of front-seat airbags for small passengers), has local applicability (a study about drinking water contamination from flooding would get more coverage in communities subject to overflowing rivers), or is of broad public interest (anything that could raise or lower cancer risks, for example). Inventors at one company developed an airport scanner that detects concealed plastic or metal weapons on passengers. Whenever there is a publicized airline incident thought to involve weapons, the inventors prepare themselves for a flurry of media inquiries regarding airport security technologies.

Timing and accuracy are important in releasing information to the public through the media. Make sure that the information you are releasing is mature enough to be credible and defensible. One scientist independently released his highly preliminary findings about potential electromagnetic hazards of cell phones right before the Christmas season. The research had not been reviewed by others, and it was presented without explanation of its preliminary nature. The resulting media coverage prompted enough controversy that his company faced potential lawsuits from cell phone manufacturers angry about losing business.

Prepare Messages and Materials Carefully

Planning is critical for successful media interaction. When preparing to talk with a reporter, understand what might be asked and consider in advance how to respond. Table 16-1 gives examples of questions to help you prepare interview responses.

For an interview, have two or three short, crucial messages that you want to leave with the audience. According to Rene Henry, who led communications for the mid-Atlantic

Table 16-1. Questions to ask before an interview

Background questions	What is the reporter's name, organization, and phone number?
	What stories has the reporter previously covered?
	Who generally reads, sees, and/or hears the publication or program?
Logistics questions	Where and when will the story appear?
	What is the deadline for the story?
	Where will the interview take place?
	How long will the interview take?
	How long will the story be?
	Does the reporter verify the accuracy of specific quotes attributed to the person being interviewed?
Topical questions	What is the story's theme?
	What topics does the reporter want to cover in the interview?
	What types of questions will be asked?
	Has the reporter done any background research?
	Does the reporter want to receive background material before the interview?
	Who else has been interviewed?
	What did they say?
	Who else will be interviewed?

Adapted from work by Vince Covello (CDC et al. 2003).

States region of the U.S. Environmental Protection Agency after years of private service, sound bites broadcast from interviews average 7.2 seconds; quotes from print materials average 1–2 seconds (Henry 2000). In developing these messages, consider what the audience most needs to know, most wants to know, and is most concerned about. (See Chapter 9 for more details on message development.) Find ways to "bridge" the conversation to these points even if you are not asked questions that directly reflect them. This way, you not only answer the reporter's questions but also focus on the most critical information. Hyer and Covello (2005) offer more than 30 bridging statements that may be useful when speaking.

Take advantage of advice and training provided by public affairs specialists in your organization or through consultants. Many organizations use professional trainers with news media backgrounds when a finding of high public interest is about to be released. These trainers put technical and managerial staff who will be spokespersons through extensive rehearsals, giving them realistic practice in explaining the significance of findings and in responding to

> [Scientists who go into a media interview unprepared] will be required to present the equivalent of a one-draft dictated statement without preparation or notes to an audience of a few thousand to several million people, most of whom know nearly nothing about the topic. In view of the task, the arrogance of some scientists seems inappropriate; terror would be more justified.
> —Robert McCall (1988, p. 87).

challenging questions. For more information on selecting an appropriate spokesperson for various situations, see "Choose the Appropriate Spokesperson" in Chapter 15.

Offer to provide photos or video footage appropriate to the medium. Remember that for television, pictures, rather than words, usually determine what the audience remembers.

Know Where to Draw the Line

Reporters often will ask questions outside the scope of your expertise. When you are asked a question for which you do not know the answer, it is perfectly acceptable to say, "I don't know," or give the reporter another source who will know the answer. Avoid speculating about a hypothetical situation. Avoid guessing, especially in matters that involve quantities. For example, when asked how many people in your community may be affected by a certain risk, one response is to give a range based on various conditions.

Be aware of the limits of your position in representing your organization on a particular issue. One technical expert declined to be interviewed on radio with the head of a nationally known civil liberties organization. The reporter was asking them to discuss public acceptance aspects of a new technology that the expert was developing. Sensing a potentially contentious debate that held little value for his project, the technical expert declined that particular format. "I'm an engineer, not a constitutional lawyer," he explained to the reporter. He did, however, agree to do other, selected media interviews that focused on the need for the technology and its applications.

When being interviewed, do not feel obligated to answer every question. Never say anything you do not want to see in the paper or on the air. There is no such thing as "off the record," despite what we see in the movies. Do not say "no comment," especially in a crisis communication situation; many audiences will interpret that as an assumption of guilt on the part of the organization. Examples of better responses are:

> When being interviewed, do not feel obligated to answer every question.

- I don't know, but we're working to learn the answer.
- I'll have an update at [a specific time] or [when we learn X].
- I wasn't involved in that aspect, but I can tell you about. . . .
- This isn't about blame, it's about saving lives [property, the environment]. Here's what we're doing. . . .

Put Your Message in Terms That the Reporter's Audience Can Understand

When providing printed or visual information to media representatives, or when being interviewed, remember that you are not speaking to your peers but to a general audience. Many of them know far less than you do about the risk in question. Guide journalists in interpreting the results of studies. Replace all technical jargon with terms and concepts to which the public can relate. It may be helpful to pretend that you are talking with your neighbors or to a relative who does not know what you do.

People need a "yardstick" to evaluate a new or unfamiliar risk. For example, in reported studies of electromagnetic fields, exposures often are compared with levels received from home appliances. (But be cautious about comparisons. See "Principles for Comparing Risks" in Chapter 6.) And get to the point: What is the bottom-line message about the risk that people should know? Audiences typically want to know two things: How does this affect me and my family, and what can I do about it?

> Audiences typically want to know two things: How does this affect me and my family, and what can I do about it?

Prepare press kits to give media representatives. The kits should contain factual, explanatory background information on a particular topic or event. Reporters may use the information as reference material to help them put the story in context or as self-education about technical details. A typical situation in which press kits are used is an event that media representatives have been invited to attend, such as a major release of information about a local or regional risk. Press kits, electronic or hard copy, can be offered to media representatives who are doing a story on a topic that involves your organization, especially if the topic is complex. Hard copies may be best for materials that, if digital, would represent very large files. Some reporters do not accept large files as attachments, to avoid viruses and malware, and because of server space restrictions.

A kit might include some of the following materials: a fact sheet, a published article from a magazine or trade publication, a question-and-answer sheet, a press release, a photo depicting a visual aspect of the topic, a list of contacts, and a business card. The materials should use nontechnical, straightforward language; the closer the language to what the reporter will write or produce, the better. Keep the press kits short. Busy reporters will ignore press kits crammed with technical journal articles, official reports, or promotional materials that have been doctored with the organization's "spin."

Put the Risk in Perspective

In providing printed, visual, or oral information, tell how new findings build on or contradict previous studies. This kind of "big picture" approach helps audiences evaluate the risk. A way to help audiences evaluate the magnitude of the risk is to differentiate between relative and actual risk. Cristine Russell, Special Health Correspondent to the *Washington Post*, uses the following example: A finding that a drug poses nine times greater risk of cancer (its relative risk) is misleading without explaining that the cancer has an *actual risk* of one in a million of occurring in the first place (Russell 1993).

Similarly, Russell encourages communicators to distinguish between individual and societal risk. Is the risk a public health problem, or is it significant only for a localized or specialized population with certain characteristics?

To make its risk coverage even more useful, the British media conglomerate BBC News developed the following reporting guidelines for its journalists (Harrabin et al. 2003). Risk communicators may find them helpful in preparing messages:

- What exactly is the risk? How big is it? Whom does it affect?
- How has the risk been measured? How big is the sample?
- Who funded the research? How reputable is the source?
- If you are reporting a relative risk, have you made clear what the baseline risk is? (For example, a 100% increase in the problem that affects one person in 2000 will still only affect one in 1000.)
- Have you asked how safe is this rather than is this safe?

- If a scientist or a victim is taking a view that runs against majority scientific opinion, is that clear in the report and in the casting of the discussion and subsequent questions?
- Have you told the audience how to find more information?
- Can you find a comparison to make the risk easier to understand?
- Have you given the audience information to put the risk in context? (For example, women who stop taking the pill during a pill scare face worse risks from either abortion or childbirth.)
- Is the scale of reporting in proportion to the extent of the risk? Will our reporting increase or decrease risks in society?
- Can we use a story about a specific risk as a springboard to discuss other related risks (for example, train safety vs. road safety)?

Recommend that the reporter interview other people who see the risk from a different angle: those who cause, manage, benefit from, study, or prevent it. Including such a variety of people is especially useful in a press conference. This holistic approach gives audiences a clearer picture of the risk by illuminating more dimensions and points of view.

For preparing medical risk messages, it may be helpful to review the criteria from an organization known as HealthNewsReview.org, funded by the Foundation for Informed Medical Decision Making. The website is dedicated to improving the accuracy of news stories about medical treatments, tests, products, and procedures, thus helping consumers evaluate the evidence about new ideas in health care. The organization independently rates specific news stories on medical issues based on criteria defined on its website. Adapting the criteria from the risk communicator's point of view, you should be able to answer the following questions in your messaging:

- Costs of the intervention?
- Benefits and harms of the treatment/test/product/procedure?
- Quality of the evidence?
- Independent sources?
- Conflicts of interest?
- New approach compared with existing alternatives?
- Availability of the treatment/test/product/procedure?
- Novelty of the approach?

Respect the Reporter's Deadlines

Return calls as soon as possible. Provide additional oral or printed information, video clips, photos, and reviews of materials when you say you will. Most reporters will identify their deadlines but ask if you are unsure. You may only get one chance to provide information. If you do not respond on time, the story will appear or air without your input.

Maintain Ethical Standards of Disclosure

When talking with media representatives, disclose any proprietary interest or other potential conflict of interest. This applies especially to scientific findings or risk-reduction technologies with direct personal benefit to you or your organization. The Jacobs Institute of Women's Health describes the increasing use of self-serving information releases that actually are advertising, such as highlighting premenstrual syndrome and menopause programs, two examples that have been used to generate revenue for health institutions (Lebow and Arkin 1993).

An example of ethical disclosure is identifying a research sponsor that could be seen as biasing the results, such as studies on lung cancer sponsored by the tobacco industry. If you fail to disclose an aspect of the work that is later shown as covertly benefiting you, your organization, or your sponsor, you and your work risk losing credibility with both the news media and the public.

Take Action When Inaccurate or Misleading Material Is Published or Aired

Reporters almost never ask a source to review a story before it is aired or published, though you can encourage reporters to fact-check a story with you before it runs. You do have the right to ask the reporter what the gist of the story will be. You can also ask to have your quotes read back to you during an interview.

> Avoid complaining about the journalists' writing style, omission of superfluous details, or elements that do not change the main message of the story.

It is also possible to alert reporters about factually inaccurate, incomplete, or misleading reporting after it has appeared. Use this option judiciously. Avoid complaining about the journalist's writing style, omission of superfluous details, or elements that do not change the main message of the story. Keep your end goal in mind: to get a more correct story next time, not to badger the reporter for an apology.

Most journalists appreciate being alerted to inaccuracies; they want to be viewed as credible. However, do not expect an automatic correction; it is the decision of the media representative, the editor, or producer, and, sometimes, it is the media organization's policy.

Evaluate News Media Coverage

Regardless of how you engage with the news media, if you are including media coverage as part of your risk communication efforts, be sure to evaluate its usefulness. For traditional news media channels, Hyer and Covello (2005) suggest the following process-oriented metrics:

- Number of newspaper stories, radio and TV spots, and website mentions that carried your information, and prominence of the information presented
- Extent to which the information was accurate, edited appropriately, and conveyed your messages without distortion
- Information conveyed by partners reinforced the same messages

Use this information to modify your news media strategy while it is in process, or for future efforts. For example, you may realize that you need to focus media and partner interactions more closely on key messages. You may ask your spokespersons to address new misperceptions or gaps that cropped up in initial media coverage. If you see that the news media have touched on a new "slant" to the story that you had not anticipated, you may wish to capitalize on that by reaching out to additional media outlets, such as trade publications, that specialize in that topic.

Other information on evaluation can be found in Chapter 10. Specific guidelines on evaluation of social media channels are covered in Chapter 19. Evaluation of public health campaigns is described in Chapter 23.

USING TECHNOLOGY

Organizations are increasingly using the Internet and other technology tools to make their interactions with reporters faster and more efficient. Many organizations have entire multimedia sections on their websites for media professionals and the public to access. The World Bank, for example, provides video and radio news releases, B-roll (extra footage that helps tell a story), a photo library, and downloadable public service announcements (PSAs). It also contains a password-protected online media briefing center that contains embargoed news (advance news that is requested not to be reported until a certain time) for accredited journalists.

> Many organizations have entire multimedia sections on their websites for media professionals and the public to access.

The following sections describe some of the more common technology tools used when working with the news media. For more on using technology in risk communication, see Chapter 18.

Distribution Services

Many organizations distribute their news releases using subscription-based web services that make news available to journalists. For risk-related news, many of these services are in the science and technology realms. The leading such service is EurekAlert!, operated by the American Association for the Advancement of Science. EurekAlert! is an online press service with which journals, research institutions, universities, government agencies, corporations, and others can distribute science-related news to reporters and news media. EurekAlert! also archives its press releases for the public. As with many online press services, you must register on the site as a public information officer to submit a news release. Only those who have registered as a reporter or freelancer can access embargoed news for use in a news story.

> Many organizations use e-mail lists to distribute their news releases on a regular basis. Make sure that each release contains an e-mail link to a contact person and the organization's website address.

Another web service, Newswise, also specializes in research results and news from research institutions worldwide. Newswise works similarly to EurekAlert!, but without prepublication access to journals.

Some reporters also use services that match reporters and sources. With the subscription service Profnet, part of PR Newswire, a reporter sends a topical query to the service at no charge, and Profnet compiles the queries and dispatches them via e-mail to institutional subscribers, who then contact the reporter directly with their information that fits the query. Another service called Help a Reporter Out (HARO) works the same way as Profnet, but it does not charge institutional users.

Many organizations use created and self-subscribed e-mail lists to distribute their news releases on a regular basis. Make sure that each release contains an e-mail link to a contact person and the organization's website address. To capture the attention of busy reporters, use a descriptive subject line in the e-mail (not "Press Release from ABC Company" but "New Discovery Reduces Health Risk from Asbestos"). Link to online background information and photos when available. Include an "unsubscribe" option.

Video and Audio News Releases

Video and audio news releases provide broadcast-ready information. The Broadcast Media and Technology Center of the U.S. Department of Agriculture (USDA), for example, produces more than 90 video news releases annually on stories including biotechnology, water quality, food safety, and other issues. The news releases usually air on nationally syndicated programs for rural audiences, as well as on commercial television stations. The center also makes its news releases available to the public as streaming media files on its website. USDA's radio news stories and news conferences, more than 2000 annually, are available to radio stations via telephone dial-up service and MP3 audio files on the USDA's website. This way, broadcasters can use the Internet to directly access nightly radio feeds that have high-quality sound.

One note of caution is that some reporters shun video news releases as manufactured propaganda. In keeping with journalistic objectivity, they would rather explore all the angles and issues, not just what the producing organization wants to convey. On the other hand, media outlets with more restricted budgets, such as in smaller markets, may welcome video news releases as high-quality products without the hefty production price tag. As with any communication method, it is important to determine whether video news releases will be used and whether the costs are worth the return.

When using video news releases, keep the following considerations in mind:

- Make sure that your organization's infrastructure can handle the additional bandwidth required by high-quality video files. Your information technology professionals may require you to set up an online newsroom that is separate from the main site to prevent crashes. Alternatively, consider using an existing video-sharing site such as YouTube or Vimeo to host your videos and to ensure wider access.
- Keep video clips for media brief. A good rule of thumb, according to Tim Roberts from Wieck Media, is to separate clips into 15- to 20-second bites. Journalists are more likely to use video that does not require extra editing on their part.
- Contact your target outlets to let them know when a new video clip has been added online. Or, include the web link in press releases.

Public Service Announcements

PSAs are advertisements that serve the public interest. PSAs educate and raise awareness about significant social issues in a way that will change attitudes and behaviors and create positive social change. PSAs are not intended to promote a commercial product, brand, or service, and are usually put out by government and nonprofit organizations. PSAs are common in care communication, such as anti-obesity campaigns, and crisis communication, such as the Federal Emergency Management Agency telling people how to get help after a flood. TV and radio are the most common media outlets, but newspapers, magazines, and websites may also accept PSAs.

Most PSAs run as a community service at no charge by the media, but some run in purchased time and space. Some nonprofit organizations and government agencies, such

> Most PSAs run as a community service at no charge by the media, but some run in purchased time and space.

as the Office of National Drug Control Policy and the Centers for Disease Control and Prevention, purchase media time and space for some of their PSAs. This gives them more control over placement and scheduling.

PSAs were more common in the past, when the Federal Communication Commission imposed stricter requirements for broadcasters to demonstrate that they were operating in the public interest. Today, with deregulation, most stations are setting their own standards on what constitutes fulfillment of their public service programming responsibilities to their local communities. Some stations run their own community-affairs programming instead of PSAs. Others air public-service-type messages from paying advertisers and use on-air promotions featuring their own network TV stars. Stations may run PSAs in the early morning hours when audiences are almost nonexistent, thus meeting the letter of the law if not the spirit. To counter this trend of diminishing PSAs, some organizations are delivering more content through video news releases and B-roll to the same markets that used to run PSAs.

Nevertheless, PSAs may be one strategy in an overall risk communication program. Sometimes, a local ad agency may develop a PSA for your organization at no charge; local media may also lend their development support. The Advertising Council supplies dozens of free, high-quality PSAs; risk-related topics range from fire safety to childhood asthma to homeland security. In creating PSAs, it is important to tailor the message to audiences you are trying to reach and restrict the content to one takeaway message.

Understand the various media's policies, lead times, and required formats. If the PSA relates to a specific event, make sure that it gets to the stations at least 2 weeks in advance, to allow them time to schedule it.

Evaluate the results based on the broadcast statistics from the media outlet and the contact phone number and/or website you have provided in the PSA. The PSA Research Center is a good source of additional information, including articles and case studies by PSA topic.

Telebriefings

Telebriefings use a conference call format that replaces or augments the traditional press conference. Similar to a conference call, they can be conducted anywhere and do not require participants to travel to a news conference location. Sometimes the telebriefing is broadcast via streaming audio on the Internet. The Centers for Disease Control and Prevention (CDC) has conducted more than 40 telebriefings each year since 2001, on public health topics including anthrax, smallpox, West Nile virus, severe acute respiratory syndrome, and many others. The CDC sends an e-mail message to its listserv of media and provides a listen-only line to public health and Congressional contacts, giving them a toll-free number to call and a time for the briefing. CDC Media Relations hosts the briefing with a group of experts, and reporters ask questions following a brief opening statement. CDC posts transcripts of its telebriefings on its website for reporters and the public.

When considering a telebriefing, plan accordingly for capacity. The CDC, for example, pays for 100 phone lines to handle the deluge of callers during and after each telebriefing.

Social Media

The rise of social media is influencing how journalists report risk and other topics in two important ways. First, journalists are delivering their news via more social media platforms, thus targeting more niche audiences. Second, journalists are using social media channels to find and develop story ideas.

Here is what journalists themselves are saying. A national media survey found that up to three quarters of U.S. journalists who responded use social media tools in reporting, including Facebook, blogs, Twitter, Wikipedia, and LinkedIn (Middleberg and McClure 2011). They use these tools to research individuals or organizations, participate in conversations, monitor sentiment or discussion, keep up on issues or topics of interest, and find story ideas and sources. Up to half have work-related Twitter accounts, a third have professional Facebook pages, and others had their own blogs or podcasts.

> We don't break things on air anymore. We break them on Twitter and Facebook and then fill in the details on the show.
> —A broadcast journalist (Middleberg and McClure 2011).

How can your organization take advantage of this phenomenon to communicate about risks in the news media? Engage in the journalists' own social media channels to get a feel for what they cover and are interested in. This can also help you direct your story ideas to the most appropriate journalists. Use your own blog or other social media channels to get journalists' attention. Fifty-six percent of the journalists who responded to the 2011 survey said that they quote bloggers in their stories. Post comments to establish yourself as an expert.

Create profiles on social networking sites and use relevant tags, or key words, so that journalists can find you easily when searching these sites. Take care, however, when pitching story ideas to media bloggers. Understandably, online journalists get annoyed when they get pitches that do not relate to their blog topics. Choose blogs carefully and create customized pitches for each.

Remember that the personal touch is still important. According to the 2011 survey, journalists still prefer traditional communications and relationship building when researching stories.

For more on how organizations can use social media to communicate about risks, see Chapter 19.

CHECKLIST FOR WORKING WITH THE NEWS MEDIA

When planning to work with media representatives, assure that:

☐ The role and mission of news media are understood and have been factored into risk communication goals as appropriate.
☐ Productive relationships have been developed with key members of the media who are most likely to cover specific risk issues.
☐ A plan has been prepared for when to approach the media and/or what to do if events prompt them to approach you.
☐ Appropriate social media channels are used to engage journalists.

(continued)

CHECKLIST FOR WORKING WITH THE NEWS MEDIA (*continued*)

When working with media representatives, assure that:

☐ Risk messages and materials have been carefully prepared and tailored for specific media.

☐ Reporters are given the information they need and you want to convey, in language aimed at your target audiences.

☐ Questions outside the scope of your expertise or role in the organization are referred to others.

☐ Reporters' deadlines are known and respected.

☐ Ethical standards of disclosure are understood and maintained.

☐ Appropriate action is taken when inaccurate or misleading material is printed or aired.

REFERENCES

Adams, W. C. 1986. "Whose Lives Count?: TV Coverage of Natural Disasters." *Journal of Communication*, 36:113–122.

Atkin, C. K., S. W. Smith, C. McFeters, and V. Ferguson. 2008. "A Comprehensive Analysis of Breast Cancer News Coverage in Leading Media Outlets Focusing on Environmental Risks and Prevention." *Journal of Health Communication*, 13(1):3–19.

CDC (U.S. Centers for Disease Control and Prevention), Agency for Toxic Substances and Disease Registry, Oak Ridge Institute for Science and Education, and the Prospect Center of the American Institutes of Research. 2003. Emergency Risk Communication CDCynergy (CD, February 2003). http://www.orau.gov/cdcynergy/erc/ (accessed January 23, 2013).

Combs, B. C. and P. Slovic. 1979. "Newspaper Coverage of Causes of Death." *Journalism Quarterly*, 56:837–849.

Ferris, E. and D. Petz. 2011. *A Year of Living Dangerously: A Review of National Disasters in 2010.* The Brookings Institute and the London School of Economics Project on International Displacement, Washington, DC.

Greenberg, M. R., D. B. Sachsman, P. M. Sandman, and K. L. Salomone. 1989. "Network Evening News Coverage of Environmental Risk." *Risk Analysis*, 9(1):119–126.

Harrabin, R., A. Coote, and J. Allen. 2003. *Health in the News: Risk, Reporting and Media Influence.* King's Fund, London.

Henry, R. A. 2000. *You'd Better Have a Hose if You Want to Put Out the Fire: The Complete Guide to Crisis and Risk Communications.* Gollywobbler Productions, Windsor, California.

Hyer, R. N. and V. T. Covello. 2005. *Effective Media Communication during Public Health Emergencies: A WHO Handbook.* World Health Organization, Geneva, Switzerland. WHO/DCS/2005.31. http://www.paho.org/cdmedia/riskcommguide/Effective%20Media%20Communication%20Handbook.pdf (accessed January 23, 2013).

Lebow, M. and E. B. Arkin. 1993. "Women's Health and the Mass Media: The Reporting of Risk." *Women's Health Issues*, 3(4):181–190.

McCall, R. B. 1988. "Science and the Press: Like Oil and Water?" *American Psychologist*, 43(2):87–94.

Middleberg, D. and J. McClure. 2011. "How Are Media and Journalism Evolving? Insights from the 3rd Annual Middleberg/Society for New Communications Research Survey of Media in the Wired World." Presented at the 6th Annual SNCR Symposium and Awards Gala, November 4, 2011, Cambridge, Massachusetts.

Nelkin, D. 1994. "Reporting Risk: The Case of Silicone Breast Implants." Technical Risk in the Mass Media. Franklin Pierce Law Center. http://ipmall.info/risk/vol5/summer/nelkin.htm (accessed February 7, 2013).

Nelson v. McClatchy Newspapers, Inc., et al., No. 97-187-CSX, U.S. Supreme Court (1997).

O'Rourke, A. D. 1990. "Anatomy of a Disaster." *Agribusiness*, 6(5):417–424.

Russell, C. 1993. "Hype, Hysteria, and Women's Health Risks: The Role of the Media." *Women's Health Issues*, 3(4):191–197.

Singer, E. and P. M. Endreny. 1993. *Reporting on Risk: How the Mass Media Portray Accidents, Diseases, Disasters, and Other Hazards.* Russell Sage Foundation, New York.

Wallack, L.M., L. Dorfman, D. Jernigan, and M. Themba. 1993. *Media Advocacy and Public Health: Power for Prevention.* Sage Publications, Newbury Park, California.

Young, M. E., G. R. Norman, and K. R. Humphreys. 2008. "Medicine in the Popular Press: The Influence of Media on Perceptions of Disease." *PLoS One*, 3(10):e3552.

ADDITIONAL RESOURCES

Advertising Council. http://www.adcouncil.org/psa (accessed January 23, 2013).

Atkin, C. and L. Wallack, eds. 1990. *Mass Communication and Public Health: Complexities and Conflicts.* Sage Publications, Newbury Park, California.

Bertrand, J. T. and R. Anhang. 2006. "The Effectiveness of Mass Media in Changing HIV/AIDS Related Behaviour among Young People in Developing Countries." *World Health Organization Technical Report*, Series 938, 205–241.

Clement International Corporation. 1991. *Risk Communication Manual for Electric Utilities, Volume 1: Practitioner's Guide.* Electric Power Research Institute, Palo Alto, California. EPRI EN-7314.

Elder, J. P., E. S. Geller, M. F. Hovell, and J. A. Mayer. 1994. *Motivating Health Behavior.* Delmar, Albany, New York.

EurekAlert! http://www.eurekalert.org (accessed January 23, 2013).

Flynn, J., P. Slovic, and H. Kunreuther, eds. 2001. *Risk, Media, and Stigma: Understanding Public Challenges to Modern Science and Technology.* Earthscan Publications, London.

Newswise. http://www.newswise.com (accessed January 23, 2013).

Olshefsky, A. M., M. M. Zive, R. Scolari, and M. Zufliga. 2006. "Promoting HIV Risk Awareness and Testing in Latinos Living on the U.S.–Mexico Border: The Tit No Me Conoces Social Marketing Campaign." *AIDS Education and Prevention*, 19(5):422–435.

PR Newswire. http://www.prnewswire.com (accessed January 23, 2013).

Profnet. http://www.profnet.com (accessed January 23, 2013).

PSA Research Center. http://www.psaresearch.com/index.html (accessed January 23, 2013).

Rice, R. E. and C. Atkin, eds. 1989. *Public Communication Campaigns.* 2nd ed. Sage Publications, Newbury Park, California.

Rice, R. E. and C. Atkin. 1994. "Principles of Successful Public Communication Campaigns." In J. Bryant and D. Zillman, eds., *Media Effects: Advances in Theory and Research.* Lawrence Erlbaum Associates, Hillsdale, New Jersey, pp. 365–387.

Sandman, P. M. 1986. *Explaining Environmental Risk.* U.S. Environmental Protection Agency, Office of Toxic Substances, Washington, DC.

Sandman, P. M., D. B. Sachsman, and M. L. Greenberg. 1988. *The Environmental News Source: Providing Environmental Risk Information to the Media.* Risk Communication Project, Hazardous Substance Management Research Center, New Jersey Institute of Technology, Newark, New Jersey.

U.S. Department of Health and Human Services, Public Health Service, and National Institutes of Health. 1992. *Making Health Communication Programs Work: A Planner's Guide*. Office of Cancer Communications, National Cancer Institute, Bethesda, Maryland. NIH Publication No. 92-1493. http://www.cancer.gov/pinkbook (accessed January 23, 2013).

Walters, L. M., L. Wilkins, and T. Walters. 1989. *Bad Tidings: Communication and Catastrophe*. Lawrence Erlbaum Associates, Hillsdale, New Jersey.

West, B., P. M. Sandman, and M. R. Greenberg. 1995. *The Reporter's Environmental Handbook*. Rutgers University Press, Piscataway, New Jersey.

Windsor, R., N. Clark, N. R. Boyd, and R. M. Goodman. 2004. *Evaluation of Health Promotion, Health Education, and Disease Prevention Programs*, 3rd ed. McGraw-Hill, New York.

17

STAKEHOLDER PARTICIPATION

Having the audience or stakeholders interact directly with those who are communicating, assessing, and/or managing the risk can be an extremely effective way to communicate risk. For example, in 1993, when an abandoned World War I chemical warfare research facility was discovered under an upscale residential community in northwest Washington, DC, the U.S. Army Corps of Engineers took command of the situation and held nightly interactions with the community. Their community involvement is credited for preventing panic, protest, and litigation (Henry 2000).

Stakeholder participation can take many forms, such as self-help groups, focus groups, and advisory committees. Stakeholders can be involved in working through a particular risk issue, such as in a workshop. Stakeholders can also participate in how the risk is assessed or managed, for example, members of the public taking surveys or operating monitoring stations. The least effective but most often used form of stakeholder participation is the formal hearing or public meeting, for which the organization sets a time and place for the audience to present formal testimony, which is transcribed and used later in the risk management process.

Some stakeholder participation programs fail because of a lack of early and continuing involvement (Kasperson 1986). Stakeholder participation is most effective when key choices concerning the risk have yet to be made. Once an organization is locked on a course of action, participation opportunities dwindle to those that will

> Stakeholder participation is most effective when key choices concerning the risk have yet to be made. Once an organization is locked on a course of action, participation opportunities dwindle to those that will educate the audience.

Risk Communication: A Handbook for Communicating Environmental, Safety, and Health Risks,
Fifth Edition. Regina E. Lundgren and Andrea H. McMakin.
© 2013 The Institute of Electrical and Electronics Engineers, Inc., and Regina E. Lundgren and
Andrea H. McMakin. Published 2013 by John Wiley & Sons, Inc.

educate the audience. Using stakeholder participation solely to educate is more costly and time-consuming than other forms of risk communication that can be used to educate. Furthermore, many stakeholders willing to participate expect a more substantial involvement and will become hostile when they realize that their activities are limited, further constraining the risk communication effort.

Stakeholder participation is rapidly becoming the premiere way to communicate risks in consensus communication and planning for crises. Two blue-ribbon panels (the Presidential/Congressional Commission on Risk Assessment and Risk Management and the National Research Council's Committee on Risk Characterization) advocated stakeholder participation throughout the risk assessment, risk management, and risk communication processes (Commission 1997; NRC 1996). Some research indicates that the public may be more supportive of decisions that were reached through a stakeholder participation process, even if they were not personally part of that process (Arvai 2003).

Advantages and disadvantages of stakeholder participation are discussed in Chapter 10. This chapter discusses requirements for successful stakeholder participation, provides guidelines for specific types of stakeholder participation activities, and gives advice on how to choose a form of stakeholder participation.

REQUIREMENTS FOR STAKEHOLDER PARTICIPATION

To choose a form of stakeholder participation, you need to consider both organizational and stakeholder needs. Your organization must be comfortable with the way it interacts with stakeholders and vice versa. You can then evaluate the various forms of stakeholder participation for application to your situation.

Organizational Requirements for Successful Stakeholder Participation

For any form of stakeholder participation to succeed, your organization must be fully committed to it. Everyone involved with the risk assessment and risk management process—the scientists and engineers, the public health professionals, the technicians, the communicators, the public affairs specialists, the risk managers, and the organization's line managers—must believe that stakeholders have a right to be and can be involved. If anyone has reservations, those reservations will be apparent to the stakeholders and spoil any chances for meaningful interaction. Because many organizations are less than committed to stakeholder participation, effective participation in risk communication has been limited. However, in those cases in which participation has been successful, both the stakeholders and the organization have deemed the participation well worth the effort (see, for example, Aleknavage and Lyon 1997; Beierle 2002).

> For any form of stakeholder participation to succeed, your organization must be fully committed to it.

To make the effort acceptable to all involved in the risk assessment and management process, you need a clear plan. You need to know your purpose and objectives, your audience (stakeholders), your schedule, and the resources at your disposal. If these factors show that stakeholder participation would be an effective way to communicate risk, you then need a compelling reason that will convince the organization. Perhaps stakeholders have perspectives that would be particularly valuable to the risk assessment or management process. Perhaps no other method will bring about the desired results. Perhaps a regulation or policy mandates stakeholder participation. Presenting your plan and this compelling reason to the others involved with communicating, assessing, and managing the risk, and having a list of cases in which effective stakeholder participation made for a more effective risk management decision, can help convince management of the need to involve stakeholders in a meaningful way. (See Aleknavage and Lyon 1997; Arvai 2003; Beierle 2002; Imholz et al. 1990 in this chapter's reference list; and additional examples of case studies in Resources at the back of this book.)

What most organizations fear about stakeholder participation is that the organization might lose control over the risk decision. However, stakeholder participation can strengthen that decision by:

- Identifying stakeholder perceptions early in the process so that the organization does not find out after the decision is made that stakeholders are unwilling or unable to support or implement it
- Developing a consensus among all parties affected by the risk, giving the decision a weight it can achieve in no other way
- Helping to prevent conflicts such as lawsuits that can delay making or implementing the decision

In addition, inviting stakeholders to participate in the communication, assessment, or management of the risk can raise their awareness, inform them, and motivate them to action more strongly than any other form of risk communication.

Stakeholder Requirements for Successful Participation

Certain stakeholder characteristics can influence the success of interactions with the organization. These characteristics include the size and number of segments, their level of interest, and, most important, their level of hostility.

If stakeholders are widely distributed and encompass a number of diverse constituencies, each with its own perception of the risk and your organization (for example, the entire American populace), it can be extremely difficult to develop a single kind of stakeholder participation that will meet everyone's needs. Even including just one representative from each segment may require a committee with hundreds of people. Many types of stakeholder participation are limited to no more than 10 people to be effective (for example, focus groups). With more people, it becomes extremely difficult to come to any level of agreement (in the case of consensus communication) or even to have any kind of meaningful discussion (in the case of care communication or crisis communication planning). With a stakeholder group whose representatives number in the hundreds, the forms of stakeholder participation that have a chance of being successful are formal hearings, multiple meetings held around the country or region of

> Certain stakeholder characteristics—such as the size and number of number of segments, their level of interest, and, most important, their level of hostility—can influence the success of interactions with the organization.

interest, and consensus-building committees run by an expert in conflict resolution. Technology is also making large-scale involvement more feasible. See Chapter 18 for details.

Stakeholder participation can also be difficult if stakeholders are apathetic. If stakeholders truly care nothing about the risk, but your organization feels that the risk is real, you may need to raise awareness before you can effectively involve stakeholders in decision making or risk prevention.

Some people feel that silence or previous lack of involvement are signs of apathy. However, lack of attendance at meetings or lack of comments on documents may mean hostility, not apathy. Hostility is the primary obstacle to meaningful stakeholder participation. If stakeholders do not believe your organization and are angry about the way your organization is handling the risk assessment, risk management, or risk communication process, it may be difficult to develop meaningful partnerships. On the other hand, effective stakeholder participation can be one of the best ways to reduce hostility, by showing that the organization does listen to and address concerns. In a situation with hostile stakeholders, interactions with small groups to build consensus and make decisions is the best choice. The worst choice in a hostile situation is a formal hearing because (1) the layout of the meeting exacerbates the "us versus them" mentality; (2) the formal hearing process is not conducive to having organizations inform stakeholders about how their concerns will be addressed or considered in the risk decision; and (3) communication tends to be one way.

The International Association for Public Participation, a professional organization dedicated to promoting and improving the practice of public participation, provides another way to consider what stakeholders may expect from meaningful participation. The organization developed a list of core values to guide stakeholder participation efforts, such as those for risk communication:

- The public should have a say in decisions about actions that affect their lives.
- Public participation includes the promise that the public's contribution will influence the decision.
- The public participation process communicates the interests and meets the process needs of all participants.
- The public participation process seeks out and facilitates the involvement of those potentially affected by or interested in a decision.
- The public participation process involves participants in designing how they participate.
- The public participation process communicates to participants how their input affected the decision.
- The public participation process provides participants with the information they need to participate in a meaningful way (adapted from IAP2 2000).

GUIDELINES FOR SPECIFIC TYPES OF STAKEHOLDER PARTICIPATION ACTIVITIES

Once you know what your organization will approve and what stakeholders need, you can determine which type of stakeholder participation will be most effective in your situation. For each of the types of participation mentioned earlier in this chapter, there are certain things you can do to help ensure success.

The Formal Hearing

Formal hearings are those in which the risk information is presented by a member of the organization, followed by formal testimony from members of the audience; an example is a scoping hearing for an environmental impact statement. These interactions tend to be primarily one-way communication and, hence, are often not very effective in involving stakeholders in risk assessment, risk management, or risk communication. In fact, some research indicates that formal hearings may increase participants' perception of risk and decrease their perception of the sponsoring organization's credibility (McComas 2001).

However, as mentioned earlier in this chapter, there are times when the formal hearing is a useful means of stakeholder participation (for example, when stakeholders are dispersed and encompass many segments). There are also times when the law appears to require that you hold such a hearing (for example, after a draft environmental impact statement has been released for public comment). (Note, however, that your organization can usually comply with the spirit of the law by holding workshops, open houses, or other less formal and more effective types of stakeholder participation.)

> Formal hearings tend to be primarily one-way communication and, hence, are often not very effective in involving stakeholders in risk assessment, risk management, or risk communication.

If you must conduct a formal hearing, following certain guidelines can help ensure success:

- **Pick a time and place with which stakeholders have no associated negative feelings or even have associated positive feelings.** Choose a room large enough to hold everyone. Pick a time that will allow them to attend without missing work or key social functions. (For example, few people attended a hearing held on St. Patrick's Day in a town with a large Irish Catholic population, and those who did were hostile.)
- **Make sure that key decision makers attend.** Stakeholders often tend not to believe that an organization will do anything about their concerns and comments unless they see someone in authority at the meeting. This was illustrated by comments at several meetings concerning the potential dangers of hazardous-waste storage tanks. Although a number of concerns were expressed, the overriding comment was, "Why isn't the head of the organization here? We want to hear his views, and we want him to hear ours."
- **Get a good moderator.** Get one who (1) keeps the meeting on track without making anyone feel slighted, (2) can focus questions and concerns so that everyone understands them, and (3) is credible to the audience (which usually means one who is independent of the organization). We once listened to a moderator from a federal agency at a formal hearing address every member of the audience by name and provide seemingly sincere comments on their testimony. We were greatly impressed, until a woman seated in front of us leaned over to her companion and whispered, "She's so hypocritical! As if she really cares!" The moderator seemed to be doing everything right, and she still did not meet the needs of the stakeholders because, as a member of the organization, she had no credibility with the stakeholders.
- **Arrange for comments to be recorded.** In many cases, the legal purpose of a formal hearing is to record audience concerns. This recording can be done by using

a digital recorder, a court stenographer, or a video recorder. Just taking notes or writing words on a flip chart is insufficient, although both are excellent supplementary methods; they show your audience that you are listening to them and noting their concerns. However, neither is sufficient to recall exact phrasing or underlying concerns at a later date.

- **Allow time for questions and answers.** Too many meetings are structured to allow the organization and audience to speak at each other, but not to interact. Typically, the organization spokesperson makes a presentation concerning the risk, then members of the audience take turns coming to the podium or microphone and expressing their concerns. Allow time between the organization's presentation and the audience's comments so that the audience can ask questions about the risk. Answers to these questions may allay some concerns, which you will not have to deal with later.

- **Respond to audience concerns.** As soon as possible after the meeting, make sure that those who attended and their constituencies know how their comments and concerns were used in making a decision about the risk. Often, organizations publish the transcribed comments, with coding or annotation by the organization to indicate how comments were addressed. Unfortunately, few people ever see such documents, leaving most of the audience to wonder why they bothered, thus eroding your organization's credibility. It is better to contact each commenter personally, either by phone or by letter, to explain how their concerns were addressed. This can be a daunting job, given the hundreds to thousands of commenters at some formal hearings. Nonetheless, if your organization's credibility and the audience's level of hostility are important, you need to take the time and resources to do it.

Group Interactions

Group interactions generally require some sort of meeting. Much has been written about how to make meetings more effective. For risk communication purposes, there are several circumstances under which group interactions in meetings will not work:

- Data are confidential.
- The subject is trivial related to the overall risk.
- There is anger or hostility.
- The decision has already been made (Doyle and Straus 1976).

For the various types of group interactions (self-help groups, focus groups, workshops, or advisory committees), there are certain factors that help ensure success:

- **Pick a time when all members of the group can attend** and a place with which they have no associated negative feelings and, at best, with which they have associated positive feelings. Both focus groups and advisory committees should meet away from the organization if the group is independent or at the organization if it is more important to reinforce the feeling that the organization is really listening to them. For self-help groups, choose a meeting place that is conducive to sharing feelings (private, with comfortable furnishings).

- **Identify the purposes, objectives, and desired results at the beginning.** Make sure that both the group and the organization know why the group is meeting, what they hope to accomplish, any limitations to their scope or ability to make decisions, and the expected results. For example, with a group that is determining possible future land uses after the cleanup of a Superfund site, the purpose could be to explore possibilities and make recommendations to the organization. The objectives could be to meet regularly, divide the work among representatives, consult legal authorities and stakeholders, and develop recommendations based on information gathered. The results could be to present a report on their findings and recommendations to the organization by the end of a year.

- **Set clear lines of command and communication.** Who is in charge of the meetings, the organization's representative or the other members, perhaps on a rotating basis? How are decisions made, by written ballot, electronic voting, or show of hands? Do all members discuss the information and allow an elected leader to actually make the decision? Who mediates in the case of conflicts? Is consensus necessary or is it more important to capture the range of concerns and comments? Who presents the information from the group and to whom? Who presents the information from the organization and to whom? Develop a plan for how the group will function and make sure everyone involved with the group and from the organization agree to it.

- **Make sure that the organization shows support for the group.** Support can be shown by having someone from the organization whom the group respects attend the meetings; by providing resources such as funding, meeting space, and support staff (for example, to develop reports or provide guidance); or by sending letters of support to the group or outside organizations such as the press. This last suggestion will be meaningless if no one attends meetings and no resources are provided. (In situations in which stakeholders are particularly hostile, it can also undermine the credibility of the group, so consider the situation before following this advice.) If the organization does not show support, the group interaction will soon wither. If the organization does show support, the group interaction can be an extremely effective way to communicate risk information.

Researchers have also studied various group interactions to determine what makes these interactions effective. For example, researchers reviewed the work of a team of government agencies, contractors, Native American tribes, and stakeholders to evaluate potential risks to a major watershed from past releases from a nuclear facility. They found that perceptions of fairness and competence were key to process success. By fairness, they meant that everyone had the same opportunity to (1) determine the rules and the agenda, (2) speak and ask questions, and (3) access information and analysis. By competence, they meant the ability to understand language concepts and to agree on reality. Key problems that hindered the effort included the lack of attention to process rules, refusal of management to support decisions made by organization representatives, no formal dispute resolution mechanisms, and lack of mutually understood meanings of terms and concepts (Kinney and Leschine 2002).

Later in this chapter are additional guidelines for specific types of group interactions.

Self-Help Groups

The general purpose of self-help groups is to motivate stakeholders already aware of personal risky behavior (such as alcoholism or drug addiction) to prevent or address the risk in their own lives. Self-help groups can also assist stakeholders who are dealing with the aftermath of a crisis. In general, self-help groups should be led by a trained facilitator, often

> The general purpose of self-help groups is to motivate stakeholders already aware of personal risky behavior to prevent or address the risk in their own lives. Self-help groups can also assist stakeholders who are dealing with the aftermath of a crisis.

a psychologist or other health care provider. Those who are communicating risk can listen to concerns raised by the group and provide or develop information materials to answer specific questions (more information on developing information materials can be found in Chapter 13). Risk communicators can also assist the facilitator in finding ways to support the group members in managing the relevant risks.

Focus Groups

Focus groups involve stakeholder representatives who meet for a specific purpose, usually for a specified period of time (a few hours to several months), for example, to evaluate the future land uses of a government Superfund site after it is cleaned up. Focus groups have been used in risk communication to explore risk perceptions, help develop content of risk messages, pretest risk communication messages and materials, select risk communication methods, develop alternative ways to manage a particular risk, and evaluate risk communication efforts (Desvousges and Smith 1988). As we use the term in this book, focus groups serve to inform those who are communicating risk to the wide variety of opinions, as well as the prevailing opinions, within a stakeholder group on a particular subject. Used in this way, their participatory nature is somewhat limited because information is still flowing in one direction. However, because the organization thought to ask stakeholder opinions, stakeholders sometimes perceive a greater degree of involvement than some other forms of one-way communication.

> Focus groups have been used in risk communication to explore risk perceptions, help develop content of risk messages, pretest risk communication messages and materials, select risk communication methods, develop alternative ways to manage a particular risk, and evaluate risk communication efforts.
> —Desvousges and Smith (1988).

Focus group interactions require the use of a skilled moderator to probe attitudes and opinions on the specified topic. However, it is important that the moderator be independent—that is, the moderator cannot be perceived as an expert in the risk or the effort, or the meeting may turn into a question-and-answer session. Tasks or exercises are often used to focus discussion or identify perceptions (Desvousges and Smith 1988).

The key word is "focus." This applies to these groups in several areas:

- **Focus membership.** Focus groups do not function well beyond about 10 members. More than that limits the amount of time each member can speak and may intimidate some members of the group, keeping them from speaking. If the stakeholder group is large and multifaceted, try holding several focus groups with specific characteristics. For example, you might meet separately with environmental group representatives, civic group representatives, and labor representatives.
- **Focus time.** Keep the meetings short and on schedule, preferably no more than 2 hours (Desvousges and Smith 1988). Nearly everyone who might belong to your focus group will have limited time to contribute. Make their time count, and they will be more likely to find the process useful.

- **Focus effort.** As mentioned previously, make sure that all members know the purpose of the meetings and how their comments will be used.
- **Focus results.** Make sure that you act on what you hear. Do not avoid problems that surface. And once you have acted or at least have a plan to act, let the wider group know how you are responding to their concerns.

Again, having a trained facilitator at a focus group meeting can help ensure meaningful results for both the group members and those who are communicating risk.

Workshops

Workshops are similar to focus groups in that the purpose of the meeting is specific (for example, to reach a decision on which alternative to propose to manage aircraft noise in a community near a busy airport). However, the nature of the workshop is more educational than participatory in that the stakeholders will be provided with presentations and information before conducting their evaluation. These presentations serve to ensure that all members of the workshop have a common language for the topic being discussed and have a similar understanding of risk on which to build discussions. Another area that is frequently overlooked, however, is the development of a common understanding of the purpose of the workshop, the process the workshop deliberations is supporting, and how the results of the workshop will be used.

> The nature of the workshop is more educational than participatory in that the stakeholders will be provided with presentations and information before conducting their evaluation.

To develop the presentations for the workshop, follow the guidelines in Chapter 15 on speaking engagements. To develop information materials to supplement the presentations, follow the guidelines in Chapter 13. To set up the workshop, follow the general guidelines for group interactions earlier in this chapter.

Advisory Groups

An advisory committee is made up of stakeholder representatives who advise the organization about concerns, usually over a variety of subjects over an extended time. Advisory groups can be used in care, consensus, and crisis communication. Some government agencies use advisory groups in relation to environmental cleanup. For example, the Comprehensive Environmental Response, Compensation, and Liability Act (CERCLA or Superfund) strongly suggests the use of citizen advisory groups and makes provisions for technical assistance grants to support these groups. Technical assistance grants provide citizen advisory groups with the resources to hire their own risk consultants to help them understand and respond to the information in a risk assessment. In addition, the Emergency Planning and Community Right-to-Know Act establishes emergency planning activities, many of which include citizen advisory groups. Federal guidance on bioterrorism preparedness planning in 2003 also encouraged the development of advisory groups. Industry uses employee groups to advise on health and safety issues. Many industries have also developed their own citizen advisory groups to assist them in maintaining a positive relationship with local communities.

> Advisory groups can be used in across care, consensus, and crisis communication.

The U.S. Department of Defense has established Restoration Advisory Boards (RABs), advisory groups for environmental restoration processes at specific military installations. A RAB includes representatives of the military installation, the U.S. Environmental Protection Agency, state and local governments, tribal governments, and the affected local community. RAB members share community views with the installation decision makers and report information back to the community on the military's environmental restoration activities. RABs are not decision-making bodies; they advise the military installation's commanding officer.

Since 1994, the Department of Defense has established more than 300 RABs in the United States and its territories. The website of the Defense Environmental Restoration Program contains many resources about RABs, including policies, guidance, operations manuals, and a directory of RABs operating nationwide (see Additional Resources in this chapter). The Department of Energy has established similar boards (Site-Specific Advisory Boards) at various sites where they are cleaning up the legacy from the Cold War.

Advisory groups usually meet for a longer time than other types of group interactions (except for some self-help groups that go on for years), requiring a large commitment of time from stakeholder participants. This time commitment often limits who can serve as members of these groups. There are two common ways around this limitation.

The first way is to find a method to compensate members for their time. For employees meeting on safety issues, the solution may be to make group membership an equal part of their job and ensure that they are paid for their time to attend meetings and provide information back to their fellow employees. For stakeholders outside the organization, the solution may be to make arrangements with their employers for time off as well as to pay them for their time. A drawback to this solution is that some members may look upon this pay as an entitlement and fight to stay in the group even when they are no longer in a position to contribute.

> More and more, stakeholders are demanding to participate in how a risk is determined, whether and how regulations are set for particular chemicals, or in how risks are characterized in environmental cleanup decisions.

The second way to counter this limitation is to allow for a continuing rotation of membership so that no member has to spend excessive time in meetings. Members representing specific groups could serve for a quarter, a year, or a few years, depending on the frequency of group meetings. The drawback to this solution is that with continued membership turnover comes the need for continued education on the same issues, slowing the progress of the entire group.

Because of the popularity of these types of groups, much has been written as to characteristics of the group that ensure success:

- **Include members from across stakeholder groups.** In care communication efforts, this would include a member of each group at risk. For example, in developing an employee safety committee, you would need members who represent common classes of employees such as clerical, managerial, physical labor, and scientific/engineering/white collar workers. In crisis communication planning efforts, members could include representatives from all groups who will assist in emergency response efforts, both inside and outside an organization, for example, county and city governments, local firefighters, nonprofit organizations such as the Red Cross, and the news media. In consensus communication efforts, this could include representatives from each group with which consensus is sought. For example, advisory groups for environmental cleanup efforts often contain representatives of environmental groups, elected officials, business interests, regulatory agencies, Native American governments, and that nebulous group called "the public" (Serie and Dressen 1992). This latter group is the most difficult to recruit. Tactics include putting ads in the newspaper and contacting already organized groups with noncompeting agendas (for example, the local parent/teacher association might be able to identify a representative of the general public for an environmental cleanup effort but might be already deeply involved in an effort to reform the science curriculum in the education system).

- **Provide meaningful background information.** Some agencies seem to fear contaminating opinions if they share information beforehand. Advisory groups can benefit from broad overviews as well as specific technical data. They need to know agency constraints as well. It does not help the agency or its relationship with the advisory board if the group provides advice that cannot be acted upon for legal reasons.
- **Provide for independent technical advisors.** Regardless of whether the advisory group is chartered under Superfund (and thus can apply for a technical assistance grant), advisory groups can benefit from having their own expert advisor. This advisor can serve to help clarify risk information and provide additional review and perspective to the risk assessment and management process. When the advisor and the organization in charge of managing the risk agree on the interpretation of risk data, the organization's credibility is enhanced. When they disagree, the disagreement points to areas that require additional information and communication. Edward Scher of the Massachusetts Institute of Technology's Department of Urban Studies and Planning and Sarah McKearnan of the Public Disputes Program of Harvard Law School suggest that to be successful, these technical advisors must be credible with all segments of the stakeholder group, understand the process as well as the technical issues, guide but refrain from driving the interactions, and have the ability to work closely with participation facilitators (Scher and McKearnan 1997).
- **Spend time on team building.** Let the team members get to know each other so they are comfortable sharing opinions and information, and set up a process that allows you to hear all opinions, not just the most vocal ones.

Interactions Involving Risk Assessment

More and more, stakeholders are demanding to participate in how a risk is determined, for example, in how regulations are set for particular chemicals, preparations are made for public health emergencies, and risks are characterized in environmental cleanup decisions. A few years ago, many scientists, engineers, and other decision makers would have balked at the thought of a layperson conducting complex risk analyses. However, the advent of widely available risk assessment models in the form of software that can be used on a personal computer has helped make such participation a reality.

Stakeholders can get access to much of the raw data used by scientists and engineers to calculate risks. Government agencies have to make such information available, even if they do so unwillingly, as is often the case when a stakeholder group issues a Freedom of Information Act request. Industries that use hazardous chemicals must provide information on the types of chemicals and amounts to local communities when complying with the Emergency Planning and Community Right-to-Know Act. Some generic cancer registry information may also be available. With such raw data and a computer model, many stakeholder groups can calculate their own risk numbers, which, not surprisingly, may differ from those being issued by the organization in charge of assessing the risk because different assumptions and scenarios were used. When the two estimates differ by more than a small amount, the organization's credibility can be eroded, and subsequent risk communication efforts can be severely hampered.

It only makes sense, then, to assist stakeholders in their efforts to assess risk. Welcome the chance for an independent review (for that is what their assessment will be) of the risk assessment as an opportunity to improve the assessment. Meet with stakeholder groups from the beginning to identify assumptions, model limitations, and scenarios being developed. Whenever possible, incorporate their suggestions of assumptions and scenarios into your own risk assessment effort. When it is not possible (perhaps because of time or funding constraints or model limitations), explain why. Offer to provide training in the use of their chosen model. Allow them to view the data being input to your model and provide an overview of how your model functions. These steps will likely increase the similarities in the two risk assessments. However, even if the two estimates still differ, stakeholders will have a better appreciation for what you were trying to accomplish, and instead of open hostility, you are more likely to find a receptive audience to discuss the differences and to continue the risk communication effort.

Interactions Involving Decision Making

> The American public has increasingly demanded a bigger role in how decisions about risk are made.
> —Creighton (1992).

As noted by James Creighton, noted consultant on stakeholder involvement, the American public has increasingly demanded a bigger role in how decisions about risk are made (Creighton 1992). It is common to find regulatory agencies, other government agencies, and industry moving toward consulting with stakeholders before making key decisions regarding environmental, safety, and health risks. Most of the activities described in this chapter can be adopted to allow stakeholders to participate in decision making. Key considerations will be:

- **The scope of the involvement.** What does "stakeholder participation" mean for this decision? Will they provide criteria for ranking alternative ways to assess or manage the risk? Will they provide input on preferred ways to assess or manage the risk? Will they provide input on how various stakeholders might respond to different risk management strategies? Both the stakeholders and the organization in charge of managing the risk must agree on the exact parameters of involvement, or the effort will not succeed.
- **Results of the involvement.** What will happen once input has been received? The ultimate decision maker in most risk situations is the organization in charge of managing the risk, and these organizations seldom relinquish that power. Indeed, in the case of government organizations, laws and regulations prohibit the relinquishing of that power. For organizational credibility, and the success of the risk communication effort, it is vital that stakeholders understand where their decision-making power ends and the organization's begins.

The following subsections discuss specific types of stakeholder participation related to decision making involving risks. These types include facilitated deliberation and alternative dispute resolution.

Facilitated Deliberation

Facilitated deliberation usually refers to groups of people discussing issues and recommending solutions, led by a facilitator. Methods can be as casual as "Internet café" discussion groups and as formal as AmericaSpeaks electronic "21st Century Town Hall Meetings" involving thousands of citizens nationwide. For risk-related issues involving affected communities, two methods, Citizens Juries® and study circles, have been particularly useful.

CITIZENS JURIES

The Jefferson Center, in Minneapolis, Minnesota, developed and trademarked the process known as the Citizens Jury. The Citizens Jury is designed to enable citizens to engage in informed discussions, generating findings for decision makers and the broader community. Randomly selected and demographically representative panels of citizens meet for several days using a prescribed, facilitated process to examine public policy issues and present their findings. The juries hear from "witnesses" in "hearings," deliberate on complex issues, and report their findings and recommendations.

> Citizens Juries are useful for one-time decisions or input when intense discussion and deliberation occur over a condensed time period.

Citizens Juries have been used for public policy issues of local, regional, or national importance. Risk-related issues have included comparing environmental risks, land use plans, agriculture and water policy, genetically modified food (Opinion Leader Research 2003), solid waste disposal (Jefferson Center 2001), and global climate change (Jefferson Center 2002). The Institute for Public Policy Research began its own series of Citizens Juries in the United Kingdom in the 1990s (Coote and Franklin 1999), with more than 100 projects conducted in Britain since 1996. Many other countries have adopted Citizens Juries as well. The Jefferson Center disbanded in 2002, but Citizens Juries continue in various forms.

Citizens Juries have many similarities to advisory groups. They are independent of the organization that formed them. Sessions are open to observers, and the final report is in the public domain. The commissioning organization is expected either to follow the jurors' recommendations or to explain publicly why not.

Unlike most advisory boards, Citizens Juries are useful for one-time decisions or input when intense discussion and deliberation occur over a condensed time period. The planning process usually takes 3 or 4 months, but the deliberation process itself typically runs only 4 or 5 days. Jurors disband afterward, so they cannot address policy issues as they develop over time.

STUDY CIRCLES

The study circle is another deliberation and problem-solving process. Groups of 8–12 people from diverse backgrounds and viewpoints meet several times to talk about an issue, often related to a community policy or need. They usually meet over a period of weeks or months and are guided by a trained facilitator. The emphasis is on examining an issue from many perspectives. The process does not require consensus but uncovers areas of agreement and common concern, ultimately generating strategies for action.

In a large-scale study circle program, people all over a neighborhood, city, county, school district, or region meet in diverse study circles over the same period of time. All the study circles work on the same issue and seek solutions for the whole community. At the end of the round of study circles, people from all the circles come together in a large

> Study circle proponents say that the process sparks new connections between citizens and community leaders.

community meeting to work together on the action ideas. According to the Everyday Democracy (formally the Study Circles Resource Center), more than 550 communities have instituted community-wide study circles since 1989.

Risk-related topics in study circles have included community growth issues, land use, and neighborhood crime. Study circles involving nearly 400 people in Portsmouth, New Hampshire, prompted the city's Planning Board to endorse a 10-acre land purchase as public conservation habitat in 2003. Study circles of 350 people in Buffalo, New York, created programs such as "Walk and Park," whereby police officers park their cars for an hour a day to visit neighborhood businesses, and an anticrime program called "Putting the Neighbor Back in the 'Hood'."

Study circle proponents say that the process helps citizens gain ownership of the issues, discover a connection between personal experiences and public policies, gain a deeper understanding of their own and others' perspectives and concerns, create a greater ability to work collaboratively, and spark new connections between citizens and community leaders. The Everyday Democracy website contains resources to help communities institute study circles, including operations guidelines, facilitator training guidelines, evaluation forms, and best practices reports.

Alternative Dispute Resolution

Risk-related decision making often sparks conflicting points of view, sometimes leading to disputes. Alternative dispute resolution, in all its forms, represents ways to settle disputes without litigation or administrative adjudication, thus avoiding a solution that is imposed on the parties by an outsider. Research has shown that disputants often prefer alternative dispute resolution to the court process because of rapid processing, low costs, and the perception of the process being satisfying, fair, understandable, and ultimately resolving the conflict (Cook et al. 1980).

> Alternative dispute resolution represents ways to settle disputes without litigation or administrative adjudication.

Federal and state statutes have enacted resolutions for alternative dispute resolution, and many agencies have implemented their own programs (Herring 2001). Even the Internet has been used via online or electronic dispute resolution (Katsh and Rifkin 2001).

Three of the most commonly used methods of alternative dispute resolution are facilitation, negotiation, and mediation. All three use a neutral third party to facilitate an agreement. They typically share certain goals: allow voluntary participation by the disputants in a fair process, craft a creative and mutually satisfactory resolution, and enhance the parties' relationships while enabling them to save face (Renken 2002). All three methods have been used extensively in public participation processes involving risk. The following sections briefly summarize each method. Note that these definitions are fluid and somewhat interchangeable; even practitioners and professional associations disagree among themselves on what each method entails.

FACILITATION

Facilitation uses a neutral third party to help groups accomplish their work by providing process leadership and expertise. Facilitators improve the flow of information and enhance mutual understanding. The facilitator remains impartial about the substantive issues under discussion and focuses on the communication process, leading the parties and providing procedural direction. A facilitator uses skills and techniques that enable the group to clarify issues, generate ideas, prioritize goals or solutions, and solve problems. If conflicts arise, the facilitator uses process skills to help people get past their individual agendas and get on with the group task.

> Facilitators improve the flow of information and enhance mutual understanding.

Unlike other alternative dispute resolution techniques, facilitation does not usually involve caucuses (private meetings with the facilitator), except for initial meetings to launch the proceedings.

The International Association of Facilitators recommends the use of an independent facilitator (not a member of the group working on the task) under the following circumstances:

- When distrust, bias, or rivalry are present
- When participants have disparate educational, social, or economic status; are at different hierarchical levels; or are in relationships with significant power disparities
- When the task or problem is poorly defined or defined differently by various participants
- When group members all want to participate in the decision process rather than focusing only on the group process and logistics

The International Association of Facilitators offers a Certified Professional Facilitator credential and maintains a directory of certified facilitators by country and state. Certification is based on demonstrated competencies. These include the ability to create collaborative client relationships, plan appropriate group processes, create and sustain a participatory environment, guide the group to appropriate and useful outcomes, build and maintain professional knowledge, and model a positive professional attitude.

In risk communication, facilitation has been used extensively to address issues such as environmental cleanup, resource management, siting of facilities and electric transmission lines, and habitat restoration. Risk communicators typically serve in one of two roles. They may serve as the trained facilitator. They may also serve as the liaison who selects and/or recommends a facilitator and provides that person with background information and resources for the facilitation exercise.

NEGOTIATION

In negotiation, a third party helps the parties come to an agreement and may recommend a particular settlement. A long-established approach to negotiation considered it a "zero-sum" game, assuming that one party's gain is the other party's loss. But in the 1980s, Harvard Law School professors Roger Fisher and William Ury developed a highly regarded method called principled negotiation. It is based on the premise that it is possible to meet one's own needs and those of others and that conflict provides such opportunities (Fisher et al. 1991). Four principles define the method:

> In negotiation, a third party helps the parties come to an agreement and may recommend a particular settlement.

1. People problems (negative emotions, differences in perceptions, and communication difficulties) are separated from substantial issues.
2. Positions are transformed into underlying interests.
3. Many options for gain are generated before deciding.
4. Mutually agreed-upon objective criteria are developed that are legitimate, practical, reciprocally applicable, and independent of each side's will.

Negotiators can be found and evaluated in the same manner as mediators; the following section suggests a process.

MEDIATION

Mediation helps people resolve or better manage disputes by reaching agreements about what the parties will do differently in the future. Originally adapted from labor/management negotiations, mediation is often used with families, businesses, schools, and workplaces, as well as with stakeholders in public policy issues. With the involvement of a neutral mediator, the parties identify the roles of the participants and ground rules, identify and discuss the problem, identify common goals and issues, generate options, bargain, and reach agreement. Private caucuses between the parties and the mediator may be used to build support or trust, explore settlement options, or break down barriers to negotiation in a confidential setting. The agreement is put in writing and signed by representatives of the parties, and it carries the same legal weight as any contract.

> Mediation helps people resolve or better manage disputes by reaching agreements about what the parties will do differently in the future.

The Alaska Judicial Council (1999) recommends the following process for selecting a mediator:

- **Identify your mediation goals.** What do you expect the mediator to do, based on the nature of the dispute and context for resolving it? Consider your budget and time frame. Mediator organizations can help you understand which services would be best for your situation.
- **Compile a list of names.** Mediator organizations often maintain listings by state, such as the Association for Conflict Resolution Mediator Directory.
- **Contact several mediators and request information.** This could include their promotional materials, resume, references, and a sample of their written agreements. Evaluate the materials in terms of training, experience, certification, and fee structure. Do they offer an orientation session after which the parties decide whether they want to continue?
- **Interview the mediators.** Your goal is to learn more about their training, knowledge, experience, style, confidentiality policies, logistics, and cost.
- **Compare among the mediators and decide.** You may wish to suggest two or three mediators so that all parties can agree on at least one.

Several programs and associations certify mediators. Some states and agencies also have their own certification requirements.

Environmental mediation, or environmental dispute resolution, as it is sometimes called, is a subset of mediation. Environmental disputes typically involve issues such as land use, agricultural and water rights, hazardous waste, tribal–state natural resource agreements, facility siting, and growth management. In South Carolina, for example, certified mediators throughout the state help citizens resolve disputes over environmental health and natural resources. Government officials are increasingly turning to negotiators for assistance when attempting to build public consensus on permitting decisions, public project specifications, and natural resource allocations (Blacklocke 2001). In California, an environmental mediation process brought agreement on technical issues that had stalled an ecological risk assessment for Vandenberg Air Force Base for 2 years (Poncelet and Widman 2001).

Public participation and dispute resolution expert Jim Creighton cautions that environmental mediation can succeed only if the following conditions are met (Creighton et al. 1998):

- The parties have reached the point at which they no longer believe that they can impose their will on each other.
- The relative power of the interests is close enough that no one worries unduly about being exploited.
- The parties are convinced that they will achieve more through an agreement than by continuing to fight.
- The parties are well defined.
- The parties agree on the need for all of them to be at the table.
- The parties have the ability to commit their respective constituencies to the agreement.
- The agreement is binding.

An even more specific kind of environmental mediation is negotiated rule making or regulatory negotiation. It uses techniques of multiparty mediation to deal with large disputes over public policy. Representatives of stakeholder groups from industry, consumer and environmental organizations, and government agencies work with a mediator to negotiate government regulations. If they reach consensus, the agency can use the outcome as the basis for a proposed rule. The rule is still subject to public review and all other steps in the formal rule-making process.

The U.S. Institute for Environmental Conflict Resolution maintains a directory of specialists in environmental mediation.

Interactions Involving Risk Management

Risk management interactions involve stakeholders in managing the risk. In care communication involving personal risks to health and safety, those at risk are often the only ones who can manage the risk. Involving stakeholders in risk management is becoming more common in consensus and crisis communication too, for example, by having stakeholders collect and analyze environmental samples near a facility to determine

> Risk management interactions involve stakeholders in managing the risk.

whether contaminants are affecting the local area. Successful stakeholder participation in how a risk is managed is becoming more common, mostly because such interactions can be an extremely effective means of informing and raising awareness about a risk. To effectively involve stakeholders in managing the risk, certain guidelines should be followed:

- **Determine the purposes and objectives.** What do you want to gain by this interaction and how will you accomplish it? If your purpose is to inform your audience, a reasonable objective is to provide them with hands-on experience with the scientific assessment of risk. For example, they might join the scientists in running risk calculation models and explore how changes in various parameters equating to changes in lifestyles affect results. If your purpose is to raise awareness, your objective might be to show them how the risk is managed. For example, they might tour a facility under construction to identify safety features. If your purpose is to build consensus on the management of a risk, your objective might be to begin a dialogue on the issues at stake. For example, you might hold a workshop on strategies to communicate the dangers of drug addiction to teenagers. Each case will require slightly different arrangements.

- **Determine the scope of the interaction.** Exactly what will you have them do? If they will work in the field or laboratory, what will they be doing and who will be accountable for them? If they are working alongside management, who will determine what they can do within the organization and what information they are allowed to access? How long will they work? What will be the final product: a report to their constituencies, a press release, a journal article? Establishing this scope in the beginning will help you plan a meaningful interaction.

- **Make sure that those chosen to help manage the risk represent stakeholder groups.** Risk management interactions take a lot of time to set up. If the people you choose to help manage the risk do not represent stakeholders, the effort will be for nothing. No stakeholder will be interested in what they have to say unless they are credible.

- **Make sure that those chosen to help manage the risk are properly trained.** For safety reasons as well as for technical credibility, those who will be working with the risk must have the proper training. Allow them to attend the organization's training courses, or hold special sessions targeted to their needs.

- **Make sure that those chosen to help manage the risk are capable of disseminating information and that mechanisms and resources are available for them to do so.** For example, you might provide clerical assistance in typing an article by those involved in the risk management activity for a stakeholder newsletter. The main reason for conducting this type of stakeholder participation is to share risk management information. The information will be far more credible if it comes from stakeholder members who have been involved with the risk. If you disseminate the information on their behalf, it may have no greater credibility than if your organization gathered it. Make sure that the workers chosen are articulate and can write or speak well enough to present the risk to their peers in language that stakeholders will understand.

- **Show organizational support.** With this type of interaction, the organization must be heavily involved in planning, training, overseeing, and evaluating the workers. However, releasing press information, mentioning the workers by name in organization literature (if the workers agree to this), and even treating them to dinner with the organization's top management are good ways to show you appreciate their efforts. (But first make sure that the stakeholders or the workers will not see this last token of appreciation as the organization's paying off the workers.)

- **Watch out for union conflicts.** Some of the jobs assigned to these workers may fall under union rules. Discuss the importance of these workers with union representatives beforehand. If the union workers understand that the organization's future reputation and business depend heavily on the audience understanding of a particular risk, they are more likely to accept such encroachment on their areas of expertise.

Evaluating Stakeholder Participation Based on Your Situation

Table 17-1 lists advantages and disadvantages of the various types of stakeholder participation. The type of interaction you choose also depends on whether you are conducting care, consensus, or crisis communication. Table 17-2 shows which types of interaction are most effective for the three types of risk communication.

Table 17-1. Advantages and disadvantages of stakeholder participation in risk communication

Type of interaction	Advantages	Disadvantages
Formal hearings	Are easy to implement; meet minimum legal requirements for some laws; allow geographically dispersed groups to participate	Can increase hostility; give time to only vocal concerns; leave some members of the audience dissatisfied; provide insubstantial involvement
Self-help groups	Motivate audience to act; empower audience	Will be effective only if audience can affect risk; require long-term commitment from both organization and stakeholders
Focus groups	Are small scale and have well-defined purpose and time frame so they may be less intimidating to some organizations	Have limited scale and time; may not represent full audience; may not be substantial enough involvement for some stakeholders
Workshops	Educate as well as involve; gather disperse viewpoints	Require technical knowledge; require teaming between organization and stakeholders that may not be possible in hostile situations
Advisory groups	Have longer time frame so stakeholders can learn about risk; can help develop decisions	May be less effective over time; commit organization to respond; requires considerable resources (time, money, staff)
Risk assessment interactions	Provide credible review of process; increase chances of acceptable assessment	Require technical knowledge; feel too much like "challenge" to some technical experts
Decision-making interactions	Can lead to more acceptable decision; are highest form of involvement	Require organization to relinquish some control; may not be legal in some situations
Risk management interactions	Teach stakeholders about risk; empower them	Require technical knowledge; will be effective only if audience can affect risk

Table 17-2. Effective stakeholder participation for care, consensus, and crisis risk communication

Level of effectiveness	Care communication	Consensus communication	Crisis communication*
Most effective	Self-help groups; risk management interactions	Decision-making interactions; advisory groups; workshops; risk assessment interaction	Focus groups
Moderately effective	Focus groups	Focus groups; risk management interactions	Self-help groups
Least effective	Formal hearings	Formal hearings	Formal hearings

*Stakeholder participation cannot be conducted during a crisis unless the interaction has been planned months in advance. If planned, focus groups can meet quickly to help those who are communicating the risk to understand the needs of the audience and to disseminate information. Self-help groups can be used after the crisis to help the audience come to terms with what has happened.

CHECKLIST FOR STAKEHOLDER PARTICIPATION

☐ The involvement starts early and runs throughout the risk assessment and management process.
☐ The organization is committed to providing stakeholder participation. The type of stakeholder participation was chosen based both on:
  ☐ Organizational needs
  ☐ Stakeholder needs
☐ There is a written plan for the interaction.

For formal hearings:

☐ The time and place are comfortable to the audience.
☐ Key decision makers have been invited to attend.
☐ The services of a good moderator have been arranged for.
☐ Comments will be recorded.
☐ Time has been allowed for questions and answers.
☐ How audience concerns were used in the risk decision will be communicated to them after the meeting.

For group interactions:

☐ The time and place are comfortable to participants.
☐ The purpose and objectives of the group have been agreed upon.
☐ The lines of communication have been defined.
☐ The group has decided how it will operate.
☐ Someone credible from the organization will attend.
☐ The organization will provide resources.
☐ Technical support has been provided.

For self-help groups:

☐ The group is led by a trained facilitator.
☐ Additional information has been made available to meet audience needs.

For focus groups:

☐ The moderator is seen as independent.
☐ There are no more than 10 members who represent stakeholders.
☐ Each meeting lasts no longer than 2 hours.
☐ The group has an agreed-upon purpose.
☐ The group knows how its comments will be used.

For workshops:

☐ Guidelines for speaking engagements have been followed for presentations.
☐ Guidelines for information materials have been followed for those materials.
☐ Guidelines for group interactions have been followed.

For advisory groups:

☐ The organization has developed a mechanism to manage time commitments.
☐ The group includes members from across stakeholder constituencies.
☐ The group has a credible technical advisor.

For risk assessment interactions:

☐ Assumptions, limitations, and scenarios for the assessment have been related to stakeholders.
☐ Stakeholder assumptions and scenarios have been included when possible and an explanation provided when it was not possible.
☐ The stakeholders have been offered training in model use.
☐ The stakeholders have viewed the data being input to the model.
☐ The model functioning has been explained to the stakeholders.

For decision-making interactions:

☐ The scope of involvement has been delineated.
☐ Stakeholders know how their input will be used in the decision.

For risk management interactions:

☐ The purposes and objectives are clear to all involved.
☐ Scope has been determined and agreed upon.
☐ Those who are working represent the audience.
☐ Workers have received the appropriate training.
☐ Workers are capable of disseminating risk information.
☐ The organization will show support for the workers.
☐ The union has been consulted about possible conflicts.

REFERENCES

Alaska Judicial Council. 1999. *A Consumer Guide to Selecting a Mediator.* Anchorage, Alaska. http://www.ajc.state.ak.us/reports/MediatorGuide.pdf (accessed February 7, 2013).

Aleknavage, J. and B. Lyon. 1997. "Citizens and Manufacturers Work Together." *ChemEcology,* June–July:7–9.

Arvai, J. L. 2003. "Using Risk Communication to Disclose the Outcome of a Participatory Decision-Making Process: Effects on the Perceived Acceptability of Risk-Policy Decisions." *Risk Analysis,* 23(2):281–290.

Beierle, T. C. 2002. "The Quality of Stakeholder-Based Decisions." *Risk Analysis,* 22(4): 739–750.

Blacklocke, S. 2001. *Alternative Environmental Dispute Resolution in South Carolina: Emerging Opportunities to Build More Sustainable Communities.* http://www.mediate.com/articles/blacklocke.cfm (accessed February 7, 2013).

Commission (Presidential/Congressional Commission on Risk Assessment and Risk Management). 1997. *Risk Assessment and Risk Management in Regulatory Decision-Making.* Commission, Washington, DC.

Cook, R., J. Roehl, and D. Sheppard. 1980. *Neighborhood Justice Centers Field Test: Final Evaluation Report.* U.S. Government Printing Office, Washington, DC.

Coote, A. and J. Franklin. 1999. "Negotiating Risks to Public Health—Models for Participation." In P. Bennett and K. Calman, eds., *Risk Communication and Public Health.* Oxford University Press, Oxford, United Kingdom, pp. 183–194.

Creighton, J. 1992. "What Does It Take for a Decision to 'Count'?" Presentation to the U.S. Department of Energy, Richland Operations Office, Richland, Washington. Creighton and Creighton, Palo Alto, California.

Creighton, J. L., C. M. Dunning, J. Delli Priscoli, and D. B. Ayres, eds. 1998. *Public Involvement and Dispute Resolution: A Reader on the Second Decade of Experience at the Institute for Water Resources.* Institute for Water Resources, U.S. Army Corps of Engineers, Alexandria, Virginia.

Desvousges, W. H. and V. K. Smith. 1988. "Focus Groups and Risk Communication: The 'Science' of Listening to Data." *Risk Analysis,* 8(4):479–484.

Doyle, M. and D. Straus. 1976. *How to Make Meetings Work.* Wyden Books, Chicago, Illinois.

Fisher, R., B. Patton, and W. Ury. 1991. *Getting to Yes. Negotiating an Agreement without Giving In,* 2nd ed. Penguin Books, New York.

Henry, R. A. 2000. *You'd Better Have a Hose if You Want to Put Out the Fire: The Complete Guide to Crisis and Risk Communication.* Gollywobbler Productions, Windsor, California.

Herring, M. 2001. "Summary of State Alternative Environmental Dispute Resolution Institutions." http://mediate.com/articles/scsg1.cfm (accessed February 7, 2013).

IAP2 (International Association for Public Participation). 2000. "Core Values for the Practice of Public Participation." http://www.iap2.org/displaycommon.cfm?an=4 (accessed February 7, 2013).

Imholz, R. M., T. B. Hindman, and D. M. Brubaker. 1990. "Lessons Learned from Applying External Input to DOE Policy Decision Making." In *Proceedings of the International Topical Meeting on Nuclear and Hazardous Waste Management, Spectrum '90.* American Nuclear Society, La Grange Park, Illinois, pp. 12–15.

Jefferson Center. 2001. *Citizens Jury®, Metro Solid Waste.* June 18–22, 2001, St. Paul, Minnesota. Jefferson Center, Minneapolis, Minnesota.

Jefferson Center. 2002. *Citizens Jury®, Global Climate Change.* March 18–22, 2002, Baltimore, Maryland. Jefferson Center, Minneapolis, Minnesota.

Kasperson, R. E. 1986. "Six Propositions on Public Participation and Their Relevance for Risk Communication." *Risk Analysis,* 6(3):275–281.

Katsh, E. and J. Rifkin. 2001. *Online Dispute Resolution: Resolving Conflicts in Cyberspace.* Jossey-Bass, San Francisco, California.

Kinney, A. G. and T. M. Leschine. 2002. "A Procedural Evaluation of an Analytical–Deliberative Process: The Columbia River Comprehensive Impact Assessment." *Risk Analysis*, 22(1): 83–100.

McComas, K. A. 2001. "Public Meetings and Risk Amplification: A Longitudinal Study." Presented at the Society for Risk Analysis Annual Meeting, December 2–5, 2001, Seattle, Washington.

NRC (National Research Council). 1996. *Understanding Risk: Informing Decisions in a Democratic Society*. National Academy Press, Washington, DC.

Opinion Leader Research. 2003. "FSA Citizens' July: Should GM Food Be Available to Buy in the UK?" Final Report. Prepared for the Food Standards Agency, London.

Poncelet, E. C. and G. Widman. 2001. *Better Together: Mediating an End to an Ecological Risk Assessment Dispute at Vandenberg Air Force Base*. http://www.mediate.com/articles/poncelet.cfm (accessed January 22, 2013).

Renken, D. 2002. *The ABC's of ADR. A Comprehensive Guide to Alternative Dispute Resolution.* http://www.mediate.com/articles/renkenD.cfm#edn2 (accessed January 22, 2013).

Scher, E. and S. McKearnan. 1997. "Do's and Don't's from a Philadelphia Story: How to Find Helpful Expert Advice." *Consensus*, April 1997, No. 34, pp. 1, 8, and 12. MIT—Harvard Public Disputes Program, Cambridge, Massachusetts.

Serie, P. J. and A. L. Dressen. 1992. "Creating a Context for Public Confidence in Environmental Remediation Programs." In *ER '91, Proceedings of the Conference for the U.S. Department of Energy*, pp. 31–35, U.S. Department of Energy, Washington, DC.

ADDITIONAL RESOURCES

American Bar Association Section on Dispute Resolution. http://www.abanet.org/dispute (accessed January 22, 2013).

Association for Conflict Resolution. http://www.acrnet.org (accessed January 22, 2013).

Carpenter, S. L. and W. J. D. Kennedy. 2001. *Managing Public Disputes: A Practical Guide for Government, Business and Citizens' Groups*. Wiley, New York.

CR Info: The Conflict Resolution Information Source. http://www.crinfo.org (accessed January 22, 2013).

Defense Environmental Restoration Program. http://www.dtic.mil/envirodod (accessed January 22, 2013).

Everyday Democracy. http://www.everyday-democracy.org/en/index.aspx (accessed January 22, 2013).

Imholz, R. M., G. E. Rubery, and D. M. Brubaker. 1992. "Case Studies on Designing Meetings for Effective Institutional Interactions." In *ER '91: Proceedings of the Environmental Restoration Conference for the U.S. Department of Energy*, pp. 27–30, U.S. Department of Energy, Washington, DC.

International Association of Facilitators. http://www.iaf-world.org (accessed January 22, 2013).

Jefferson Center. http://www.jefferson-center.org (accessed January 22, 2013).

Lynn, F. M. and G. J. Busenberg. 1995. "Citizen Advisory Committees and Environmental Policy: What We Know, What's Left to Discover." *Risk Analysis*, 15(2):147–162.

National Association for Community Mediation. http://www.nafcm.org (accessed January 22, 2013).

National Center for Information Technology and Dispute Resolution. http://www.umass.edu/research/research-units-facilities/national-center-information-technology-dispute-resolution (accessed February 7, 2013).

National Coalition for Dialogue and Deliberation. http://ncdd.org/ (accessed January 22, 2013).

National Institute for Advanced Conflict Resolution. http://www.niacr.org/ (accessed February 7, 2013).

National Institute for Conflict Resolution. http://www.nifcr.com/ (accessed February 7, 2013).

Moore, C. W. 2003. *The Mediation Process: Practical Strategies for Resolving Conflict*, 3rd ed. Jossey-Bass, San Francisco, California.

U.S. Institute for Environmental Conflict Resolution. http://www.ecr.gov (accessed January 22, 2013).

18

TECHNOLOGY-ASSISTED COMMUNICATION

Technology gives us almost unlimited options for risk communication. Websites raise risk awareness and provide options to mitigate risks. Telebriefings and streaming audio and video convey risk information quickly to media outlets and consumers. Data-gathering and measurement devices upload content to reveal risk information for decision making. Through online virtual worlds, users practice their skills in emergency response scenarios and other risk-related activities. Electronic forums invite stakeholders to comment on risk decisions, computer models help them calculate their own risks, and group software aids decision-making processes.

Yet technology applications are not always superior to other forms of communication; in fact, they can be less useful and more expensive. Advantages and disadvantages of technology-based applications are discussed in Chapter 10. This chapter discusses how to choose technology applications based on objectives and provides advice on how to use these applications when communicating risk in the workplace and in care, consensus, and crisis communication efforts. All web addresses given here were correct at the time of publication.

See Chapter 19, Social Media, for the use of social networking sites; blogs and podcasts; microblogging; video, image, and file-sharing sites; mapping; and mobile phones in risk communication.

Risk Communication: A Handbook for Communicating Environmental, Safety, and Health Risks, Fifth Edition. Regina E. Lundgren and Andrea H. McMakin.
© 2013 The Institute of Electrical and Electronics Engineers, Inc., and Regina E. Lundgren and Andrea H. McMakin. Published 2013 by John Wiley & Sons, Inc.

CHOOSING TECHNOLOGY-BASED APPLICATIONS

As with any form of risk communication, purpose and objectives should be the key factors in choosing technology-based applications to communicate risks. Table 18-1 gives some examples.

Audience needs are another important consideration. Participants must have access to equipment and software and know how to use them efficiently. Pretesting the application with target audiences is critical. This is especially true in situations in which people will be using the technology on their own, without a person there to answer questions or facilitate the process.

A third consideration is the medium itself. Technology-based applications should be designed to use the characteristics specific to that medium. For example, websites should use navigation patterns that enable people to go immediately to particular sections of interest and drill down for more detail if desired. In computer-based training, immediate feedback about right and wrong answers should be provided to students when possible. E-mail and other alerting messages to employees during an emergency should be updated frequently, keeping workers continually up to date. On the other hand, computer-based applications should not be relied on as the primary source to communicate crisis information in situations such as natural disasters because electrical power may be lost.

> Technology tools should be designed specifically to use the characteristics of the medium, rather than merely substituting for another medium.

The following sections suggest ways to use technology to communicate about risks. For advice about how to design or adapt tools for specific situations, work with a technology specialist. Additional Resources at the end of this chapter lists helpful materials that provide more detail on each application described here.

Table 18-1. Communication objectives and corresponding technology-based applications

Risk communication objectives	Potential technology tools
Provide safety and health training	Computer-based training or course work, virtual worlds
Keep employees informed about risks, including during and after emergencies	E-mail, shared databases, shared file space, alerts on mobile devices
Provide current risk information	Websites, e-mail, listservs, telebriefings, social media, kiosks, streaming audio and video, CDs and DVDs, content delivered to mobile devices
Share information and receive feedback from interested individuals about a particular risk (participatory process)	Internet, e-mail, local area networks, shared file space, computers in central areas, designated space on social media. Tools must have a feedback feature or at least a point of contact to receive comments and discuss issues.
Compile and analyze public comments	Tailored software, designated web space
Support group decision processes	Tailored software or software/hardware combination with facilitator

WORKPLACE RISK COMMUNICATION

Risk-related communication to employees often fits into two categories: (1) ensuring that workers receive required training, and (2) keeping employees informed about current health, safety, and environmental issues, including emergency situations.

Computer-Based Training

Computer-based training is a broad term encompassing everything from self-paced DVDs to interactive, real-time multimedia training that links people and instructors in dispersed geographic locations, including online virtual worlds. Computer-based training can be a cost-effective, consistent, fast, and relatively easy way to impart information about risks and to test employee knowledge of what to do on the job. Common risk-based training topics are equipment operation, handling hazardous chemicals, fire safety, and general office safety.

> Computer-based training can be a cost-effective, consistent, fast, and relatively easy way to impart information about risks and to test employee knowledge of what to do on the job.

General Physics, for example, developed a computer-based training program based on information and guidelines contained in regulatory documents (Lobbin 1997). The software includes more than 60 scenarios that constitute an "exam bank" that can be used by nuclear plant personnel to test and evaluate their understanding of requirements for reporting nuclear "events."

Researchers at Oregon Health and Science University developed a software program to improve workplace safety and to decrease the number of on-the-job injuries. Kent Anger, associate director of the Center for Research on Occupational and Environmental Toxicology, created the program to train employees in physically demanding jobs such as field agriculture, food preparation, painting, and building maintenance. Often, these employees have limited education or limited English skills. When workers do not read English or are intimidated by computers, teaching them standard safety procedures to protect themselves can be a challenge. Anger's creative use of technology is addressing this problem.

More organizations are using live online training, where a group of people in various locations receive simultaneous training via their desktop computers, linked to an instructor. To access the training, participants use their web browsers to log onto a server, where they are automatically connected. The trainees can watch and listen, via teleconference or Internet-based phone conference, in real time as the instructor shows and demonstrates information onscreen. Trainees can practice the exercises online. The instructor can share additional documents that all trainees download to their computers with a click of a mouse. Trainees can talk with the instructor and each other via teleconference or send e-mail messages to the instructor during training. A web camera can be used to show students what the instructor is doing. Don Clark, a specialist at Pacific Northwest National Laboratory, recommends the following guidelines for live online training:

- People typically cannot stay engaged in online training for more than an hour at a time, so keep the training short or split it into several sessions.
- Trainees should be in quiet locations, ideally rooms where the doors can be shut. The background must be quiet enough so that all participants can hear each other

on speaker phone; any ambient noise is transmitted to everyone on the conference call.

- Trainees must prepare in advance not to be interrupted in their offices during training. Unlike the seclusion of a traditional classroom, employees are sitting at their desks, and colleagues may unknowingly drop in while training is in process. People who answer multiple phone lines at their desks should arrange to forward the calls or to use a different workstation so they will not constantly have to break away to answer calls.

Distance education is another application for the Internet. Distance education enables learners to obtain instruction on their own time, at their own sites. Depending on the setup, instructor and students can do projects, discuss issues, and share assignments and tests via the Internet, e-mail, various forms of shared file space, or shared multimedia spaces. One example is the Oak Ridge Institute for Science and Education, which provides a variety of courses and certifications on safety and health risk management. They use a variety of technologies including satellite uplink and downlink transmissions, video conferencing, and use of company websites and intranets.

> Distance education enables learners to obtain instruction on their own time, at their own sites.

Another example is the Walter Reed Army Medical Center in Washington, DC, which developed a web-based interactive distance learning tool to increase the clinical risk communication skills of U.S. Department of Defense health care providers. The Health-e Voice tool uses interactive simulated experiences to teach physicians how to better communicate with recently deployed veterans about medically unexplained physical symptoms, rather than disregarding them as merely psychological or stress related. The Defense Department hopes that improved clinical risk communication may alleviate unnecessary patient distress and physical health concerns, reduce frustration and tension in the doctor–patient relationship, and reintroduce patient trust in care providers and the health system.

As in any kind of training, it is important to build a strong foundation. This includes researching the applicable regulations and laws that apply, determining the organization's responsibilities in specific training areas, understanding what employees know and need to know, identifying training objectives and effective ways to impart information via technology, pretesting the training materials, and evaluating success.

> Some kinds of training are best conducted in a traditional classroom environment.

Remember that some kinds of training, in which an instructor needs to watch someone do something or students need to practice a hands-on skill, are still best conducted in a traditional classroom or field environment. Some training works well as a hybrid. For training equipment operators, for example, students could use an online course to understand the basics, and then classroom meet as a group to practice operating actual equipment under the guidance of an instructor.

When adding a new computer-based training program or replacing a traditional program with a computer-based one, trainers often must justify the increased technology cost to management by showing a return on investment. Typical benefits measured are reduced travel and labor costs, delivery to more people in a shorter time, and automation of testing and scores. Of course, computer-based training, as with all training, should also measure learning outcomes such as changes in knowledge, skills, job performance, and/ or business impact.

In developing risk-related computer-based training, additional factors must be considered:

- **Be aware of the possibility of generating or increasing hostility.** Some off-the- shelf computer-based training and some in-house developed applications have obvi- ously been created with little knowledge of risk communication principles. Watch for inappropriate risk comparisons, technical jargon, downplaying of nontechnical or opposing viewpoints, and other factors that may inadvertently generate or increase audience resentment, thereby reducing the chance that the risk communication effort will be successful. If you are developing the training course yourself, always pretest it before giving it to employees as a final product. A group of nuclear subject matter experts developed an example of good and poor risk communication for a training session being given to employees at a nuclear facility. While a team of risk communicators vetted the script, when it came time to film, the nuclear experts took a few logical shortcuts to save time and filming costs. Unfortunately, these shortcuts involved leaving out key actions employees regularly took on the job, such as being able to call a "time out" when they spotted a safety problem. When the film was first aired during a dry run of the course, the employees being trained were so fixated on the procedural inaccuracies that they could not "hear" the risk communication messages.
- **Tailor off-the-shelf courses.** While many large organizations create their own computer-based training courses, smaller organizations often rely on courses devel- oped by others. Supplement these off-the-shelf courses with information materials, presentations, and other forms of risk communication to ensure that the training fully meets the needs of the participants and reflects organization-specific information.

Organizations such as the American Society for Training and Development, the Society for Applied Learning Technology, the United States Distance Learning Association, and the Distance Education and Training Council can provide additional guidance and resources.

Informing Employees about Risks

In organizations in which most employees have access to a networked computer, certain risk-related information can be effectively and quickly disseminated. One common way is to warn employees of potential upcoming risks so that they can prepare themselves. For example, through all-staff e-mail messages, text messages, and/or instant message alerts, workers can be alerted to approaching storm fronts and told to leave work early to avoid hazardous weather.

All-staff e-mail bulletins or electronic newsletters also can be used to remind workers to follow certain safety procedures, or that a safety audit or fire drill is coming up and what they should expect. When sending all-staff messages, consider the following:

Keep the message short. As one information specialist put it, "If they have to scroll, they won't."

- **Keep the message short.** As one information specialist put it, "If they have to scroll, they won't." Try to keep the information to a size that is no larger than a standard computer screen. If you must go longer, insert topical headings in the message so that staff can get an overview quickly.
- **Tell them why they should care.** As close to the front of the message as possible, state why this message is important. In some organizations, employees receive several hundred electronic mail messages each day. Even one from the president of the company is less likely to stand out in such a crowd. Give them a reason to continue reading.
- **Give them direction.** Let employees know what is expected of them. Do you want them to take some action? Do you want them to be more aware of an issue? For what purpose? Explaining what you expect of them will help them see the value of the message as well as encourage a response.
- **Always provide a point of contact for additional information.** Because you are keeping it short, it is all the more important to provide sources of additional information for staff who want to know more.

Another way companies use e-mail is to keep employees informed after an accident or other unforeseen event has occurred; in other words, in-house crisis communication. After a chemical release, for example, a company may tell its employees what happened, define any known health or environmental consequences, and explain what is being done to rectify the situation. This kind of message usually comes from the president or other high-placed official. Often, a contact name and phone number are given so that employees may get more information.

Internal electronic communication can also be used to keep users up to date on particular issues via shared databases or intranets (in-house websites). AT&T's Environmental, Health, and Safety Engineering Center developed an integrated set of electronic tools for sharing current environmental information with its various divisions (Davis 1995). The center must help its divisions stay on top of environmental issues, including constantly changing environmental regulations worldwide.

The center used an internal customer survey and a needs analysis to develop the features of its electronic information system and the databases that support it. Users require a password to access the system. Users can download information and generate a variety of reports directly from the databases. In some cases, users can instruct the systems to fax the desired information to the user's local fax machine.

Based on the results of the survey and needs assessment, the center instituted the following computerized databases for its internal organizations:

- An online database that gives access to general environmental and safety information such as AT&T policy and practices and training opportunities.
- A single corporate chemical inventory, including a current list of approved chemicals, and a database on all chemical substances manufactured, imported, exported, processed, or used within AT&T. This database is the primary tool for showing compliance with the Toxic Substance Control Act and hazard communication guidelines. Any chemical substance not listed on the inventory must be checked for regulatory compliance before it can be approved for use by AT&T.
- A system for recording on-the-job injuries and illnesses. The system produces Occupational Safety and Health Administration reports and other customized reports for accident statistics and analysis.

- Online Material Safety Data Sheets for chemicals and hazardous materials and products used by AT&T employees in the work area.

Center representative Thomas Davis cautions that it is easy to give customers enormous amounts of information without prioritizing it or making sense of it (Davis 1995). So much information already exists that people do not have time to digest it all. It is important to prioritize information provided via computer and evaluate its usefulness to users on a regular basis.

The advantage of using computers to keep the staff informed is that everyone linked to a computer gets the message immediately and consistently, a central point of contact is given, and updates can easily be provided. One possible disadvantage is the potential spreading of sensitive information. Company officials should assume that anything they send to employees via computer can be distributed outside the company, including to the media.

> It is all too easy to give people enormous amounts of information via computer without making sense of it. Prioritize information and evaluate its usefulness to users on a regular basis.

For environmental, health, and safety issues that have a longer life, some organizations also use computers for online discussions. This approach often involves shared file space, such as Intranet sites or electronic question and answer forums. Employees can have informal discussions on particular topics, or, more formally, pose questions that are then answered by designated people in the company that everyone can see.

AT&T's Environmental, Health, and Safety Engineering Center uses an electronic folder on a shared e-mail system for two-way dialogue between the center and its internal customers. The folder contains a variety of environmental and safety information, including technical developments, new regulations and laws, recommendations, updates on progress toward environmental and safety goals, and comments and tips from employees. Unlike traditional mailing lists, which are notoriously difficult to keep up to date, the electronic folder enables users to self-select material of interest to them. The center is cautious not to place highly sensitive or restricted information on this shared folder.

Many companies use off-the-shelf or customized software for internal knowledge management. Sometimes called enterprise portals or e-rooms, they are often used for collaborating on projects, but they can also be used to share risk-related best practices and receive alerts. Many pull information from databases. With Procter & Gamble's subscription-based platform, for example, employees receive updates via e-mail or through a posting on their personalized portal pages. NASA's web-based technical questions database lists important questions that should be asked during a project design process or at a review, to identify and prevent problems later. Serving as a "mind tickler," it can be browsed, and users can submit input.

WEB-DELIVERED AND STAND-ALONE MULTIMEDIA PROGRAMS

Consumers are increasingly expecting risk information in multimedia formats. The use of digital tools makes risk information and discussions more salient, immediate, and interactive than ever. Examples include online multimedia tools, mobile platforms, web and satellite broadcasts, and emergency planning and training programs.

Online Multimedia Tools

Online multimedia tools combine video, graphics, text, animation, virtual tours, interactive tools, and other options. Users control the flow of information and can move around to various topics of interest. For example, the U.S. Centers for Disease Control and Prevention (CDC) offers a highly regarded CD series called CDCynergy, especially those related to health issues. The multimedia format makes them ideal with programs on health communication planning and education, including cardiovascular health, immunization, and communicating in emergencies (CDC et al. 2003).

> Multimedia formats are ideal for combining video, graphics, text, animation, virtual tours, interactive tools, and other options.

Many organizations provide audio and video programs online. For example, the Occupational Safety and Health Administration website offers streaming video and PowerPoint presentations on a wide range of worker safety topics from asbestos safety to workplace violence.

Many online tools are interactive and customizable. One example is the American Cancer Society's Treatment Decision Tools, interactive exercises that help patients understand and chose customized treatment decisions. The user answers questions about his or her cancer diagnosis and test results, then gets customized information about treatment options and rates of survival and recurrence. Another example is Your Disease Risk, an online tool of Barnes-Jewish Hospital and Washington University School of Medicine. On the website, users can find their risk of developing five common diseases and get personalized tips for preventing them. A number of online smoking cessation programs use customizable online tools for self-monitoring of behaviors, social support, and reinforcement timed to match enrollees' efforts to quit.

Often, these tools contain an educational component. The website of the U.S. Agency for Toxic Substances and Disease Registry, for example, includes an extensive interactive learning program about the process used to evaluate whether people will be harmed by hazardous materials from waste sites. It includes a description of how community members can get involved in the assessment process, along with interactive self-quizzes and other exercises to test the user's knowledge while going through the learning program.

> Take advantage of the interactive, multimedia nature of the medium, including audio, video, movement, sound, and interactive exercises as appropriate.

The website of the U.S. Transportation Security Administration contains a "myth busters" page that responds to specific public perceptions. Examples of rumors that the organization has addressed include officers hassling traveling children, a federal air marshal shortage, and reporters being placed on the agency's "watch list" for travelers.

More medical providers are using a variety of online multimedia tools for personalized communication. In studies, patients who could communicate with their doctors or pharmacists via a secure web connection did a better job of reducing their risk of developing cardiovascular disease (Green et al. 2008; Temple University 2006). A "virtual nurse" who explains a discharge plan to hospital patients was found to cut down on readmissions while reducing medical costs (Landro 2011). Appearing on a computer screen wheeled up to the patient's bedside, "Louise"—a name chosen by focus groups—reviews the discharge packet that the patient is holding. Patients react by using a touch screen, including asking for information to be repeated.

If you have online information that would be useful elsewhere, consider widgets. Widgets are online applications on one website that can be "pulled" and displayed by another website. CDC makes widgets available on its website for seasonal flu updates, public health statistics, and other topics of high interest. Once an organization has added a widget from someone else's site to its own site, there is no technical maintenance required because the providing organization updates the content automatically.

Consider using quick response (QR) codes to connect people to your website. QR codes, like bar codes, can be placed on literature, posters, presentations, and just about any flat object. Read with a free, downloadable QR reader, available on most smart phones; they automatically link to a location of the creator's choice, such as a website or a video on a website.

In creating web or stand-alone programs, make sure that they can be used as intended by the target audience. Take advantage of their interactive, multimedia nature, including audio, video, movement, sound, and interactive exercises as appropriate. Make sure that the tool works properly on all standard computer platforms and web browsers. Specify the program duration. It helps to include a website that contains more information about the subject of the program and a way to contact the organization for more information and troubleshooting.

> Provide indexes, summaries, and other navigational cues in computer-based information, even more than you would in a hardcopy document.

When posting video on your organization's website, make sure that the infrastructure can handle it. High-quality video files have a tendency to use huge amounts of bandwidth that can crash websites.

Mobile Platforms

Here, we define mobile platforms as a service or application that involves voice or data communication between a central point and remote locations. It includes the use of mobile phones, tablets, and other devices.

These applications are enabled by the characteristics of mobile networks and devices: near ubiquitous in many countries, locatable, connected user interface devices, often personalized, delivering computing power at low cost, integrating a range of sensors, and supporting mobility.

Many such applications are useful for public and individual care communication, particularly in the health fields. Health applications range across remote diagnostics and monitoring, self-diagnostics, management of long-term conditions, clinical information systems, targeted public health messaging, data gathering for public health, hospital administration, and supply chain management (Freng et al. 2011). At the time this book was published, there were more than 12,000 health care applications from independent developers for the iPhone and iPad. Some of these use mobile phones to distribute and transmit information about disease incidence and reporting. Some provide diagnosis and treatment via around-the-clock telemedicine.

A typical example is the T2 Mood Tracker, launched by the U.S. Department of Defense's National Center for Telehealth and Technology in 2010. It is an application for smart phones and other wireless devices to help military personnel who have been deployed to track their mood and stress levels. Service members use a touch screen with a visual scale to track their own anxiety, depression, general well-being, life stress, and posttraumatic stress. Users can correlate changes to their medication regimen or home or work environment to changes in their moods. By recording an experience at the time and place that it happens, patients can more accurately convey information about their emotional states to their health providers, thus improving the quality of treatment, the developers say.

Digital heart monitors, stethoscopes, blood glucose monitors, and other diagnostic devices can be connected or accessed remotely, sending data to a health care provider. Some include virtual doctor visits that use home monitoring devices and communication tools such as Skype.

Remote data sharing can also be used in crisis communications. After the 2011 Fukushima nuclear disaster, a crowdsourced monitoring effort was launched to obtain more information about radiation levels throughout the country. Data captured from Geiger counters were fed into central information sources, which aggregated radiation readings from government, nonprofit groups, and other sources.

Another example of crowdsourcing in crisis communication is the Regional Asset Verification & Emergency Network, which is a multilayer mapping tool that supports emergency first response in Cincinnati, Ohio. RAVEN911 uses live data feeds and intelligence gathered through Twitter to provide details such as the location of downed electric power lines and flooded roads. Authorities cooperate with other regions to implement this emergency management system to help fire departments assess the risks and potential dangers before arriving on the scene of an accident. This open source system gives emergency responders a common operating picture, to better execute time-critical activities, such as choosing evacuation routes out of flooded areas.

Those developing or using mobile applications for risk communication should consider three important issues:

- **Access.** Will all those the communicator is trying to reach have the ability to access and maintain these applications? How reliable are the service platforms, including providers of telecommunications, power, storage, security, and peripheral devices such as sensors and other diagnostic tools?
- **Maintenance.** Who will update the applications when needed? How will users know that updates are available?
- **Privacy.** Is gathering personal information, including health status and location, legal and ethical in the application used? Are users made aware of what data can be collected and shared about them? How is information security assured? Can users adjust privacy settings or maintain anonymity? A good place to find privacy checklists is GSMA, a trade organization representing mobile operators worldwide. GSMA has published universal privacy principles and design guidelines for mobile applications that collect and use personal information (GSMA 2011, 2012).

Web and Satellite Broadcasts

Telebriefings, web seminars (also called webinars), webcasts, live streaming video and audio, and satellite broadcasts are increasingly used for communication to select or large populations. Like traditional mass media, they quickly reach many people at the same time, but often with the added benefit of being interactive, enabling discussion among participants.

Telebriefings are similar to press conferences or updates, but they use a conference call format. The advantage of telebriefings is that they can be conducted anywhere as they do not require participants to travel to a news conference location. The organizers plan the briefing in advance, often to address urgent or time-sensitive issues, and invite people to call in. Sometimes telebriefings are broadcast in streaming audio on websites, and people have the option of calling in to ask questions. For toll-free lines, participation is usually controlled in advance by using a listserv, e-mail list, or password-protected area on the Internet. Organizers can send electronic presentations or other materials in advance to participants, or put them on a website.

The CDC has conducted more than 40 telebriefings each year since 2001 on public health topics including anthrax, smallpox, West Nile virus, severe acute respiratory syndrome (SARS), and many others. The organization posts transcripts of its telebriefings on its website.

Webcasts and streaming video are audio and video programs that are converted into files for hearing and/or seeing on a website. They can be live or recorded and archived for on-demand viewing later. Live broadcasts are advertised in advance, then people log on by clicking on a website link that becomes active when the webcast occurs. Viewing requires access to a standard media player program and sufficient bandwidth connection to hear and see the program clearly. Participants can be given access to audio files, transcripts, and related reference materials.

Good examples of web-delivered audio and video abound. The American Cancer Society's Cancer Survivors Network® makes good use of the multimedia nature of the Internet by including downloadable

> Webcasts and streaming video can be live or taped and archived for on-demand viewing later.

talk shows, interviews with transcripts, and webcasts, along with its chat and discussion board areas and other web-based resources. The U.S. Food and Drug Administration (FDA) website includes archived 15-minute news programs for health professionals called "FDA Patient Safety News." The programs cover products recently approved by the FDA, product recalls and safety alerts, and tips on protecting patients and preventing medical errors. Viewers can see video segments, find more information on each story, and report problems with products through an online link.

Satellite broadcasts are good for reaching sites around the country simultaneously. Viewers congregate at downlink host sites that have the proper coordinates to link to the broadcast. Many satellite broadcasts are interactive. The CDC, for example, broadcasts programs simultaneously in the United States and Canada via satellite, Dish Network, webcast, and web conference. Panels of experts answer questions posed by viewers throughout the broadcast via fax, e-mail, and telephone.

For computer-delivered multimedia programs, consider the following guidelines:

- Clearly state user access instructions in introductory materials or on the Internet.
- Specify the program duration. For live programs, note the time zone.
- List the requirements to view or hear the program. It is easiest for the user to click on a link to see whether his or her system is configured properly to handle the program. Consider having various bandwidth speeds to accommodate various users.
- Provide a solution, such as a technical support person standing by or a person to e-mail, in case participants have problems.

- Tell people where to get more information about the topic, online and elsewhere.
- For telebriefings, clearly introduce the speakers at the beginning. Make sure that all participants (including questioners) state their names and affiliations.
- Use a password for online access if you want to limit the information to specific audiences.
- Consider making a very short questionnaire available at the end of web-based programs for user evaluation and program planning.

TRADITIONAL ELECTRONIC FORUMS

Many organizations also use more traditional electronic forums for risk-related discussions, including e-mail lists, listservs, and newsgroups.

E-mail lists and listservs enable subscribers to receive and sometimes send messages. Traditional mailing lists are often one way, with organizations sending information to subscribers, although a contact name is often provided for inquiries. For example, the CDC has more than two dozen mailing lists on topics ranging from ambulatory care to preventing chronic disease. The Food Safety Network has several e-mail lists that provide current public risk perception information about food-related issues, generated from journalistic and scientific sources worldwide and distributed daily to subscribers from academia, industry, government, the farm community, journalism, and the public. The Leukemia and Lymphoma Society uses an online live chat feature on its website to convey expert information. Users click on a link and are instantly connected to someone who can answer their questions in real time.

Some listservs are set up as a discussion forum, where any subscriber can send a message that goes to all other members, to which anyone can respond. The Society for Risk Analysis, for example, runs separate listservs on risk analysis and risk communication. The listserv for the European Risk Communication Network fosters discussion between researchers and practitioners in Europe about best practices for risk communication.

The International Society for Infectious Diseases operates PROMed, an international electronic reporting system to speed identification of major diseases of plants, animals, and humans. No government approval is needed to post, and moderators are experts in a wide range of medical and epidemiological issues. It includes formal reports, lay reports from the news media, and anecdotal reporting from subscribers. The system was responsible for the first-reported existence of SARS internationally and helped track its spread (Green et al. 2007).

Some listservs archive past postings. Risk World, an online commercial site, contains several online discussion groups on risk-related topics, including risk analysis, risk management, and technology.

Usenet is a very general source of discussion forums accessible via the Internet, consisting of newsgroups with names that are classified hierarchically by subject. People post articles or messages to these newsgroups. In some newsgroups, the articles are first sent to a moderator for approval before appearing in the newsgroup. Many Usenet sites are commercial entities, but universities, research labs, or other academic institutions also operate Usenet sites.

Because of their "viral" propagation with little or no controls, e-mail, listservs, and newsgroups have significant potential to affect risk perceptions. One study looked at the power of Usenet to affect risk perceptions about NASA's plutonium-powered space probe, Cassini. A study of Usenet messages about the probe from 1995 through 1999 showed that six people originated messages that ultimately evolved into more than 8000 messages by more than 900 authors (Rodrigue 2001).

> Because of their "viral" propagation with little or no controls, social media and other online discussion forums have significant potential to affect risk perceptions.

Additional guidance on using e-mail, list-servs, and newsgroups to communicate about risk includes the following:

- **Focus the topic.** E-mail lists are proliferating, and readers are being asked to choose which they will join based on content and information. At the same time, more users are blocking spam or filtering out information from lists on a particular topic. So, for example, instead of hosting a list for employees related to all safety risks at your company, you might want separate lists on laboratory safety, equipment operations safety, and office safety to better meet audience needs.
- **Establish ground rules.** Clearly state the kind of e-mail behavior you expect. Many lists have rules against posting personal information, advertisements for products or services, inflammatory or obscene language (sometimes called flaming), and attachments, which can overload some mail programs and also carry viruses.
- **Clearly state whether the list is moderated.** Many lists have a host who ensures that messages arrive and are sent to all list subscribers appropriately. Sometimes these hosts act as moderators, screening postings to ensure that nothing proprietary, inflammatory, redundant, or off the subject is sent to the subscribers. Especially for lists communicating risk information, let all subscribers know in advance whether a moderator will be used, what rules the moderator will use to screen information, and how to appeal if a decision inadvertently censors important information. When the moderator does reject a message, make sure that the subscriber is aware of the reason and has the opportunity to rephrase the message in more appropriate terms. If subscribers post information that does not appear, with no explanation, resentment will increase, and participation will fall off.

INTERACTIVE MULTIMEDIA PROGRAMS IN PUBLIC PLACES

User-navigated computer programs contained in kiosks are another option for risk communication. Many of these programs are similar to interactive CDs; the user selects information at will and can skip around to various topics. Many use touch screens for easier user interfacing than with a mouse. Nevertheless, the degree of sophistication varies widely. Kiosks may range from a self-contained computer program to those that connect to the Internet or other networks. Computer kiosks can let users browse the Internet, send e-mail photos, send video, conduct business transactions, view virtual reality exhibits, and print information.

A kiosk in a public place, such as a community center, university, library, or health center, can be one component of a broader communication plan. Being unstaffed, kiosks can add efficiency and reduce labor costs by freeing employees to do other things. Through questions, computer programs can tailor information to the user's demographics, experience, literacy level, and other

> Kiosks are especially useful for populations without widespread computer access.

characteristics. Kiosks are especially useful for populations without widespread computer access.

Staff from a federally funded climate change research program created an educational, interactive touch-screen program that runs on a kiosk in a community center in Alaska (Figure 18-1). The kiosk uses a custom-built multimedia program running on a standard desktop computer. The program consists of interviews of Alaskan community members, primarily Iriupiat tribal elders, with additional animations and material from program scientists. The program is designed to answer questions about the research program and describe how the climate has changed over time.

In the video clips, a local whaling captain speaks about how he has seen the sea ice become thinner over time, and how this makes the spring whale hunt more difficult and more dangerous for the hunters. One elder speaks about how it freezes later each year, which affects when people can put their nets through the ice. Scientists speak about what they are measuring and what causes climate change. Residents and visitors have been very excited about the program; an elaborate tribal celebration with dancing and singing was held when the kiosk was first unveiled.

Kiosks have been shown to be effective in health interventions, when risks are communicated but the underlying goal is to encourage behavior change. Researcher Armando Valdez designed a successful kiosk-based information program designed to promote breast cancer screening among low-income, low-literacy Latinas in California (Valdez 2002). Extensive formative research had revealed misconceptions and information gaps among the target audience that were directly addressed in the 10-module program. Video clips showed Latina women speaking about their experiences, getting mammograms, and giving advice about early screening. The program also directly

Figure 18-1. Touch-screen program in a kiosk in Barrow, Alaska. The program, which uses video clips and animation, describes climate change and research being conducted in the surrounding North Slope of Alaska. (Source: Mike Ebinger, ARM Climate Research Facility Education and Outreach.)

addressed common misconceptions such as "Breast cancer is an automatic death sentence," "Breast cancer risk decreases with age," and "Putting off treatment gives the cancer a chance to get better."

The kiosk used a standard desktop computer, a headphone jack for privacy or for the hard of hearing, and a thermal printer. Using the touch screen, women were asked a few introductory questions about their spoken language, age, and personal screening history, then given access to various customized modules, which they could choose at will. Every module emphasized the importance of taking action to get early screening. Women were also given the option of printing out information about where to get a mammogram, including clinics offering low-cost or free services.

Results showed increased knowledge and, most importantly, a very high rate of behavior change. Fifty-one percent of women got mammograms after viewing the program, as opposed to other interventions, which typically produce results ranging from 20% to 40%. The kiosk format was effective when other intervention methods—videos, public service announcements, and print media—were not working well for the target population.

In 1997, the University of Michigan's Health Media Research Laboratory developed its "Health-O-Vision" software and placed 100 interactive, touch-screen health kiosks in shopping malls, supermarkets, medical centers, libraries, community centers, and other high-traffic settings. The kiosks are designed to convey a range of topics, including smoking cessation and prevention, cancer screening, bicycle helmet safety, immunizations, cardiovascular disease prevention, alcohol problem detection, and sexually transmitted disease prevention. The kiosk is designed to look like a television, allowing users to select and interact with risk factor "channels." Each kiosk is linked via the Internet to a central data collection system at the University of Michigan's Health Media Research Laboratory, allowing data collection on usage and satisfaction. Public health and medical specialists, computer programmers, graphic artists, and Hollywood Screen Writers Guild writers joined forces on the project, which was funded by proceeds from the state tobacco tax. More than 400,000 people use the kiosks each year (Strecher et al. 1999).

When designing kiosk programs, keep these things in mind:

- Understand how a kiosk fits into your overall communication strategy.
- When estimating the cost, include not just the hardware and software but also the network connections, implementation, maintenance, and upgrades.
- Understand what your audiences want and need to know, and design the information accordingly.
- Test the information with the intended audience for usability. The Alaska kiosk program, for example, was developed and tested for more than a year before it was installed in 2003.
- Make sure that there is a way for users or administrators to report any problems, fix technology glitches, and update information as necessary.
- Build in ways to evaluate effectiveness.

TECHNOLOGY IN CARE COMMUNICATION

When risk managers, technical and public health professionals, members of the public, and other individuals need to find current information about a given risk, the Internet has become a starting point. Though the Internet is unregulated and thus subject to abuse and commercial exploitation, many credible organizations are represented online. Table 18-2

Table 18-2. Some organizations that provide online risk-related information

Source	Content highlights
Government organizations (examples)	
Centers for Disease Control and Prevention (U.S.) http://www.cdc.gov National Center for Health Marketing: http://www.cdc.gov/communication	Searchable by health topic. Contains strategies and materials for entire intervention campaigns for a variety of health issues, for example, antibiotic use, diabetes, hepatitis C, skin cancer, and immunization. Research-tested intervention programs are searchable by age, race, and setting (for example, urban, school, workplace). Many have their own websites and/or are downloadable. English and Spanish, with parts of the site in multiple languages.
National Cancer Institute (U.S.) http://www.nci.nih.gov	Downloadable, research-tested intervention programs and a risk communication bibliography. English and Spanish.
Occupational Safety and Health Administration, U.S. Department of Labor http://www.osha.gov	Workplace safety and health.
National Institutes of Health (U.S.) http://www.nih.gov	Medical and behavioral research under the U.S. Department of Health and Human Services. Represents more than 20 institutes and centers, including the National Cancer Institute.
U.S. Agency for Toxic Substances and Disease Registry http://www.atsdr.cdc.gov	Information on hazardous substances and sites in the United States
U.S. Department of Homeland Security http://www.dhs.gov	Threat advisories, emergency planning, contacts, and resources. Includes the Federal Emergency Management Agency for emergency planning and disaster assistance.
U.S. Environmental Protection Agency http://www.epa.gov	Searchable list of environmental protection topics, laws and regulations, resources, and contacts by state.
U.S. Food and Drug Administration http://www.fda.gov	Information on U.S. Food and Drug Administration activities and regulated products. Searchable by demographic category (for example, consumers, patients, health professionals, industry, press).
University research centers (examples)	
Department of Engineering and Public Policy, Carnegie-Mellon University http://www.epp.cmu.edu	The Risk Analysis and Risk Communication Section of this department conducts research, including behavioral decision making and policy. Researchers pioneered approaches including mental models and internationally recognized risk rankings.
Environmental Risk Analysis Program, Cornell University Department of Communication and Center for the Environment http://environmentalrisk.cornell.edu	Resources for citizens and policy makers about environmental risks.

Table 18-2. *Continued*

Source	Content highlights
Harvard Center for Risk Analysis http://www.hcra.harvard.edu	Decision science applied to risk analysis, including public response to risk.
Johns Hopkins Bloomberg School of Public Health http://www.jhsph.edu/risksciences	The Risk Sciences and Public Policy Institute focuses on environmental policy to improve public health. The School's Center for Communication Programs focuses on public health interventions.
Professional associations, societies, and nonprofits (examples)	
American Cancer Society http://www.cancer.org	Information about various cancers, treatments, statistics, and research. Includes treatment decision tools, interactive exercises that help patients understand and choose customized treatments
American Industrial Hygiene Association http://www.aiha.org	Occupational and environmental health and safety issues.
Cancer Research UK http://www.canceresearchuk.org	Britain's largest cancer research charity. Cancer news, patient information, discussion forum, searchable database of clinical trials, podcasts, and research grant opportunities.
International Association for Public Participation http://www.iap2.org	Advancing public participation in decision making and policy making. English, Spanish, and French.
International Consumer Product Health and Safety Organization http://www.icphso.org	Health and safety issues related to consumer products manufactured and marketed in the global marketplace. Online newsletter, conference proceedings.
International Union Against Cancer http://www.uicc.org	Global news alerts, cancer documentary film festival.
National Safety Council (U.S.) http://www.nsc.org	Resources for protecting life and health.
Society of Environmental Toxicology and Chemistry http://www.setac.org	Science-based environmental quality.
Society for Risk Analysis http://www.sra.org	Multidisciplinary, international focus on risk analysis, including risk assessment, characterization, communication, management, and policy. A subgroup, the Risk Communication Specialty Group (http://www.sra.org/rcsg), focuses on risk perception, public participation, mass media coverage of risk, trust and credibility, social influence, and evaluation. It includes a listserv for information sharing.
Society for Technical Communication http://www.stc.org	Resources for technical communicators worldwide.

(*continued*)

Table 18-2. *Continued*

Source	Content highlights
Databases (examples)	
ATSDR's Hazardous Substance Release and Health Effects Database http://www.atsdr.cdc.gov/hazdat.html	U.S. hazardous sites, searchable by state and contaminant.
National Library of Medicine http://www.nlm.nih.gov	Health-related databases of publications and resources, including MEDLINE and PubMed.
National Priorities List (U.S.) http://www.epa.gov/superfund/sites/npl/index.htm	U.S. Environmental Protection Agency database of Superfund sites, searchable by state.
World Health Organization http://www.who.int/en	Health statistics searchable by country, including chronic and infectious diseases, risk factors, causes of death, life expectancy. Also a database of publications on health topics, many downloadable.
Commercial organizations (example)	
Risk World http://www.riskworld.com	Information on the analysis and management of health, environmental, financial, and technological risks. Includes press releases, news, book reviews, and links to other resources.

shows some examples of risk-related information online. This is not intended to be a comprehensive list but to show the types of organizations that provide health, environmental, safety, and risk communication information online.

Many publications give guidelines and standards for properly preparing information to be viewed on the Internet; see the resource list at the end of this chapter. When designing websites to communicate, a few points should be emphasized:

- **Establish your organization's credibility.** Anyone can put risk-related advice on the Internet, and it is difficult for users to know whom they should trust, especially when they receive conflicting information. Establish your credibility in the "About" section by describing your organization, its role, and your accountability. It helps to have testimonials from others. The Department of Engineering and Public Policy at Carnegie-Mellon University, for example, has a "What Others Say" page that includes quotes from notable third parties endorsing the department's programs. Include contact information (e-mail, phone, and postal mail) for people to get in touch or ask questions.

> Establish your credibility in the "About" section of your website by describing your organization, its role, and your accountability.

- **Put your risk information in context and qualify it.** When including risk information, tell how it was prepared and how accuracy is ensured. Include any qualifying statements about the limitations of risk estimates. Medical professionals, for example, caution that online cancer risk calculators often neglect to state whether the risk is from getting a disease or dying from it, how the risk compares with those of other people or other cancer risks, and the level of uncertainty inherent in the estimate (Woloshin et al. 2003). Say when the risk information was last updated. Make sure that all downloadable papers, presentations, and other information include publication dates and are traceable back to their sources.

> Make sure that all downloadable papers, presentations, and other information include publication dates and are traceable back to their sources.

- **Be ethical.** Specify the purpose of the site, including any commercial purposes and advertising. Identify any potential conflicts of interest or biases. Explain how the privacy and confidentiality of any personal information collected are protected.
- **Collaborate with credible others.** View the Internet as a way to collaborate, especially for skeptical or hostile audiences. Provide links to related regulatory agencies, environmental or civic groups, universities, and other such sites. Common Internet courtesy is to request such linking beforehand and to provide reciprocal linking if possible (you link to them, and they link to you). Check with your organization first before linking. Some government agencies and industrial organizations have policies about linking to certain other sites to avoid the appearance of endorsing them.
- **Conduct usability testing.** Content is king, so make sure that users can find the information they want and need, in language to which they can relate. Make sure that content, including graphics and downloaded files, comes up within seconds on a variety of computer platforms and bandwidth speeds. Make sure that the site's structure, navigation, and search function are clearly and logically organized to let users find what they need, know where they are, and get back to where they were.

Also consider the guidelines for the visual representation of risk (Chapter 14) as well as the development of information materials (Chapter 13).

TECHNOLOGY IN CONSENSUS COMMUNICATION

Computer technology can be used effectively in risk communication that involves groups or individuals in a decision-making process. This section summarizes some of the most common tools and guidelines for each.

Websites

In consensus communication, websites can describe a risk and its options for mitigation and invite public input that can be used to craft a decision. The U.S. Department of Energy and other federal agencies, as part of public involvement on environmental analyses, often put draft publications online for review and comment. Stakeholders can use a feedback form on the website to submit comments. The U.S. Army Corps of Engineers built a webpage for the Great Lakes Navigation Team to share information with stakeholders and local officials on risks associated with coastal infrastructure. The site includes notices to alert stakeholders to upcoming meetings, presentations made at past meetings, and additional information to facilitate understanding and involvement in the decision-making process.

In 2012, the U.S. Environmental Protection Agency (EPA) launched NEPAssist (http://nepassisttool.epa.gov/nepassist/entry.aspx), a mapping tool designed to facilitate the environmental review processes and project planning under the National Environmental Policy Act (NEPA). The web-based tool shows environmental assessment indicators for locations that the user specifies, drawing environmental data dynamically from EPA's geographic information system databases and web services. For example, users can enter any location of interest in the United States and choose to see information about that area such as hazardous wastes, air emissions, water discharges, toxic releases, Superfund sites, demographic data, flood hazard zones, wetlands, and other data sources. Users can visualize features geographically, such as the outline of an aquifer on the map, in context with other features—a difficult task to find in sometimes-obscure environmental impact review documents. This tool was designed to streamline the review process for decision makers and stakeholders, raising important environmental issues at the earliest stages of project development.

These types of applications can be effective in consensus communication for several reasons:

- They are accessible from a variety of locations, allowing stakeholders to interact at home, work, school, or other settings.
- They break what would otherwise be insurmountable amounts of information into manageable chunks that provide insight into the larger picture.
- They offer the ability for stakeholders to manipulate the data in ways that are most meaningful for them.

To be successful in communicating risk information in a consensus process through a website, then, make sure that your application can meet these criteria. See Chapter 17 for additional information on working with stakeholders.

Local Area Networks, Extranets, and Bulletin Boards

Community-level decision making may benefit from the use of a local area network (LAN). A LAN enables electronic communication among users within an organization or area, such as a neighborhood. All those who are hooked up to the LAN can communicate with each other. Users of a LAN are not necessarily on the Internet. A site may use a LAN to place documents for comment, meeting notifications, announcements, and multimedia. The U.S. Department of Agriculture's Broadcast Media & Technology Center, for example, uses a LAN to carry messages, press conferences, and taped events throughout the Washington, DC, metro area. To span a larger geographic area, two or more LANs can be linked to form a wide area network.

Extranets are external intranets. By setting up an extranet, a company can allow selected people outside an organization to connect via the Internet to information that is normally internal to the company. The extranet can be set up to maintain security for sensitive or proprietary company information. Companies typically use extranets for business communication and commerce among employees, suppliers, customers, and other business partners. However, extranets also can be used for consensus-type communication such as community-wide workgroups. An extranet can be used to give participants access to internal information such as documents and databases. A feedback mechanism can be included for making comments or requesting additional information.

An electronic bulletin board or blog can be internal to an organization, on a LAN, or on the Internet. Newsgroups, or collections of people with a common interest, talk to each other via bulletin boards or blogs that focus on specific topics. Everyone who logs onto the bulletin board or blog can see all the other messages in the discussion and can jump into the discussion as well. For those who have a good way to get the current access, bulletin boards or blogs are a good way to get the current "pulse" on an issue, share information, address rumors, and correct misinformation.

Some of the same guidelines apply to LANs, bulletin boards, blogs, and extranets as to e-mail lists. Make sure that your audience knows the ground rules, let them know

> Electronic bulletin boards are a good way to get the current "pulse" on an issue.

whether a moderator will be used and define that person's role, and focus the topic of the discussion to ensure that appropriate information is being communicated in a way that meets audience needs.

Tracking and Analyzing Comments and Responses

Software programs are often used to track stakeholder comments and issues and an organization's or agency's responses. The idea is to track who commented, when, what they said, and what the response was. One government organization tailored a standard database software application to record information associated with its public briefings during the public consultation process (McMakin et al. 1995). Database fields included date, commenter's name and affiliation, commenter's location, comment summary, comment category (topic), recorder (note taker at a meeting), and follow-up actions and dates. The fields could be searched, and summary reports are generated. Database input and maintenance time depend on the number of briefings, comments, and responses.

The Regional Municipality of Ottawa-Carleton in Ottawa, Ontario, has adapted for public consultation commercial software originally designed for managing sales and telemarketing contacts (McMakin et al. 1995). The municipality is a regional government responsible to 750,000 citizens for transportation, environmental services, health, social services, and planning. The municipality adapted the software to make a several-thousand-person mailing list available to municipality staff with criteria for when and how to contact citizens regarding policy and program consultation. The software also keeps track of who was contacted and can track public comments. The database outputs addresses to mailing labels and reports. The municipality is putting the database on a community-based net system so that all constituency groups in a particular area can access the database. The net system also enables people to send e-mail to the municipality.

Two online collaboration tools specifically for NEPA reviews were praised by the Council on Environmental Quality in 2012. Both make the NEPA process more efficient, including shortening the time needed to process and analyze public comments received through online submission. They also enable collaborative development, and online publication and storage, of NEPA documents. The first is the Planning, Environment, and Public Comment (PEPC) system (http://parkplanning.nps.gov/) developed by the National Park Service. The site lets users see and select from all the NEPA documents open for review and comment that pertain to U.S. national park sites. For each project, the public can see the plan process, meeting notices, and document lists and can comment online. Many parks now are using the system as the primary method for submitting and receiving comments, with some also accepting comments through "traditional" methods such as mail, fax, and e-mail. Those responsible for NEPA compliance use PEPC for project management—to structure, streamline, and track the compliance process. The other online collaboration tool is the Electronic Modernization of NEPA (eMNEPA), a suite of web-based tools and databases used by the U.S. Forest Service (http://www.fs.fed.us/nepa/nepa_home.php).

The key feature of these systems from a risk communication standpoint is transparency. Both those tracking and analyzing the comments and individual commenters can see the process, how each comment relates to the decision, and final resolutions. If you decide to make use of similar systems, ensure that they too are transparent to your audience and that your organization has procedures in place to ensure that information is appropriately reviewed before being posted online.

Facilitating Group Decision Making

The increase in community and technical advisory panels, as well as public and technical workshops, means that many viewpoints must be fairly considered and balanced, especially to reach consensus. This process is made more difficult when viewpoints are polarized and issues are complex.

> Studies show that situations in which a group must reach consensus work best when computer-mediated communication is combined with face-to-face interaction.

An increasing number of software and hardware combinations are available to help facilitate group decision processes. These technologies help with brainstorming ideas, presenting and weighing group members' viewpoints, visualizing "what-if" scenarios of processes or systems, ranking or rating items, voting, and reaching consensus. One way to use these systems is to have all group members in a meeting room, or in diverse geographic areas but in a shared web area, where they type comments via computer terminals, and comments are redisplayed on an electronic whiteboard or on each other's screens. If the comments are anonymous, participants experience more freedom to verbalize and criticize others' ideas without the fear of reprisal (Jessup et al. 1990; Valacich et al. 1992). Though this process may cause greater conflict within the group, the conflict tends to be substantive rather than interpersonal, and decision making may be enhanced (Watson et al. 1988).

One caution from research studies is that situations in which the group must reach consensus appear to work best when combined with face-to-face interaction. More social interaction than just working individually on computers is needed to reach agreement, especially when the group is given restricted time periods (Hiltz et al. 1986; Siegel et al. 1986). One such study showed that the highest decision quality was achieved with a two-phase arrangement. Anonymous computer-mediated communication was used for brainstorming ideas, and face-to-face interaction was used for evaluation and reaching consensus (Olaniran 1994).

CH2M Hill, an environmental engineering firm, created a software application that they use to help groups reach consensus (McMakin et al. 1995). They use the software in conjunction with a nominal group technique, a structured method for discussing and evaluating issues. The process involves developing criteria that capture the salient points of the issues, assigning weights to the criteria (each participant does this), conducting a statistical analysis of the weights, and discussing the results as group. This process can be repeated several times, in which the group members often move closer together in their views as they carve out common ground. This process requires not only mastery of the software that documents the statistical analysis but also excellent facilitation skills to capture and clarify viewpoints while keeping the process moving.

A national program to clean up the U.S. Department of Energy's former sites used computer-mediated communication in its public participation process to rank criteria, vote, and conduct other group activities. The software they used displayed visual results, such as bar charts. Group members used remote touch pads for recording input and voting. Agency representatives report that the software has saved hours of discussion time that would ordinarily be required by traditional methods (McMakin et al. 1995).

Professional facilitator and dispute resolution expert Jim Creighton offers some recommendations for using technology in group decision making (Creighton and Adams 2002):

> Focus the process on the user and the purpose, not the technology.
> —Jim Creighton (Creighton and Adams 2002, p. 180).

- **Put the collaboration first, rather than the technology.** Technology does not magically transform diverse people, especially adversaries, into collaborative partners. A group decision process must be expertly designed to involve all participants, resolve disputes, and achieve the goals of the collaboration.
- **Match the technology to the process.** Information briefings from a trusted source may require only e-mail or intranet communication, whereas conflict resolution may require more elaborate technologies that demonstrate that all points of view have been heard and recorded.
- **Count the costs and benefits.** Costs can include software and equipment, technicians to set up and troubleshoot it, transaction charges for telecommunications, and leader and/or facilitator labor. But remember to balance that with likely cost reductions: less travel by participants, less time spent in meetings, and a more efficient decision-making process.

TECHNOLOGY IN CRISIS COMMUNICATION

Most of the same technology tools used in care and consensus communication can also be used in crisis communication. This section describes some of the basic principles for using these methods during crises. Chapter 21 of this book includes more information about technology-assisted communication in emergencies. Chapter 19, Social Media, describes the use of smart phones and other interactive technologies in crisis communications.

Websites, Wired, and Wireless Technologies

Websites that are updated frequently can be good resources in helping the public know what to do in case of an emergency and how to respond when one occurs. The websites of the U.S. Department of Homeland Security, U.S. Federal Emergency Management Agency, and the American Red Cross, among others, tell citizens what to do in case of various threats. Websites for localized threats can be more specific.

Websites can also be used to share information about threats and what is being done to counter them. For example, InciWeb reports current information on U.S. wildland fires. The U.S. Forest Service, with other agencies, developed this online information and alerting system. Users can sort incidents by name of the fire or state and see percent containment, acres affected, photos, and news articles about each fire. They can also register to receive RSS feeds on specific fires.

Many state and local government agencies are creating their own social networking sites for crisis communications. For example, the Virginia Department of Emergency Management has a YouTube channel in partnership with Google to reach Virginia citizens with emergency-related information and public service announcements from the governor.

Wired and wireless telecommunications are increasingly being used to convey critical information and updates in crisis situations. Several communities and states have the so-called reverse 911 telecommunication systems that phone citizens in targeted geographic areas with a recorded message to notify them about specific threats. Reverse 911 also includes a number that citizens can call to hear prerecorded information, such as emergency evacuation procedures. Emergency communication systems send information instantly to subscribers' mobile devices (cell phones, pagers, tablets) as to e-mail systems. Government organizations are increasingly using these emergency communication systems to keep their first responders, employees, and citizens informed.

> Reverse 911 telecommunication systems phone citizens in targeted geographic areas with a recorded message to notify them about specific threats.

Ironically, as much as we rely on technology in emergencies, it can be the first thing to fail. In recent years, various crises worldwide have crashed websites, overloaded phone lines, rendered cell phone towers inoperable, and triggered regional electrical blackouts. Organizations responsible for communicating risks should have backup communication plans that account for failure of standard public communication methods. The Oregon chemical weapons depot, for example, distributed battery-powered tone alert radios to citizens in surrounding communities. Some organizations put their emergency risk communication plans, including a list of people to contact in emergencies, on CDs that run on battery-powered laptop computers.

Emergency Planning and Training Tools

Emergency planning and exercises have their own technology tools, such as software programs, that are useful for planning and implementing crisis communications. One example, used by the Federal Emergency Management Agency and its partners, is the Standard Unified Modeling, Mapping, and Integration Toolkit, known as SUMMIT. SUMMIT is a modeling and simulation software environment that enables analysts, emergency planners, responders, and decision makers to access integrated suites of modeling tools and data sources for planning, exercise, or operational response. SUMMIT has a library of scenarios for a number of threats to support scenario development and exercise evaluation. For example, SUMMIT has a scenario that links contaminant plume modeling to medical effects, infrastructure effects, medical needs, and hospital bed shortfall models. Any changes in the plume release will automatically and realistically be reflected in the other parts of the model. Users can visualize building damage and other conditions that occur after a disaster. They can analyze the "what-if" trade-offs that are important in effective response. Since 2010, SUMMIT has been used in small- and large-scale exercises to accelerate scenario planning, provide scientifically grounded scenario data, and enhance the realism and common operating picture.

Virtual reality websites, which are immersive synthetic environments, are also useful for crisis communication planning. They enable you to create a persona, in the form of a physical avatar, and interact with others and three-dimensional objects online in virtual worlds. Virtual worlds often also offer resources such as blogs, wikis, instant messaging, and sharing user-created objects to connect with others on topics of interest. Research has shown advantages for the use of virtual worlds for educational and training purposes. By allowing learners to interact with other avatars in a safe, simulated environment, it is possible to decrease student anxiety, increase competency in learning a new skill, and encourage cooperation, collaboration, and conflict resolution (Hansen 2008).

The most established virtual world, Second Life, is a three-dimensional environment that is "inhabited" by millions of people worldwide. Emergency responders have used Second Life and a commercial software development toolkit known as OLIVE to simulate real-life disaster and emergency incidents (Figure 18-2 and Figure 18-3). The training exercises have included government sectors and specialty fields such as medical and transportation to create small- and large-scale responses from multiple agencies. Users practice elements of crisis communication, incident command, and resource management. Simulation-based training with avatars can fill gaps in traditional training techniques for first responders. Trainers can replicate emergencies in the locations where they are likely to occur without disturbing the public (Figure 18-4). Once built, these scenarios can be practiced numerous times over the Internet with participants in various locations.

As virtual worlds grow in size and prominence, real-world governments and other organizations are increasingly using them to pursue their missions such as educating the public or enforcing federal laws. The Emergency Management Nexus, for example, was one of the first massive multiplayer virtual environments specifically designed for use by government personnel. Emergency Management Nexus supports workplace education

Figure 18-2. Mass casualty response scene in OLIVE, software for creating virtual worlds. Sixty avatars were logged into this simulation, which was used for emergency response training. (Source: SAIC; used with permission. The emergency preparedness and medical features of the OLIVE technology were developed by Forterra Systems with funding from the U.S. Army Telemedicine and Advanced Technology Research Center.)

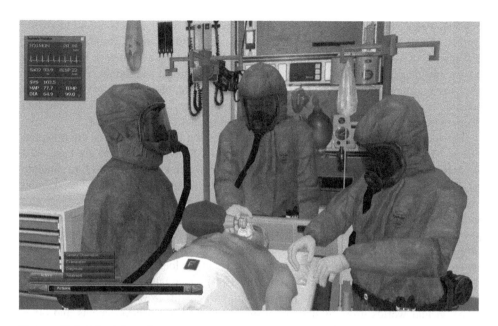

<u>Figure 18-3.</u> Virtual reality emergency room exercise. A physician and two nurses "treat" a virtual patient who has been exposed to sarin nerve agent. The physician and nurse avatars are controlled by an actual emergency department physician and nurses who are going through a chemical hazard response scenario; the victim is played by a role player. The victim's vital signs are generated by a physiology computer model and displayed in the "bedside monitor" display in the upper left hand corner of the screen. (Source: SAIC; used with permission.)

training and collaboration through the use of virtual classrooms, training facilities, and collaborative workspaces. It aims to enhance worker productivity and reduce costs. The National Institutes of Health Tox Town on Second Life highlights "chemical environmental health concerns and toxic chemicals where you live, work, and play." The National Oceanic and Atmospheric Administration's Earth System Research Lab enables people to experience some of the scientific research center's earth and weather simulators, including an interactive weather map.

The SciLands Virtual Continent, a Second Life section devoted exclusively to science and technology, includes government agencies, universities, and museums. They have regular meetings in Second Life where they share ideas, help each other, and plan future projects. The Information Resources Management College at the National Defense University in Washington, DC, has a Second Life island that includes a virtual replica of the college's Crisis Management Center at Fort McNair in Washington.

The Federal Consortium for Virtual Worlds is a group of federal government employees and contractors that are exploring the use of virtual worlds in government. They set standards, share best practices and policies, create shared repositories, and network. They often hold their meetings simultaneously in real and virtual locations.

When choosing to create or participate in a virtual world, consider the costs and benefits, and whether your target audience or stakeholders all have access. Also consider the broadband requirements and the needs for security and firewalls.

(a)

(b)

<u>Figure 18-4.</u> Stanford Medical Center, real (a) and virtual (b), for emergency response training. (Source: SAIC; used with permission.)

CHECKLIST FOR TECHNOLOGY-ASSISTED COMMUNICATION

☐ The particular type of technology is determined to be appropriate and useful to meet the communication objectives.

☐ Tools are designed to use the special characteristics of the medium, not just as a substitute for other forms of communication.

☐ Computer-based information is prioritized and organized so users can easily find what they need.

☐ When possible, technology applications are pretested with the intended audiences before implementation.

☐ The participants have access to the equipment and software needed to use the tools.

☐ The participants are willing to use the tools and know how to use them effectively, or training and/or facilitation are provided.

☐ Proprietary or sensitive information is not used on networked computer systems.

☐ Information provided on the Internet is updated as appropriate to maintain its usefulness.

☐ In public participation situations, technology tools include feedback systems to support a two-way communication process.

☐ In crisis communication, a backup plan for communication is in place in case of electricity loss or other disaster.

REFERENCES

CDC (U.S. Centers for Disease Control and Prevention), Agency for Toxic Substances and Disease Registry, Oak Ridge Institute for Science and Education, and Prospect Center of the American Institutes of Research. 2003. Emergency Risk Communication CDCynergy (CD, February 2003). http://www.orau.gov/cdcynergy/erc (accessed January 21, 2013).

Creighton, J. L. and J. W. R. Adams. 2002. *Cyber Meeting: How to Link People and Technology in Your Organization*. Xlibris Corporation, Philadelphia, Pennsylvania.

Davis, T. S. 1995. "Communicating Environmental, Health, and Safety Information to Internal and External Audiences." *Journal of the Society for Technical Communication*, 42(3):460–466.

Freng, I. A., S. Sherrington, D. Dicks, N. Gray, and S. Chang. 2011. "Mobile Communications for Medical Care: A Study of Current and Future Healthcare and Health Promotion Applications, and Their Use in China and Elsewhere." University of Cambridge and China Mobile.

Green, B. B., A. J. Cook, J. D. Ralston, P. A. Fishman, S. L. Catz, J. Carlson, D. Carrell, L. Tyall, E. B. Larson, and R. S. Thompson. 2008. "Effectiveness of Home Blood Pressure Monitoring, Web Communication, and Pharmacist Care on Hypertension Control." *Journal of the American Medical Association*, 299(24):2857–2867.

Green, M., J. Zenilman, D. Cohen, I. Wiser, and R. Balicer. 2007. *Risk Assessment and Risk Communication Strategies in Bioterrorism Preparedness*, NATO Security through Science Series-A: Chemistry and Biology. Springer, Dordrecht, Netherlands.

GSMA. 2011. "Mobile Privacy Principles: Promoting a User-Centric Privacy Framework for the Mobile Ecosystem." GSMA, London, United Kingdom. http://www.gsma.com/publicpolicy/wp-content/uploads/2012/03/gsmaprivacyprinciples2012.pdf (accessed January 21, 2013).

GSMA. 2012. "Privacy Design Guidelines for Mobile Application Development." GSMA, London, United Kingdom. http://www.gsma.com/publicpolicy/privacy-design-guidelines-for-mobile-application-development/ (accessed January 21, 2013).

Hansen, M. M. 2008. "Versatile, Immersive, Creative, and Dynamic Virtual 3-D Healthcare Learning Environments: A Review of the Literature." *Journal of Medical Internet Research*, 3(10):e26. http://www.jmir.org/2008/3/e26 (accessed January 21, 2013).

Hiltz, S. R., K. Johnson, and M. Turoff. 1986. "Experiments in Group Decision Making: Communication Process and Outcome in Face-to-Face Versus Computerized Conferences." *Human Communication Research*, 13:225–252.

Jessup, L. M., T. Connolly, and J. Galegher. 1990. "The Effects of Anonymity on Group Decision Support System Group Process with an Idea-Generating Task." *MIS Quarterly*, 14:312–321.

Landro, L. 2011. "Don't Come Back, Hospitals Say." *The Wall Street Journal*, June 7.

Lobbin, F. 1997. "10 CFR 50 Event Reporting Computer-Based Training Program." *Nuclear Plant Journal*, 15(5):45.

McMakin, A. H., D. L. Henrich, C. A. Kuhlman, and G. W. White. 1995. *Innovative Techniques and Tools for Public Participation in U.S. Department of Energy Programs*. PNL-10664, Prepared for the U.S. Department of Energy, Pacific Northwest National Laboratory, Richland, Washington.

Olaniran, B. A. 1994. "Group Performance in Computer-Mediated and Face-to-Face Communication Media." *Management Communication Quarterly*, 7:256–282.

Rodrigue, C. M. 2001. "The Internet in the Social Amplification and Attenuation of Risk." Presented to the 26th Annual Natural Hazards Research and Applications Workshop, July 15–18, 2001, Boulder, Colorado.

Siegel, J., V. Dubrovsky, S. Kiesler, and T. W. McGuire. 1986. "Group Processes in Computer-Mediated Communication." *Organizational Behavior and Human Decision Processes*, 37: 157–187.

Strecher, V. J., T. Greenwood, C. Wang, and D. Dumont. 1999. "Interactive Multimedia and Risk Communication." *Journal of the National Cancer Institute Monographs*, 25:134–139.

Temple University. 2006. Low-Income Patients Reduce Heart Risk Via Internet 'Visits'. http://www.temple.edu/temple_times/4-20-06/telemedicine.html (accessed February 7, 2013).

Valacich, J. S., A. R. Dennis, and J. F. Nunamaker. 1992. "Group Size and Anonymity Effects on Computer-Mediated Idea Generation." *Small Group Research*, 23:49–73.

Valdez, A. 2002. "Innovative Multimedia Cancer Education Interventions for Latinas." Presented at the 130th Meeting of the American Public Health Association, November 9–13, 2002, Philadelphia, Pennsylvania.

Watson, R. T., G. DeSanctis, and M. S. Poole. 1988. "Using Group Decision Support Systems to Facilitate Group Consensus: Some Intended and Unintended Consequences." *MIS Quarterly*, 12:463–478.

Woloshin, S., L. M. Schwartz, and A. Ellner. 2003. "Making Sense of Risk Information on the Web." *BMJ (Clinical Research Ed.)*, 327:695–696.

ADDITIONAL RESOURCES

American Society for Training and Development. http://www.astd.org (accessed January 21, 2013).

Distance Education and Training Council. http://www.detc.org (accessed January 21, 2013).

Eng, T. R. 2000. *Wired for Health and Well-Being: The Emergence of Interactive Health Communication*. U.S. Government Printing Office, Washington, DC.

McGovern, G. and R. Norton. 2002. *Content Critical: Gaining Competitive Advantage through High-Quality Web Content.* Financial Times Prentice-Hall, Pearson Education Limited, Edinburgh Gate, United Kingdom.

Rice, R. E. and J. E. Katz, eds. 2000. *The Internet and Health Communication: Experiences and Expectations.* Sage Publications, Thousand Oaks, California.

Society for Applied Learning Technology. http://www.salt.org (accessed January 21, 2013).

Spyridakis, J. H. 2001. "Guidelines for Authoring Comprehensible Web Pages and Evaluating Their Success." *Technical Communication,* 47(3):301–310.

Tsagarousianou, R., D. Tambini, and C. Bryan, eds. 1998. *Cyberdemocracy: Technology, Cities and Civic Networks.* Routledge, New York.

Usability.gov. National Cancer Institute's website of resources for "usable, useful and accessible websites and user interfaces." http://www.usability.gov/ (accessed February 7, 2013).

U.S. Distance Learning Association. http://www.usdla.org (accessed January 21, 2013).

U.S. Food and Drug Administration website, video programs called "FDA Patient Safety News." http://www.fda.gov/psn (accessed January 21, 2013).

Van Duyne, D. K., J. A. Landay, and J. I. Hong. 2002. *The Design of Sites: Patterns, Principles, and Processes for Crafting a Customer-Centered Web Experience.* Addison-Wesley, Reading, Massachusetts.

19

SOCIAL MEDIA

A more recent method of communicating risk information comes in the form of social media. Social media enables individuals to connect to each other online and to share information, videos, photos, and comments using easy-to-publish tools. The visible discussions that result can include experiences, both negative and positive, about any particular issue, in real time. According to Charlene Li and Josh Bernoff of the social technology research firm Forrester Research, which regularly analyzes online activities, people use social media to connect, take charge of their own experience, and get what they need from each other rather than from traditional institutions (Li and Bernoff 2008).

> The development of new media in today's culture calls for a strategic model of information diffusion that alters the classic top-down model of organizations relaying material to interested parties.
> —Lucy Leiderman (2012, p. 1).

Social media can be a valuable component of risk communication. Not only does it add more distribution channels, but it also gives organizations almost instant and continuous feedback on what people want to know about risks and what they are concerned about. It also enables organizations to respond quickly as situations change. More than that, it allows organizations to become part of a community, which may increase the organization's credibility and trust with those it serves. Used properly, social media can supplement or replace more expensive and nondynamic solutions, such as video, conference calls, overbooked conference/meeting rooms, and print media.

Risk Communication: A Handbook for Communicating Environmental, Safety, and Health Risks,
Fifth Edition. Regina E. Lundgren and Andrea H. McMakin.
© 2013 The Institute of Electrical and Electronics Engineers, Inc., and Regina E. Lundgren and
Andrea H. McMakin. Published 2013 by John Wiley & Sons, Inc.

Much research has centered on using social media for crisis communication. However, social media can be used effectively for care communication and has a great potential to facilitate consensus communication. The advantages and disadvantages of using social media are discussed in Chapter 10. Chapter 18 discusses other uses of technology-assisted communication, such as websites, virtual reality settings, online forums, and multimedia programs. At the time this edition was published, social media included social networking sites; blogs and podcasts; microblogging; video-, image-, and file-sharing sites; mapping; and use of mobile phones. However, the social media landscape is constantly changing, and new applications are only a thought away. Because of the adaptability of social media, this chapter focuses on the general principles associated with the methods before delving into the three main uses of social media for risk communication: sharing content, engaging with stakeholders, and monitoring changes in perceptions, as well as laying out some guidelines for specific types of social media. Finally, the chapter discusses techniques for evaluating the effectiveness of social media efforts.

GENERAL PRINCIPLES ON PARTICIPATING IN SOCIAL MEDIA TO COMMUNICATE RISK

Social media allows the two-way communication that is necessary for successful risk communication. However, use of the methods requires a shift in thinking from more traditional ways of communicating risk. For one thing, the audience is more in charge of risk communication in social media settings. As public health communication expert Craig Lefebvre puts it, in the networked world, practitioners creates messages that reach audiences, but audience members also talk to practitioners, and, just as importantly, with each other (Lefebvre 2007). Social media gives the audience unprecedented control— in what they read, in how they respond, and in whether they choose to act. Lefebvre recommends that practitioners engage this world through collaboration, sharing, and interactivity.

Another way social media differs from more traditional methods of communicating risk is the sense of community. While some sociologists and educators have expressed concern that our growing reliance on technology will mean less personal interaction, the Pew Internet and American Life Project found that two-thirds of adult users of social networking sites are there to connect with family and friends (Smith 2011). Organizations are tapping into this sense of community to help communicate risks. For example, the U.S. Army advocates official use of social media, even in combat zones, to inform the public, keep families connected, and address inaccurate reports elsewhere. It uses Facebook to share information with Family Readiness Groups, providing a one-stop interactive source of reliable information (Office of the Chief of Public Affairs 2011).

Part of that sense of community involves the audience sharing knowledge and recommendations with each other. Health care professionals, risk managers, and emergency response agency staff lament the effects of "crowdsourcing," where people who would have once turned to experts in private sector industries or government agencies for answers now turn to their online connections for information about personal health issues, environmental stewardship, and emergency situations. Crowdsourcing, however, has the potential for staff augmentation, creating ambassadors armed with information who can spread the word much faster and with greater credibility than some understaffed agencies.

As one agency official put it, social media can also be about empowering the unaffected to help reach the affected (Tinker 2009).

The use of social media for risk communication can provide other benefits as well, such as increased confidence in the organization, greater understanding of audience needs, and faster response during emergencies. For example, by having a social media presence, the Red Cross found that it received fewer media calls during an emergency, gained access to other social networks when needed, and developed better situational awareness (Tinker 2009). In addition, social media can serve as a valuable alternative for crisis communications, when phone lines or other infrastructure may be damaged. In 2008, Hurricane Gustav took down communication systems, including the Community Emergency Response Team's call notification system. Because mobile broadband was still available, they were able to send messages through Facebook. Similarly, the victims of the 2011 Japan earthquake and tsunami used Twitter to call for help when phone lines were down (Lindsay 2011).

Because of these and other factors, risk communicators need to consider the characteristics of the audiences they hope to reach and barriers within their own organizations before using social media as a method to further care, consensus, or crisis risk communication.

Determining Audience for Social Media

While understanding the needs of the audience is important for any type of risk communication, social media poses a unique situation, for the audience participates in content development and message transmission. Risk communicators need to consider the following:

- **Can the intended audience access social media?** At the end of 2011, 70% of the total households in developed countries had Internet access, compared with only 20% of households in developing countries. The United States ranked twenty-seventh among the top 50 Internet-using countries, with 78% of the population using it (ITU 2012).
- **How does the intended audience use social media?** Studies show that people online fall into four categories: those who create content, those who comment on others' content, those who collect information, and those who like to join groups for interaction (Li and Bernoff 2008). If your primary audience spends considerable time online as commenters or information collectors, blogs and podcasts might be a wise choice for at least one type of social media engagement. If they tend to be joiners, then having a presence on social networking sites such as Facebook might be a good choice. Another way to determine where to engage is to discover where the risk you are interested in is being discussed online. For example, Google alerts allow you to track key phrases as they appear across the Internet. Are most of the links to social networking sites, blogs, video- or file-sharing sites, or some combination?

For additional information on how various segments of the American population use social media, see the Pew Research Center's Internet and American Life Project (http://pewinternet.org/). Chapter 8 of this book provides additional information on understanding your audience.

Organizational Barriers to Social Media Adoption

Chapter 4 discusses constraints to risk communication, including those constraints imposed by an organization. Because of the relative newness of social media as a risk communication tool and because of the perceived lack of control over risk information in this newer tool, many organizations hesitate before sharing risk information. A survey of more than 500 agencies in 2009 found that the biggest barriers to using social media for crisis communications were lack of staff time or ability, inability to understand the tools, and the culture of the organization (Tinker 2009). Our work with organizations considering using social media to communicate risks found additional barriers, such as concern about tarnishing the organization's image, the potential to perpetuate misinformation, information security, and inability to sustain the investment.

Lack of Time or Ability

It is a rare organization that never faces staffing constraints. Chapter 4 includes a section on how to overcome such constraints in general. For social media, partnering with volunteers or other organizations can be the key, particularly for crisis communications, where such partners may not be as overwhelmed in an emergency. For example, for care communication, CaringBridge, a popular web service that informs family and friends about people going through a serious illness and recovery, partners with a variety of organizations, including health care providers, faith-based communities, insurance providers, and employee-assistance programs. Partner organizations receive training and information on how to share CaringBridge with those they serve.

Another way to increase staff time or ability is to harness a wider swath of the organization. A panel of public relations specialists who had delved into social media for their government and private sector organizations for care and crisis communications suggested that workers need three sets of skills to effectively engage in social media: information-gathering skills, information-packaging skills, and conversational skills. Finding this skill set may take you outside the traditional public relations or communications functions in your organization (Lesperance et al. 2010) and expand the number of staff who can support the effort.

Inability to Understand the Tools

While social media usage continues to expand, some demographics have yet to fully embrace it. Lack of adoption personally often translates to lack of support professionally. But even a frequent user at home may struggle to see how social media can be used for risk communication. One way to overcome this barrier is to identify the power users in your organization. Does anyone access social media for their club or association? Has anyone developed blogs or file-sharing accounts for personal use? If all else fails, tapping into interns from high schools or colleges may give you a ready reference for social media use.

A misunderstanding of social media can also get in the way of using it to its full potential. If social media is viewed as only another distribution channel, efforts will fail. Risk communication in social media is about audience members searching for content of most interest to them, creating additional content, and sharing it with other like-minded individuals. It is not so much about the organization pushing out information and hoping someone learns from it but the audience members shaping information to suit their needs. This dynamic interaction may take additional thought when planning risk communication.

Culture of the Organization

Organizations that maintain strict control of their risk information may find using social media daunting. One expert panel suggested starting with pilot programs to overcome such a barrier (Tinker 2009). Social media activities should also be part of the organization's overall risk communication plan, with similar safeguards of information and a focus on audience needs. Additional information on overcoming organizational culture can be found in Chapter 4, Constraints to Effective Risk Communication.

Concern about Tarnishing the Organization's Image

The free flow of information on social media and the possible use of nonpublic relations staff or even volunteers to manage it cause some managers to wonder whether the organization's image will be tarnished by using social media to communicate risk. One way to lessen this concern is to develop guidelines or policies for social media participation. Guidelines cover when the organization's name can be mentioned; how to gain permission to post certain types of organization-related content; reminders about confidentiality, intellectual property rights, and privacy issues; and when to forward information to the organization's legal, public relations, or communications departments. In addition, posted guidelines for public interaction (no defamatory comments, no profanity, etc.) can be shared with the audience participants. Note, however, that some organizations involved in care communication situations such as drug development and medical equipment research have been advised that additional insurance may be needed to venture into social media. Contact your organization's legal counsel to ensure that you are following appropriate regulations and restrictions.

Potential to Perpetuate Misinformation

Another concern about crowdsourcing and audience interaction is the potential for misinformation and rumor to spread. For example, in reviewing response to the damage of the Fukushima Nuclear Facility following the earthquake and tsunami off the shores of Japan in 2011, the Center for Biosecurity found that social media did perpetuate misinformation, but that even the news networks were following the online conversation and using it as a source for stories (Center for Biosecurity 2012). On the other hand, research has also indicated that the occurrence is far less likely than originally feared. For example, the Congressional Research Office found that while some information on social media following the Fukushima disasters was unintentionally inaccurate (for example, passing along requests for help long after a victim was rescued), the information was generally good (Lindsay 2011). Public health researchers in Toronto found a similar result when they looked at Twitter posts related to the swine flu pandemic of 2009. Less than 5% of the posts were misinformation. The same research found, however, that only 1.5% of the posts were links to government- and health-related agencies, which would once have been considered the most reliable resources for such information (Chew and Eysenbach 2010). This finding points to one of the best ways to prevent misinformation from spreading—be the one to share the correct information and act quickly to respond when misinformation begins circulating. The only way to do that is to engage in social media.

Information Security

Many organizations worry that in the highly interactive environment of social media, participants could deliberatively or inadvertently reveal information that should be kept secure or proprietary. This could include, for example, trade secrets, personal information about customers, or emergency response details that could hinder a criminal investigation.

Organizations also worry about sabotage and data leaks, where adversaries could use social media channels to introduce malicious software or get "backdoor" access to company data.

These are valid concerns, and organizations need to determine how much risk they are willing to incur. A few organizations have decided not to engage in social media at all, though they may have a website. Others screen blog comments before posting them, explaining in their policy that the blog is moderated. Some turn off the comment feature on their YouTube videos or monitor and delete inappropriate comments on their Facebook pages. Some organizations block employee access to certain social media sites. One company limited its employees to LinkedIn because documents cannot be uploaded to it and users cannot see peoples' profiles unless they are in an approved network.

The bottom line is that organizations must develop strategies to balance the risks against the benefits of social media. This should include clearly communicating usage policies to their own staff.

Inability to Sustain the Investment

Unlike other forms of risk communication, social media requires a sustained investment. Where a report is generally issued once and a television program is produced once, social media requires continuous interaction, often several times a day. In addition, the engaged audience craves instant information, especially where health and safety are concerned. The Seattle Police Department, for example, found that people valued timeliness of information even over accuracy when it came to risks. As long as the information was kept updated, the audience was willing to forgive the department if the initial information was wrong (Lesperance et al. 2010). In addition, after the Fort Hood shootings in 2009, Army spokespersons found their press conferences commandeered to clarify what was moving at the speed of light through social media channels (Office of the Chief of Public Affairs 2011).

One way to help alleviate this concern is to develop social media policies and plans for the organization. Knowing your audience, selecting the best social media tools to reach that audience, and keeping engaged can be better than casting a wider net that is more difficult to maintain. Research at the North Atlantic Treaty Organization (NATO) Defense College laid out three phases for using social media:

1. Gather resources, including the support of management, the cooperation of all staff involved, and hiring of the right talent.
2. Develop by optimizing existing communication mechanisms, finding ways to attract attention from the appropriate audience, and optimizing content for the audience.
3. Manage and maintain by constantly updating content and actively engaging in the conversation (Leiderman 2012).

When warranted, activities can be scaled up to meet increased needs. For example, during the H1N1 epidemic of 2009, the Centers for Disease Control and Prevention used viral marketing techniques to communicate the risks, including offering e-mail updates, running a webcast for people to ask questions, providing a widget that linked to more information, and rapidly updating its website as more information became available.

Be wary, however, of immediately adopting every new tool. Research the tool thoroughly to ensure that it meets the audience and your organization's needs. For example, the user agreements for some tools include language that would give the tool developer unacceptable access and even ownership of content developed by your organization. Read all agreements closely to ensure that the effort will further your goals and not run afoul of organizational requirements.

SHARING CONTENT VIA SOCIAL MEDIA

Sharing content is the closest social media practice to more traditional forms of risk communication, but even here, engagement is the key. This is true for care, consensus, and crisis communications. For example, Bruce Lindsay, analyst for the Congressional Research Service, found that social media sites were the fourth most popular source for locating emergency information. Organizations used social media to issue warnings, help families locate missing members, and raise funds for relief efforts (Lindsay 2011).

Content can be shared via any social media tool, but how it is shared and how much can be shared varies as does the audience it reaches. According to the NATO Defense College, Facebook has the widest reach, but Twitter has the fastest dissemination (Leiderman 2012). Always remember, however, that your audience will expect you to interact. If you post an article to a blog and send a microblog note, be prepared to respond to comments.

Best practices include the following:

- Most social networking sites have templates, so there is no need to pay someone to design your page. However, **think carefully about the kinds of information you want to share** and use graphics and color to your advantage to get your points across. Beware of too much text and jargon. Look at "branding" your sites to match your organization's other communication mechanisms. Be sure to coordinate messages across all communication mechanisms.
- **Share information that your audience cares about** (Lindsay 2011). This information may include more personal stories from those who faced similar risks, specific actions audience members can take to lower their risks, or ways to prevent risks to loved ones.
- Social networking sites thrive on community, which means interaction. **Give people something to do at your site**: take a quiz, play a game, watch a video, share personal stories on risks they have faced, or chat with a risk expert.
- On some sites, it is something of a coup online to have hundreds of "friends," people you allow to connect to your site. Other people can also find your site through their friends who are linked to you. However, not everyone who requests to be included on your list of friends will have the same philosophy about the risk you are communicating. **Determine how you will decide to accept connections** and share the approach with all involved in the effort.
- Most social networking sites allow other people to leave comments, which are visible to all visitors. Sometimes the comment section is the most interesting part of a page, and the most controversial. When you create your page, **set your preferences to approve any comments before they are posted**, so you will not be surprised to find colorful language, slander, and other offensive posts popping up where you least expect them.
- **Make it easy for people to pass along the content you provide**. "Buttons" provide the option to share information instantly on a variety of social media sites. You can also provide code to allow audience members to paste your information into their sites verbatim, increasing the exposure to your message.

Finally, remember that if you do not share it, others will. For example, when government agencies struggled to keep up with requests for information during the 2007 California wildfires, the local public broadcasting system radio teamed with Google to develop map-based guides to evacuation routes, fire status, and other information based on reports from the government agencies (Lesperance et al. 2010). When that effort was such a success, the agencies involved requested the radio team to keep up the site for the duration of the emergency. However, spontaneous efforts armed with less accurate information could do more harm than good. Be first, be accurate, and be the expert on your risk, even in emergencies.

ENGAGING WITH STAKEHOLDERS

> As social media becomes more a part of our daily lives, people are turning to it during emergencies as well. We need to utilize these tools, to the best of our abilities, to engage and inform the public, because no matter how much federal, state, and local officials do, we will only be successful if the public is brought in as part of the team.
> —Craig Fugate, Federal Emergency Management Agency, press release, August 2010.

Social media makes it easier to engage with stakeholders surrounding a risk than ever before. Where attendance at public meetings has grown sparse in some cases, stakeholders often have less trouble posting comments online, from home and at times more convenient to them. Indeed, research at the NATO Defense College found that the use of social media can increase transparency and public involvement. The research singled out social networking sites like Facebook because the content is controlled by the user and it is available worldwide (Leiderman 2012).

Engagement can take several forms:

- **Responding to comments on your own content.** Ideally, each comment would receive some sort of response, even if it is merely "thank you for posting." On contentious issues where comments may number in the hundreds, responding to key topics or issues may be necessary for the time available. Information can be clarified; additional sources of information can be provided. Misinformation can be countered respectfully with correct information from credible experts.
- **Asking for the public to provide feedback on particular issues.** Be prepared for a diverse set of opinions. Make sure that people understand how their input will be used. Occasionally, a comment will appear completely off topic, use offensive language, or be highly critical of the organization or message. What will sometimes happen is that the community will self-regulate—that is, outlying comments will result in responses from within the audience in support of the organization or approach.
- **Allowing the audience to post content associated with a particular risk.** Provide visible guidelines on the type of content welcome, review and approve posts to ensure that they meet guidelines, and thank posters for their willingness to help.

Engagement can also mean commenting on other people's content. When commenting on content not your own, remember the following guidelines:

- **Verify your organization's policies.** Is there an "official" commenter for your organization? Does anyone have to review or approve the comment before it is posted? Can employees post using their titles and organizational affiliations?
- **Contact blogs and other sites that seem to offer appropriate information and have a number of followers.** You can generally tell the popularity of a blog by how many comments each post receives or how many subscribers it has, but do not forget the many silent readers. Look also for blogs that are repeated or quoted to determine how often they are being read. Post comments or contact the owner to offer content for future posts.

MONITORING CHANGES IN PERCEPTION VIA SOCIAL MEDIA

Just as social media allows the audience unprecedented control of information, it also offers organizations unprecedented access to their audiences. When the first edition of this book was written, audience analysis required labor and resources for interviews, research on how to contact people and who to contact, and time, sweat, and sometimes tears. Social media provides an easier way to interact with your audience, ask questions, and learn. The Red Cross, for example, uses social media to listen to what people need following disasters. The local Washington DC emergency management agency monitored Facebook to manage risks during the first Obama inauguration.

Monitoring changes in perception can be as simple as noting trends in comments on content over time or as elaborate as using computer-based tools to analyze content and run reports. For example, researchers at Southeastern Louisiana University used Twitter to monitor for flu outbreaks, following the rises in patterns of people complaining of flu-like symptoms. They were able to predict the outbreak faster than the Centers for Disease Control and Prevention (CDC), which relies on clinician reporting for its predictions (Southeastern Louisiana University 2010). Medical researchers in Ottawa reported that Internet searches for listeriosis peaked almost a month before officials announced an outbreak (Wilson and Brownstein 2009). Other examples of care communication tools for monitoring social media include Google's Flu Trends (http://www.google.org/flutrends), which tracks search term use associated with influenza worldwide and shows how searches increase ahead of the number of cases reported; and Health Map (http://www.healthmap.org), which tracks emerging diseases across the globe. Researchers caution, however, that these tools are less effective for areas with limited Internet access (Wilson and Brownstein 2009).

For cases where active monitoring of specific comments is not feasible (for example, during a crisis), tools are being developed that will report on trends in social media content. These trends can help identify where changes are needed in tactics (for example, by opening shelters in emergencies) and whether messages being sent through other methods are reaching the intended audience.

Another way to monitor changes in perception is to see how often key words or concepts are showing up on blogs and how the terms are being used. You can also use a portal site that lists a variety of blogs. For example, Technorati.com searches and organizes millions of blogs and other user-generated online content.

Reviewing blogs frequented by the intended audience for risk communication can also yield insights into how perceptions are changing. The tone of a blog, its popularity, and its responses hint at the level of concern about a particular risk. The number of responses and their tendency to agree or disagree show how segmented your audience might be.

> In today's social media era, bad news circulates minutes after it happens. Organizations no longer have the luxury of waiting for a few hours or days before responding to a crisis.
> —Amiso George, in George and Pratt (2012, p. 33).

The literacy of those responses provides insight into reading level and fluency.

While such monitoring can bring a wealth of information, risk communicators need to be aware of potential pitfalls. One is the issue of privacy. Twitter and Facebook have been credited by many sources as helping fuel the Arab uprisings in 2010 and 2011 (for example, see George and Pratt 2012), but repressive regimes can use social media to spy as well. If your monitoring is obvious to audience members or becomes known to them, they may well ask questions. How much information is being kept? How long will it be kept? How will it be used? Some of these questions may already be answered in your organization's policies and procedures. For example, the Congressional Research Service advises that a privacy impact assessment is required for any revised or new information technology system developed through funding by the federal government. The Department of Homeland Security developed such an assessment for sending information via social networking in 2010 and more generally for social media in 2011 (Lindsay 2011).

Another issue is cost. Tools are available to gather and parse the millions of posts issued each day through microblogging sites like Twitter. Some tools cost money to use. In addition, some social media site owners are beginning to realize that the data they possess can form another income stream and are charging for usage. Look into all aspects before deciding what is best for your organization.

GUIDELINES FOR SPECIFIC TYPES OF SOCIAL MEDIA

Each type of social media can be used to communicate risk information, but different types lend themselves well to specific uses in care, consensus, and crisis communications, as shown in Table 19-1.

Social Networking Sites

Social networking sites remain one of the most heavily used types of social media. As of September 2012, Facebook had recorded more than 937 million users. Approximately 184 million lived in North America. Internet World Stats estimated that 50% of North America, 38% of Australia/Oceania, and 29% of Europe was active on Facebook (http://www.internetworldstats.com).

Users of social media networking sites have pages that list their interests, personal information they care to share, and the ability to send short posts on what they are currently doing (status updates). Individuals generally connect through liking or "friending" each other. Most social media networking sites also allow companies, organizations, and government agencies to host pages, and individuals can friend or like those pages to have access to more information. Social networking sites such as Facebook have numerous pages on care communication issues like drunk driving, child abuse, HIV/AIDS, drug abuse, and depression. As people share opinions through comments on status updates or posts to pages, consensus can be built around a particular issue, supporting consensus communication. Facebook has also been used for emergency communication. For example,

Table 19-1. How social media can be used for various types of risk communication

Type of social media	Type of risk communication		
	Care communication	Consensus communication	Crisis communications
Social networking sites (for example, Facebook, Google Plus)	Provide compelling content about risk Share stories from those at risk Gather feedback on communication messages and protective actions the audience can take Allow those at risk to counsel and support each other Address misinformation and rumors	Gather insight on appropriate risk management and risk communication approaches Provide live feeds to stakeholder involvement meetings Solicit input on locations for meetings	Provide updates on emergency response activities and protective actions the audience can take Counter misinformation Gather information on those at risk
Blogging and podcasts (for example, Blogger, WordPress)	Provide compelling content about risk Share stories from those at risk Train caregivers to provide additional support to those at risk	Gather feedback on risk management approaches Provide recordings of stakeholder involvement meetings	Coordinate plans before emergencies
Microblogging (for example, Twitter)	Point to compelling stories on social networking sites, blogs, or websites Update stakeholders on actions taken Gather feedback on messages or issues	Update stakeholders on involvement activities Solicit comments on risk communication or risk management approaches	Provide updates on emergency response activities and protective actions the audience can take Counter misinformation Gather information on those at risk
File-sharing sites (for example, YouTube, Pinterest)	Share interviews with those at risk to inspire action Share pictures of consequences to encourage taking protective actions Make statistics more meaningful through graphics (see Chapter 14 for more information)	Share video or slides from stakeholder involvement meetings	Share plans before emergencies

(*continued*)

Table 19-1. *Continued*

Type of social media	Type of risk communication		
	Care communication	Consensus communication	Crisis communications
Mapping	Show how risk has changed over time Pinpoint locations for more help in taking protective actions	Solicit input on locations for meetings Show locations of activities on maps Show alternatives to environmental changes	Show spread of risk (for example, fire, flood, or pandemic) Provide directions to protective actions (evacuation routes, emergency shelters) Identify infrastructure or persons at risk
Mobile phones	Develop an application that charts protective actions (for example, diabetes management, weight loss control) Provide updates on research related to the risk	Send reminders of stakeholder involvement meetings Allow voting on issues	Provide updates on emergency response activities and protective actions Gather information on those at risk

the International Atomic Energy Agency (IAEA) shared updates on the Fukushima disasters through its Facebook page (Lindsay 2011).

Books and articles abound on how to effectively use Facebook for various types of communication activities. See the section in this chapter on sharing content for additional guidelines specific to risk communication.

Microblogging

Microblogging shares tiny snippets of information, often 140 characters or less, among users who have chosen to connect as friends, family, professional colleagues, or fans of a particular activity. Information jumps from one set of networked people to another, with the potential to be seen by thousands of people, or very few, depending on how often the specific post is shared. For example, in 2010, residents of the Boulder, Colorado, area used Twitter to keep abreast of the Fourmile Canyon Fire. Content included photos, contact information for volunteer organizations, and requests for prayers.

In general, the key to communicating risk information via microblogging is to share information that will be read and shared by ever-widening groups of people potentially at risk. To determine what exactly made a post memorable enough to share, researchers provided a website that allowed Twitter users to anonymously grade posts (tweets) as worth reading, neutral, or not worth reading. Based on the results, only 36% of tweets were worth reading. Questions were about three times more likely to be judged worth reading compared with "presence maintenance" tweets like "hullo, twitter." Best gambits were questions with a unique hash tag (a way to reach a preset group of people already interested in that type of information), information sharing, self-promotion with content such as links to blog posts, and even random thoughts. Not-worth-reading reasons included being boring, repeating old news, being unclear, having too little content, and using too many hash tags or symbols. Users also did not like whining or hearing about local news when they are not local. Users preferred posts that were personal, honest, transparent, and concise (André et al. 2012). See Chapter 9 for more information on crafting risk communication messages.

Blogging and Podcasts

Blogs are online commentaries that share the opinions of their creators and those who want to respond. They range from personal journals to sophisticated publications with readership levels at or above the level of mainstream media. Blogs cover any topic with which people have deep expertise and a desire to express it.

The U.S. Transportation Security Administration is a good example of an agency not normally thought of as friendly or transparent that, nevertheless, has an effective, moderated blog. The blog receives more than 4000 unique visitors weekly and includes candid conversations on everything from traveler watch lists to the latest security screening technology. It even has an "off-topic-comments" area for bloggers who want to comment more broadly than the scope of the current post topics. Blogs sometimes include podcasts, or podcasts may stand alone on websites. A podcast is a digital audio file, often in mp3 format, made available for download on the Internet via a feed. It often takes the form of a radio show, and many traditional broadcasters have begun to use it as an alternative delivery source for their audio content.

Though podcast publishers, known as podcasters, offer links for direct download or streaming of their audio content, a podcast is distinguished from other digital media by its ability to be syndicated, subscribed to, and downloaded automatically, using a feed-reader application, also known as a podcast aggregator or pod catcher.

On its website, the World Health Organization has dozens of podcasts on health issues worldwide. Users can subscribe to podcasts and send feedback about them. Some podcasts also have transcripts online. Past podcast topics have included health recovery efforts after the 2008 Cyclone Nargis in Myanmar, tuberculosis control, and health impacts of climate change. The CDC has numerous podcasts on health topics, some in Spanish, ranging from college health and safety to emergency preparedness.

How can you use the power of blogs and podcasts to share information?

- Identify existing bloggers who cover your risk and add them to your media list for distribution of press releases and other media information.
- Post comments on blogs, offering risk information and sharing in the debate. Be careful not to point readers overtly to your organization, which is seen as a breach of etiquette and could derail your efforts.
- Monitor other peoples' blogs on your risk issue of interest. When a blog written by senior public health practitioners criticized the World Health Organization's response to pandemic influenza, an official posted a comment in reply, explaining the organization's stance on the issue.
- Start your own blog. Use informal language, offer fun facts, and make sure that you post at least twice a week. Daily is optimal.

See the content sharing section of this chapter for more information.

Video-, Image-, and File-Sharing Sites

A number of social media sites allow users to post their own content for comment by others. YouTube hosts videos; Flickr shares photographs; Pinterest shares photos and graphics; and sites like docstoc and Slideshare.net share presentation materials. Many organizations charged with communicating risk are using these sites to share risk information. For example, eyeSIGHT INTERNATIONAL, an advocacy group that strives to collectively work with leading World Health Organization partners in the effort to prevent avoidable blindness worldwide, has "its story in pins" on the Pinterest site showing its services, facilities, and volunteers to encourage usage and support.

To make the most effective use of these sites for risk communication, consider the following:

- **Be aware of permission requirements.** You may need to get written permission from those you photograph before including the information online. If your organization is not the original photographer, you may also need permission from the photographer to post the picture. Likewise, any music on videos must be in the public domain, have been bought for the purpose, or have permission on file for its use. Pages that do not adhere to permission requirements can be removed by the social media site and in some cases have been sued by the owner of the material.
- **Do not worry about Hollywood quality for video or high resolution for still photographs.** Optimize for the web and make sure that the information is useful to your audience. For example, Army Chief of Staff General George W. Casey, Jr., took a flipcam on his travels and posted his interviews on YouTube in his ChiefCam (Office of the Chief of Public Affairs 2011).
- **Ensure that anything you post would be suitable for your audience.** That means no overly graphic posts for care (for example, car crashes to advocate seatbelt use or avoiding texting while driving) or crisis communications (for example, cadavers in rubble after an earthquake or tornado). It also means that information should be put in context to avoid misunderstandings.
- **Make sure that you retain ownership of what you post.** Some social media sites "own" all posted material and can even resell it elsewhere without asking for additional permission from those who post.

Mapping

Maps have proven especially useful for crisis communications, easily showing the spread of the risk, safe evacuation routes, and shelter locations. However, university researchers in geography and disaster management in Germany found that maps could be used to improve risk perception, help people more coherently frame risks, and establish greater credibility for the communicating organization (Dransch et al. 2010).

Maps can be created using many online tools or through partnerships with online map developers such as Google. For example, CrisisCommons (http://crisiscommons.org/) describes itself as a global community of volunteers from technology, crisis response organizations, and government agencies, along with citizens working together to build and use technology tools to help respond to disasters and improve resiliency and response before a crisis. The movement was active in using mapping technology to help support response efforts during the 2010 earthquake in Haiti.

The key, once again, is interaction. Provide ways for users to drill down for more information, enlarge sections of most interest to them, and perhaps even add information to improve the representation. See Chapter 14 for more information on using visuals to communicate risk information.

Mobile Phones

Mobile phones have become an important way to communicate risk information, particularly in crisis situations. According to the International Telecommunication Union, at the end of 2011, there were nearly 6 million mobile phone subscriptions worldwide, corresponding to a global penetration of 86% (http://www.itu.int/ITU-D/ict/). The Pew Internet and American Life Project estimates that, as of March 2012, 46% of American adults own smart phones. The research also indicated that two-thirds of American adults use their cell phones to find information exactly the moment they need it, whether through calls to services or through Internet access. Nearly 20% of American adults report using their cell phones to get help in emergencies (Zickuhr 2012).

Mobile phones are being used as conduits to share risk information, for care, consensus, and crisis communication. For example, smart phone applications allow users to track calorie intake to manage diabetes or weight loss, create an emergency response plan, and monitor air quality near them. In addition, in 2012, the Federal Emergency Management Agency partnered with the Federal Communications Commission and wireless carriers to send geographically targeted text-like alerts to the members of the public who agreed to participate in the system. The alerts relay president messages, AMBER alerts, and imminent threat alerts to mobile phones using a broadcast technology that will not get backlogged during emergencies when wireless voice and data services can become congested.

To share information over mobile phones, keep information short and optimized for a smaller screen. This may include having a separate website designed for mobile phones and updated frequently. You can also partner with software firms to develop your own application for risk information.

EVALUATING SOCIAL MEDIA EFFECTIVENESS

Evaluation is often the most overlooked aspect of risk communication efforts, and social media is no exception. Chapter 20 discusses evaluation of risk communication efforts in general. For social media, however, many of the traditional metrics no longer make sense. For example, research at the University of California Riverside Center for Internet Retailing suggests three key areas to consider for social media evaluation: brand awareness, brand engagement, and word of mouth (Hoffman and Fodor 2011).

For care communication, measurements might include the number of unique visits to a blog, the number of times your agency is associated with the issue on a microblogging site or social networking site, and the number of followers you gain on similar sites. For consensus communications, you might measure the number of comments received from users, the diversity of those users, and whether comments tend to converge on an opinion. For crisis communications, the number of individuals reached and the actions they reported taking may be important factors.

Others have found that tracking how far a message spreads is a more valuable metric for the engagement over social media. For example, how many times is a message retweeted on Twitter or how many links does a blog post receive on other blogs or websites? Also important is the growth over time. Has your following increased over a month, a quarter, a year? These types of performance metrics can help determine the effectiveness of your social media work and determine whether the time and cost invested have been worthwhile.

CHECKLIST FOR SOCIAL MEDIA

☐ Audience access to and usage of social media have been considered in choosing how to engage.
☐ Staff have been identified with time and ability to engage the audience using social media.
☐ The organization has guidelines for its use of social media, and such use is consistent with industry regulations.
☐ Social media use is part of a wider risk communication plan.

When sharing content:

☐ Sites share consistent information that is attractively presented.
☐ Content consists of information the audience cares about.
☐ Content is easy for the audience to share.
☐ Mechanisms are in place to screen comments and determine appropriate people to accept as "friends."

When engaging with stakeholders:

☐ Guidelines are in place to deal with misinformation or difficult users as well as users posting content.
☐ Organizational policies for engagement have been followed.

When monitoring changes in perception:

☐ An approach has been agreed upon including how to ensure privacy for individuals.
☐ Information gained is being used to inform further risk communication efforts.
☐ The type of social media most effective for the risk communication effort is being used, and best practices are being followed.

REFERENCES

André, P., M. S. Bernstein, and K. Luther. 2012. "Who Gives a Tweet? Evaluating Microblog Content Value." Presented at the Association for Computing Machinery Conference on Computer Supported Cooperative Work, February 11–15, 2012, Seattle, Washington.

Center for Biosecurity. 2012. *After Fukushima: Managing the Consequences of a Radiological Release.* University of Pittsburgh Medical Center, Baltimore, Maryland.

Chew, C. and G. Eysenbach. 2010. "Pandemics in the Age of Twitter: Content Analysis of Tweets during the 2009 H1N1 Outbreak." *PLoS One,* 5(11):e14118.

Dransch, D., H. Rotzoll, and K. Poser. 2010. "The Contribution of Maps to the Challenges of Risk Communication to the Public." *International Journal of Digital Earth*, 3(3):292–311.

George, A. M. and C. B. Pratt, eds. 2012. *Case Studies in Crisis Communications: International Perspectives on Hits and Misses*. Routledge, New York.

Hoffman, D. L. and M. Fodor. 2011. "Can You Measure the ROI of Your Social Media Marketing?" In K. Partridge, ed., *Social Networking*. The Reference Shelf, The H. W. Wilson Company, New York, pp. 62–74.

ITU (International Telecommunications Union). 2012. *World Telecommunication/ICT Indicators Database*. http://www.itu.int/ITU-D/ict/ (accessed January 28, 2013).

Lefebvre, R. C. 2007. "New Technology: The Consumer as a Participant rather than Target Audience." *Social Marketing Quarterly*, 13(3):31–42.

Leiderman, L. 2012. "Policy Making in 140 Characters of Less: NATO and Social Media." Research Paper No. 77, NATO Defense College, Rome.

Lesperance, A. M., M. A. Godinez, and J. R. Olson. 2010. *Social Networking for Emergency Management and Public Safety*. PNNL-19601, Pacific Northwest National Laboratory, Richland, Washington.

Li, C. and J., Bernoff. 2008. *Groundswell: Winning in a World Transformed by Social Technologies*. Harvard Business Press, Boston, Massachusetts.

Lindsay, B. R. 2011. Social Media and Disasters: Current Uses, Future Options, and Policy Considerations. 7-5700, Congressional Research Service, Washington, DC.

Office of the Chief of Public Affairs. 2011. *U.S. Army Social Media Handbook*. U.S. Army, Online and Social Media Division, Office of the Chief of Public Affairs, Washington, DC.

Smith, A. 2011. "Why Americans Use Social Media." Pew Internet and American Life Project. http://pewinternet.org/Reports/2011/Why-Americans-Use-Social-Media.aspx (accessed January 28, 2013).

Southeastern Louisiana University. 2010. "Twitter Used to Predict Flu Outbreaks." Science Daily at http://www.sciencedaily.com (accessed January 22, 2013).

Tinker, T. 2009. *Special Report—Expert Round Table on Social Media and Risk Communication during Times of Crisis: Strategic Challenges and Opportunities*. American Public Health Association, Washington, DC.

Wilson, K. and J. S. Brownstein. 2009. "Early Detection of Disease Outbreaks Using the Internet." *Canadian Medical Association Journal*, 180(8):829–831.

Zickuhr, K. 2012. "Three-Quarters of Smartphone Owners Use Location-Based Services." Pew Internet and American Life Project. http://pewinternet.org/Reports/2012/Location-based-services.aspx (accessed January 28, 2013).

ADDITIONAL RESOURCES

Agency for Health Research and Quality. *Social Media Standards and Policy (Example)*. http://www.ahrq.gov/news/socmedia/socstandpol.htm (accessed January 23, 2013).

American Red Cross. 2010. *Social Media in Disasters and Emergencies*. http://www.redcross.org (accessed January 23, 2013).

Centers for Disease Control and Prevention. 2011. *The Health Communicators Social Media Toolkit*. http://www.cdc.gov/socialmedia/Tools/guidelines/pdf/SocialMediaToolkit_BM.pdf (accessed January 23, 2013).

Centers for Disease Control and Prevention. *CDC Social Media Tools, Guidelines & Best Practices*. http://www.cdc.gov/SocialMedia/Tools/guidelines/ (accessed January 23, 2013).

Department of Health and Human Services. *HHS Center for New Media*. http://newmedia.hhs.gov/ (accessed January 23, 2013).

IV

EVALUATING RISK
COMMUNICATION EFFORTS

Every risk communication effort can benefit from being evaluated. Evaluation can help a current effort achieve its purpose and objectives and make future efforts more successful.

> From a public health perspective, one should be no more willing to expose the public to an untested message than to an untested drug.
> —Baruch Fischhoff (1989, p. 115).

20

EVALUATION OF RISK COMMUNICATION EFFORTS

Every risk communication effort should undergo some sort of evaluation. Whenever possible, evaluations should be conducted during as well as at the end of a project; the former tells you the changes to be made to reach your objectives; the latter tells you what you should change in future efforts. Evaluation can also be conducted in the middle of an effort to provide midcourse corrections. However, the thoroughness and timing of the evaluation will depend on your objectives and purpose (for example, evaluation of crisis communication may have to wait until the worst is over), funding and resources, and organizational constraints.

WHY EVALUATE RISK COMMUNICATION EFFORTS?

Evaluating risk communication efforts takes time and resources. Given the fact that both are limited for many risk communication efforts, why should you bother with evaluation?

Information from the evaluation can be used to refine risk communication policies, procedures, and practices. Most organizations involved in risk communication efforts communicate risk more than once. Information gained from one effort can be applied to strengthen future efforts.

Evaluation can also serve to prove that laws are being followed. As noted in Chapter 3, many risk communication efforts are in some way responses to a law or regulation. Showing a regulatory agency that you have evaluated your efforts can help to prove that you are complying with both the letter and the spirit of the law.

Risk Communication: A Handbook for Communicating Environmental, Safety, and Health Risks,
Fifth Edition. Regina E. Lundgren and Andrea H. McMakin.
© 2013 The Institute of Electrical and Electronics Engineers, Inc., and Regina E. Lundgren and
Andrea H. McMakin. Published 2013 by John Wiley & Sons, Inc.

> Because few risk managers really want to be shown how badly they are doing, evaluations are often more readily conducted to illustrate good performance than in situations where the results will highlight intractable problems.
> —Peter Bennett and Sir Kenneth Calman (1999, p. 257).

In addition, evaluation can serve to prove to your organization's management that risk communication efforts are valuable. If results show that efforts have met organizational goals, management will be more likely to continue funding. If, on the other hand, the results show that efforts have failed to communicate risk, the information gained in the evaluation should identify where improvements are needed.

Carnegie-Mellon researcher Granger Morgan and his colleagues offer several arguments for using the mental models approach to risk communication (see Chapter 2 for more information), but these arguments could easily apply to any risk communication evaluation effort:

1. "You'd never design a new product on the basis of an engineer's best guess. You'd insist on careful empirical design and testing. The same standard should apply to risk communication."
2. "Why balk at spending an amount of money on getting the message right that is a tiny fraction of the stakes riding on correct public understanding?"
3. "We wouldn't release a new drug without adequate testing. Considering the potential health (and economic) consequences of misunderstanding risks, we should be equally loath to release a new risk communication without knowing its impact" (Morgan et al. 2002, p. 180).

THE MEANING OF SUCCESS

In recent years, evaluations have come under fire as not yielding enough information to improve risk communication activities. Although crafting the evaluation is critically important to its success, certain factors about the risk communication effort make that effort easier to evaluate. Edwin Zedlewski, previously the Acting Deputy Assistant Director for Research and Evaluation for the U.S. National Institute of Justice, helped develop the Evaluability Assessment as a way to identify programs that are likely to yield evaluations that maximize the organization's return on investment. By adopting this approach, organizations at all levels can save considerable time and money.

The approach starts by determining which programs can sustain a rigorous outcome evaluation. The determination takes 1–5 days and is guided by questions such as the following:

- Are program components stable as opposed to still evolving?
- Can logical and plausible connections be traced between a program's activities and its intended outcomes?
- Are there enough cases or observations to permit statistically robust conclusions?
- Can the program's effects be isolated from other related forces operating in the community?

If answers to all four questions are yes, then the risk communication program may be a good candidate for a thorough evaluation (Zedlewski 2006).

The question remains, however, as to how a risk communication program measures its activities. How can you tell whether risk communication efforts have been successful. Success for risk communication efforts relates back to your risk communication plan: Did you meet your objectives? Assuming that you chose the right objectives to begin with, if you met those objectives, you succeeded.

Additional evaluation factors will give you useful information to refine future risk communication efforts, particularly when using the methods of information materials, face-to-face communication, visual representation of risk, mass media, and technology-assisted applications.

David Dozier and colleagues at the University of Maryland, who studied excellence in communications and public relations programs, suggest that, at a minimum, evaluation should measure awareness, knowledge, opinions, and behaviors of the target audience before and after the risk communication program was implemented (Dozier et al. 1995). They also encourage the evaluation of outcomes, not process.

Risk communication experts Weinstein and Sandman (1993) recommend that the following be considered to measure the success of risk communication efforts:

- Does the audience understand the content of the communication?
- Does the audience agree with the recommendation or interpretation contained in the message?
- Do people facing a higher level of the risk perceive the risk as greater or show a greater readiness to take action than people exposed to a lower level of the risk?
- Do audience members facing the same level of risk tend to have the same responses to this risk?
- Does the audience find the message helpful, accurate, and clear?

When it comes to technology-assisted communication, some measurements include presence, influence, and audience reach. Other measurements of social media that are applicable to care communication, suggested by practitioner Craig Lefebvre, include the following:

> Success for risk communication efforts relates back to your risk communication plan: Did you meet your objectives? If you met your objectives, you succeeded.

- **Education.** Are you educating people about issues and problems that are relevant to them?
- **Engagement.** Are you engaging people in positive and meaningful ways?
- **Entertainment.** Is there an entertainment value to your offerings?
- **Empowerment.** Do people believe and feel empowered as a result of their experiences with your programs and services?
- **"Evangelism."** Do you take advantage of opportunities to let your customers and audiences become advocates of your programs in their discussions with others?

For stakeholder participation methods, another set of factors must be considered. Early work at Tufts University suggested the following factors (Rosenbaum 1978):

- **Accessibility.** Did the risk communication effort increase the audience's opportunities to obtain relevant information, air views before decision makers, and hold officials accountable?

- **Fairness.** Were all views given equal consideration in the decision-making process?
- **Responsiveness.** Did the risk communication effort foster recognition of public views on decisions?

More recent work suggests that evaluating stakeholder involvement efforts may be far more difficult. What appears to be success from one side may be abject failure from the other. For example, if litigation was avoided, some organizations may mark the effort as a success, whereas stakeholders leave the tables feeling disenfranchised and still in the dark as to the risks they face. For example, risk communication luminaries Susan Santos of Focus Group and Caron Chess of Rutgers University used two different approaches to evaluate citizen boards advising the Department of Defense on environmental cleanup issues. The more theoretical approach, which considered some of the issues addressed earlier in this chapter (such as fairness), would have ranked the efforts rather low, whereas, using stakeholder perceptions, the efforts were deemed successful from both a process and an outcome point of view (Santos and Chess 2003).

Other factors to consider depend on your particular situation, including resources, organizational requirements, and audience needs. Did you accomplish the most with the funding and resources available? Would a significant change have been made with additional funding or more staff? Did the fact that all materials had to be approved by seven layers of management delay production or lessen the organization's credibility with the audience? Did a recent election affect the way your audience views a particular risk, and should you have predicted that and planned for it? Table 20-1 lists possible additional factors for care, consensus, and crisis communication.

TYPES OF EVALUATIONS

Once you have decided to evaluate the risk communication effort, you must determine what type of evaluation you need. Kasperson and Palmlund (1989) developed a set of factors to consider when determining how to evaluate risk communication efforts,

Table 20-1. Evaluation factors to consider for care, consensus, and crisis communication

Care communication	Consensus communication	Crisis communication
Did the audience change to less risky behavior?	Were all segments of the audience represented in building the consensus?	Have all members of the audience been alerted to the risk?
How long did the behavior change last?	Does the audience understand enough about the risk to make decisions?	Does the audience understand enough about the risk to make decisions?
Have all members of the audience been alerted to the risk?	Was a consensus reached about the decision?	Did the audience change to less risky behavior?
Does the audience understand the risk well enough to make decisions?	Can the decision be implemented?	Was consistent information given regarding the risk?

including the objectives of the evaluation, the choice of evaluators, timing, training and monitoring of evaluators, how the audience is to be involved, boundaries of the evaluation, and how success is measured.

First, determine the objectives of the evaluation effort. Why are you conducting the evaluation? Possible reasons include determining how the current effort is going so that you can revise it, determining what to improve in future efforts, demonstrating to management the results of your program, and proving compliance with regulations. These reasons apply to care, consensus, and crisis communication.

Another factor is who conducts the evaluation. Who will evaluate your efforts? Possible evaluators include those involved in the risk communication efforts, upper management, an outside organization that specializes in such evaluations, and your audience. Which you choose will depend on organizational requirements and the situation. Table 20-2 lists the advantages and disadvantages of these evaluators.

Another factor to consider is timing. When should the evaluation be conducted to best get the information you want? Choices include before the effort begins (that is, evaluating the plan), during the effort, and after the effort. Evaluations conducted during the effort should be timed so that there has been enough activity conducted to provide the information needed and so that no activity will be compromised by rescheduling staff time and resources for the evaluation. Evaluations conducted after the effort should generally be conducted soon enough so that facts are still clear in people's minds. Evaluations may be conducted later if retention is a factor to be evaluated.

Training of evaluators should also be considered. Do the evaluators know what to look for, and are they capable of getting the kind of information you need? If you are using inexperienced evaluators, have someone experienced in risk communication programs train them in what to look for and how to go about it. Even if you are using experienced evaluators, you will need to acquaint them with your situation. One project manager with whom we worked always had his risk communication messages evaluated by a team of

Table 20-2. Advantages and disadvantages of using various evaluators

Evaluator	Advantages	Disadvantages
Risk communication staff	Staff intimate with program and risk communication practices; approvals generally not needed; staff trusted by organization	May lack credibility with regulating agencies or audience; may have difficulty being objective
Upper management	Positive interactions can increase support; less costly than outside evaluations; staff intimate with organization's constraints	Negative interactions can decrease support; may lack credibility with regulatory agencies or audience
Outside organization	Staff intimate with risk communication practices; may have more credibility with regulators and audience; may be more objective	May be more costly than in-house evaluations; may not understand organization's constraints
Audience	Interactions can increase support; audience understands own needs best; highly credible to audience	May be difficult to get approvals; does not understand organization's constraints or risk communication practices

technical experts to make sure that the technical information was correct. With them, he also sent descriptions of the laws he was trying to comply with, his expected audience, and other background information along with the message so that the evaluators could judge the effort in context. He found that this helped to eliminate suggestions and comments that were unreasonable given his situation.

You will also want to monitor the work of the evaluators to ensure that they are collecting their information without hindering your efforts and that you receive their input in time to use it.

Another consideration is how the audience is to be involved. How and when should the audience be involved in the evaluation effort? The audience can be involved in several ways. They can serve as the evaluators themselves, particularly if you have an interactive advisory committee or focus group. They can also serve as the research subjects of a questionnaire or survey.

Also consider the boundaries of the evaluation. Boundaries to consider are access to proprietary data (especially if an outside group will be conducting the evaluation) and access to the audience. Organizational requirements should be considered. Setting a time limit for the evaluation may also be necessary.

> Who will evaluate your efforts? Possible evaluators include those involved in the risk communication efforts, upper management, an outside organization that specializes in such evaluations, and your audience.

The final factor to consider is how the success of the risk communication effort should be judged. How will the evaluators determine whether your efforts have been successful? As noted earlier in this chapter, they should evaluate your efforts based on your plan. Did you meet your objectives? If they are evaluating the plan itself, they should evaluate it relative to your reasoning for its content.

Based on these factors and your ultimate purpose and objectives, you can design an evaluation effort that will bring you the information you need.

CONDUCTING THE EVALUATION

Little literature has been devoted to evaluation in risk communication as opposed to other aspects of the field (examples include Kasperson and Palmlund 1989; Regan and Desvousges 1990; Santos and Chess 2003; and Weinstein and Sandman 1993). However, Julie Downs, Carnegie-Mellon University, recommends three types of evaluation for what we call care communication (Fischhoff et al. 2011). The first, formative evaluation, is part of the planning process. During this stage, communicators conduct research to understand the target audiences and how they view the risk. Focus groups (discussed in this book's Chapter 17), interviews (Chapter 15), and the mental models approach (Chapter 9) are useful tools in this stage. This firm understanding of the audiences and their likely responses form the basis for the communication plan and information materials.

The second type of evaluation is a process evaluation, meaning assessing how well the risk communication program is carried out, ensuring that each step has maximum impact. This approach is particularly important when many people are involved in carrying out the plan so that its implementation is as consistent as possible. A process evaluation can help explain why a communication effort had the effects that it did, and let others know what to expect if they follow a similar strategy.

The third type of evaluation is assessing the outcome—the extent to which the communication effort achieved your purposes and objectives. Because your purposes and objectives will most likely involve how your audience will react to your efforts (changing behavior, gaining awareness), your evaluators must consider your audience's reactions to determine whether your efforts have been successful.

> Because your purposes and objectives will most likely involve how your audience will react to your efforts, your evaluators must consider your audience's reactions to determine whether your efforts have been successful.

The best ways to determine audience reactions are surveys and interviews. Surveys may be conducted by mail, phone, electronic mail, or in person, depending on the situation. Consider your resources and the availability and disposition of your audience. For example, David Chrislip and Carl Larson, in their research that led to their 1994 book, *Collaborative Leadership: How Citizens and Civic Leaders Can Make a Difference*, developed a measurement device that could be used for consensus communication activities. The survey, which covers five dimensions (context of the effort, design of the effort, stakeholder skills and attitudes, consensus process, and results), asks respondents to mark statements as true, more true than false, more false than true, or false. The instrument has been shown to accurately correlate with success of the effort, as judged by comparisons with results of other instruments (Chrislip and Larson 1994).

Interviews can be conducted separately with representative members of your audience or in focus groups. Occasionally, however, time or organizational constraints make it impossible to use surveys or interviews. For example, some government organizations must receive permission from the federal Office of Management and Budget before conducting any survey of more than 10 people. Alternative methods to evaluate audience reactions to risk communication efforts include the following:

- **Reviewing risk communication plans.** Evaluators can look at your plan and evaluate it against the audience's, organizational, and regulatory requirements. For example, regulators often evaluate an organization's community relations plan for Superfund sites to ensure that the community will be kept informed of and involved in activities.
- **Reviewing specific messages for content.** If the evaluators are sufficiently schooled in the theories, principles, and practices of risk communication, they may be able to point to problems by reviewing the information you are disseminating. For example, J. Harrison Carpenter of Michigan Technological University found that the way technical terms are defined in text can result in potential manipulation of the content (either to paint too rosy a picture or to create fear). He goes on to present a possible tool for classifying terms and their definitions that might be used to test risk communication messages to ensure appropriate presentation of information (Carpenter 1997).
- **Reviewing entire efforts for such factors as continuity of content, timing, and follow-through.** How the risk is communicated is often as important as the content of the risk messages themselves. Evaluators may be able to determine patterns that should be changed or maintained. For example, evaluators might review the timing, scope, and content of health care campaign messages to ensure appropriate coverage across time and audiences.

Researchers for the U.S. Nuclear Regulatory Commission offer three low-cost ways to evaluate efforts: (1) reading the local newspaper to see if coverage matches key messages and facts, (2) having a colleague observe interactions with the audience and see how the audience responds, and (3) asking the audience for feedback at the end of meetings (Persensky et al. 2004).

Regardless of which technique is employed, evaluations are generally conducted in a similar manner. Michael Regan and William Desvousges (1990), in their risk communication evaluation handbook for the U.S. Environmental Protection Agency, suggest five steps to this evaluation process:

1. **Clarify the risk communication goals and objectives.** You must know what you are trying to accomplish before you can evaluate how well you did it. See Chapter 7 of this book for additional information on setting the purpose and objectives.
2. **Determine information needs for evaluation.** What kinds of information will you need to prove that you are meeting your objectives? For example, if your objective is to raise awareness of an issue, you might want to conduct a survey at the beginning of your risk communication effort to understand the current level of awareness among your intended audience. After conducting the risk communication effort, you would do another survey to see how awareness had changed.
3. **Collect the information.** As part of your risk communication effort, you would gather the information you had identified as needed for evaluation purposes.
4. **Analyze the data.** You would then look at the information you had gathered to identify any trends or difficulties.
5. **Draw conclusions.** You would determine what might be changed to enhance any positive trends or to resolve any difficulties.

They further suggest that this activity be included in the original risk communication plan to ensure that timing, resources, and information gathered are appropriate.

See Chapter 19 for information on evaluating social media efforts, and see Chapter 23 for evaluating public health campaigns.

CHECKLIST FOR EVALUATING RISK COMMUNICATION EFFORTS

☐ What constitutes success in my risk communication efforts has been determined.

The following were determined before the evaluation:

☐ Evaluation objectives
☐ Evaluators
☐ Timing
☐ Training of evaluators
☐ Monitoring of evaluators
☐ Audience involvement
☐ Possible boundaries
☐ Judgment of success

My purpose in evaluation is:

- ☐ Refining practices
- ☐ Determining whether efforts should continue
- ☐ Proving compliance with the law

To evaluate the risk communication efforts, evaluators will:

- ☐ Conduct audience interviews
- ☐ Survey the audience
- ☐ Review risk communication plans
- ☐ Review specific messages for content
- ☐ Review the effort for continuity of content, timing, and follow-through

Evaluators will follow this process for the evaluation:

- ☐ Clarify the risk communication goals and objectives
- ☐ Determine information needs for evaluation
- ☐ Collect the information
- ☐ Analyze the data
- ☐ Draw conclusions
- ☐ Share results with the risk communication team

REFERENCES

Bennett, P. and K. Calman. 1999. *Risk Communication and Public Health.* Oxford University Press, New York.

Carpenter, J. H. 1997. "Define and Conquer: Technical Definitions and the Rhetoric of Risk Communication." Proceedings of the Fourth Biennial Conference on Communication and Environment, State University of New York-Syracuse, New York.

Chrislip, D. D. and C. E. Larson. 1994. *Collaborative Leadership: How Citizens and Civic Leaders Can Make a Difference.* Jossey-Bass, San Francisco, California.

Dozier, D. M., L. A. Grunig, and J. E. Grunig. 1995. *Manager's Guide to Excellence in Public Relations and Communication Management.* Lawrence Erlbaum Associates, Mahwah, New Jersey.

Fischhoff, B. 1989. "Helping the Public Make Health Risk Decisions." V. T. Covello, D. B. McCallum, and M. T. Pavlova, eds., *Effective Risk Communication: The Role and Responsibility of Government and Nongovernment Organizations.* Plenum Press, New York, pp. 111–116.

Fischhoff, B., N. T. Brewer, and J. S. Downs eds. 2011. *Communicating Risks and Benefits: An Evidence-Based User's Guide.* Chapter 3: "Evaluation." U.S. Department of Health and Human Services, Food and Drug Administration, Washington, DC.

Kasperson, R. E. and I. Palmlund. 1989. "Evaluating Risk Communications." V. T. Covello, D. B. McCallum, and M. T. Pavlova, eds., *Effective Risk Communication: The Role and Responsibility of Government and Nongovernment Organizations.* Plenum Press, New York, pp. 143–158.

Morgan, M. G., B. Fischhoff, A. Bostrom, and C. J. Atman. 2002. *Risk Communication: A Mental Models Approach.* Cambridge University Press, New York.

Persensky, J., S. Browde, A. Szabo, L. Peterson, E. Specht, and E. Wright. 2004. *Effective Risk Communication: The Nuclear Regulatory Commission's Guidelines for External Risk Communication.* NUREG/BR-0308, U.S. Nuclear Regulatory Commission, Washington, D.C.

Regan, M. J. and W. H. Desvousges. 1990. *Communicating Environmental Risks: A Guide to Practical Evaluations*. U.S. Environmental Protection Agency, Washington, DC. EPA 230-01-91-001.

Rosenbaum, N. 1978. "Evaluating Citizen Involvement Programs." S. Langton, ed., *Citizen Participation Perspectives*. Lincoln Filene Center for Citizenship and Public Affairs, Tufts University, Medford, Massachusetts, pp. 82–86.

Santos, S. L. and C. Chess. 2003. "Evaluating Citizen Advisory Boards: The Importance of Theory and Participant-Based Criteria and Practical Implications." *Risk Analysis*, 23(2):269–280.

Weinstein, N. D. and P. M. Sandman. 1993. "Some Criteria for Evaluating Risk Messages." *Risk Analysis*, 13(1):103–114.

Zedlewski, E. 2006. "Maximizing Your Evaluation Dollars." *National Institute of Justice Journal*, (254), July 2006.

ADDITIONAL RESOURCES

Desvousges, W. H. 1991. "Integrating Evaluation: A Seven-Step Process." A. Fisher, M. Pavlova, and V. Covello, eds., *Evaluation and Effective Risk Communications Workshop Proceedings*. U.S. Environmental Protection Agency, Washington, DC, pp. 119–123. EPA/600/9-90/054.

Desvousges, W. H. and V. K. Smith. 1988. "Focus Groups and Risk Communication: The 'Science' of Listening to Data." *Risk Analysis*, 8(4):479–484.

Kline, M., C. Chess, and P. Sandman. 1989. *Evaluating Risk Communication Programs: A Catalog of "Quick and Easy" Feedback Methods*. Rutgers University, Cook College, Environmental Communication Research Program, New Brunswick, New Jersey.

Santos, S. L. 1990. "Developing a Risk Communication Strategy." *Management and Operations*, November:45–49.

Smith, V. K., W. H. Desvousges, A. Fisher, and F. R. Johnson. 1987. *Communicating Radon Risk Effectively: A Mid-Course Evaluation*. U.S. Environmental Protection Agency, Office of Policy Analysis, Washington, DC.

V

SPECIAL CASES IN RISK COMMUNICATION

Since the initial publication of this book, certain situations have challenged risk communicators, even beyond the normal challenges of communicating environmental, safety, and health risks to audiences who may be apathetic (in care communication), frustrated (in consensus communication), or fearful (in crisis communication).

Two of the most challenging cases for risk communicators are unexpected emergencies and international communication. The terrorist attacks of September 11, 2001, in the United States, and the subsequent public health emergencies like anthrax and mad cow disease, have shown that emergency risk communication requires different strategies and tactics to be successful. In addition, the globalization of commerce, communication, and crises has made it critical that risk communicators understand the international nature of risks and how to communicate across cultures. In addition, public health campaigns, while a mainstay of risk communication efforts, must adapt to both crises and the need to reach international audiences.

In emergency risk communication, it's your job to provide the public with information that allows them to make the best possible decision within nearly impossible time constraints.
—Linda Sokler, Managing Regional Director of American Institutes for Research, Prospect Center, 2002.

21

EMERGENCY RISK COMMUNICATION

This book uses a paradigm that divides risk communication into care, consensus, and crisis communication. That paradigm is based on a number of factors such as agreement on the magnitude of the risk between scientific experts and those at risk, the level and type of involvement by audiences or participants, and the urgency of the risk. In this chapter, we make a distinction between a crisis and an emergency. A traditional definition of a crisis is a turning point that will decisively determine an outcome. Medical practitioners once spoke of a crisis as a critical point in a disease. Thus, a crisis follows a process, even if it is an undesirable one, and is not altogether unexpected. An emergency, on the other hand, has traditionally been defined as a sudden or unforeseen situation that requires immediate action. Emergencies are random, they often do not follow predictable processes, and they are unexpected. By these definitions, the sudden rupture of an underground oil tank that has been leaking for some time would be a crisis, whereas a bioterrorist attack would be an emergency. Other recent examples of emergencies include disease outbreaks (pandemic flu; botulism; severe acute respiratory syndrome or SARS; West Nile virus), natural disasters, and terrorism-related events.

Risk communicators have long dealt with crises such as industrial accidents (hazardous releases internal to a facility, environmental releases) and routine disease outbreaks (flu seasons, measles). Indeed, crisis communication, as our paradigm

> A crisis follows a process. An emergency is a sudden or unforeseen situation that requires immediate action. The sudden rupture of an underground oil tank that has been leaking would be a crisis, whereas a bioterrorist attack would be an emergency.

Risk Communication: A Handbook for Communicating Environmental, Safety, and Health Risks,
Fifth Edition. Regina E. Lundgren and Andrea H. McMakin.
© 2013 The Institute of Electrical and Electronics Engineers, Inc., and Regina E. Lundgren and
Andrea H. McMakin. Published 2013 by John Wiley & Sons, Inc.

implies, is a distinct branch of risk communication, with its own strategies and tactics. What is different about risk communication in an emergency? At times, communication principles, strategies, and tactics in emergencies fit within the boundaries of crisis, care, and even consensus communication. Nevertheless, those communicating risk must understand the unique characteristics of emergency risk communication so they can plan for the unexpected and communicate during and after an emergency.

UNDERSTANDING EMERGENCY RISK COMMUNICATION

Emergencies have their own unique characteristics that affect how risk communication is put into practice. Table 21-1 shows some of these characteristics. The following text explains these characteristics in more detail.

Table 21-1. Characteristics of emergency risk communication

What is different	The result	Practices
Purpose	Communicators must explain, put in context, correct misperceptions, give options for action, empower people to make decisions, move people to recovery, and help them attain a new level of readiness.	Use principles of care and crisis communication as appropriate.
Sense of urgency, rapid rate of change	Decisions must be made within a narrow time frame, with an uncertain outcome, to reduce risks that are still unknown and to rapidly recover from an event that is still unfolding.	Recognize that communication may be confusing, contradictory, and subject to change as the event evolves. Preplanning can reduce confusion.
Disrupted logistics	Normal or preplanned communication channels and actions may not be available, such as electrical power, cell phones, Internet connections, and transportation to scenes.	Use preplanning and be flexible during the event to identify alternatives.
Potential for large numbers of ill or injured people across wide jurisdictions	Lines of authority cross for responders, family and friends demand immediate information, and the health care infrastructure can be severely stressed.	Team with a wide variety of agencies and organizations and look for creative communication alternatives.
Intense media attention	Reporters are seeking and reporting information nonstop.	Designate and train spokespeople, but prepare others to speak as well.
Emotional response	People may experience a range of strong emotions including fear, anger, panic, denial, laying blame, solidarity with others, desire to help victims, and need for personal control. All these may affect how people respond to a risk.	Develop and implement communications that account for these responses, including giving people reasonable and appropriate actions to take.

Table 21-1. *Continued*

What is different	The result	Practices
Incomplete or unknown information	Misunderstanding facts about the risk may affect people's response. Uncertainty may increase fear and panic.	Address factual misperceptions in planning and during an emergency. Explain what is known at the time and what is unknown, saying that it is preliminary. Say what you are doing to find out more. Concede errors and modify previous statements as more information comes to light.
Involvement of multiple organizations, sometimes with competing agendas, including possible documentation for law enforcement investigation.	Sources may issue conflicting information, leading to confusion and reduced credibility.	Get buy-in of organizations during emergency planning. Understand agency roles, jurisdictions, and preservation of evidence/criminal/regulatory follow-up. Draw clear lines of authority and responsibility, and make sure everyone understands the roles.
Security and privacy issues	Some information, such as victims' names, cannot be released.	Explain the kind of information that cannot be released and why. Say whether it will be released later and under what circumstances.
Backlash	After the emergency is over, people may seek those to blame.	Evaluate deficiencies. Take responsibility for things that were your organization's fault. Explain what is being done differently now.

Emotions and Public Actions

People's reactions during an emergency can be intense and complex. Fear may prompt a debilitating response, with people acting in extreme and sometimes irrational ways to avoid a perceived or real threat. Although research shows that in some natural disasters people draw together and ordinary folks become heroes, some emergencies like bioterrorism attacks may engender more panicked responses because the agent is unseen and at first unknown. For example, after the sarin attacks in the Tokyo subway, thousands of worried people who thought that they might be affected overwhelmed the medical system for a short time (Green et al. 2007).

On the other hand, people may feel apathy or hopelessness, that nothing they do will help. They may experience denial, leading to avoiding warnings, not believing the threat is real, or not believing it applies to them. Psychological factors associated with hurricane warnings are a good example (Revkin 2011). Some people ignore mandatory evacuation orders because they feel that staying to protect their property and belongings is more important than any personal harm they may suffer. Those who have lived through previous hurricanes may have a false sense of familiarity that leads to taking new warnings less seriously: "I've lived through these before." And others have a false sense of control—the feeling that they can get out safely if conditions become dangerous, sandbag their house against flooding, and so on. Attitudes such as these can result in not taking action until the last minute or until it is too late.

Despite these negative responses, a national survey after the September 11, 2001, attacks in the United States showed that people want honest and accurate information about terror-related situations. They want this information even if it makes them worried, angry, or fearful. They want the facts, want to know whom to blame, and want to feel solidarity with fellow citizens (Fischhoff 2002).

> Americans do not want leaders to hide their own response to tragedies. Risk communication consultant Peter Sandman likes to use the example of former New York City Mayor Rudy Giuliani. When asked about the number of casualties just hours after the World Trade Center attacks, Giuliani simply answered, "More than we can bear."

They do not want leaders to hide their own responses to tragedies. Risk communication consultant Peter Sandman likes to use the example of former New York City Mayor Rudy Giuliani. When asked about the number of casualties just hours after the World Trade Center attacks, Giuliani simply answered, "More than we can bear." Giuliani's impact in the days that followed resulted not just from his calmness, competence, and compassion, Sandman says, but from the fact that these traits were accompanied by Giuliani's readily detectable pain, which enabled people to identify with him.

Sandman also argues that communicators should not try to "over-reassure," or convince, people that there is nothing to be afraid of. People may rightly be afraid. Instead, acknowledge and accept that the fear is legitimate, then tell people what is being done and what they can do.

This concept of giving people reasonable actions to take is very important in an emergency, especially ones involving public health. You want people to be concerned and vigilant, and to take reasonable precautions. Through actions, people share control of the situation, and, in some cases, they can keep it from getting worse. Having a constructive role engages people in a common mission and provides a sense of control. People can do things to help themselves, victims, and emergency responders. They can also be prepared to do things to minimize the risk of more bad things happening.

Sandman suggests giving people a choice of three actions matched to their level of concern: a minimum precaution, a maximum response, and a recommended middle response. For example, for drinking water safety, a public health official might

> Having a constructive role engages people in a common mission and provides a sense of control.

give three options: use chlorine drops (minimum), buy bottled water (maximum), or boil water for 2 minutes (the recommended middle response). Another way of looking at this is: you must do X, you should do Y, and you can do Z.

Credibility and Trust

The need to establish credibility and trust is a common theme in risk communication. Ideally, trust is built over time and is the result of ongoing actions, listening, and communication skills. In an emergency, it helps if the responding organizations have already built trust in the affected communities. If they have not, or if people have not come in contact with them enough to build trust, there are still things the organization can say and do to build trust in an emergency.

> Information is the most valuable commodity during emergencies or disasters and helps in generating visibility and credibility.
> —S.A. Barrantes et al. (2009, p. 12).

Research has shown that several factors affect trust: caring and empathy, dedication and commitment, competence and expertise, honesty and openness, fiduciary responsibility, confidentiality, and equity (Slovic 1999; Thomas 1998). Trust and credibility are eroded when there is disagreement among experts; lack of coordination among risk management organizations; lack of listening, dialogue, and public participation; an unwillingness to acknowledge risks; not disclosing information in a timely manner; and not fulfilling risk management responsibilities (Chess et al. 1995; Covello et al. 1989).

A number of examples in recent years underscore the importance of establishing trust in emergency risk communication. After the 2010 BP oil spill in the Gulf of Mexico, Chief Executive Tony Hayward undermined trust when he stated that underwater oil plumes did not exist, that the environmental impact would likely be very modest, and that sickened cleanup workers were not affected by oil fumes but by some other cause, perhaps food poisoning. He showed an utter lack of empathy when he stated that he wanted his life back, failing to acknowledge the loss of life and livelihood facing those killed in the original well blowout. Beijing was widely criticized for its initial cover-up of SARS cases, which surfaced in southern China in 2002 and killed more than 800 people worldwide before subsiding in 2003. In Hong Kong, which suffered 299 deaths, the health department first learned about the emergency through media reports. To its credit, China has since instituted a 24-hour online monitoring and consultation system to gather information and answer medical questions to prepare for future outbreaks.

Similarly, the British government was criticized about covering up facts in the early stages of the mad cow disease epidemic that began in the 1980s in the United Kingdom. Perhaps learning from that experience, when the disease surfaced in the United States in 2003, government agencies and industry groups immediately began communicating with the public about actions to recall meat, trace the affected cow's history, and ensure the safety of the food supply.

> The British government's big mistake, at the time of that [mad cow disease] epidemic, was to cover up facts and hide statistics. Official secrecy led to increased anxiety. The British lesson is clear: If more facts are revealed, consumers will feel safer, and the industry is less likely to suffer permanent damage.
> —*The Washington Post*, December 28, 2003.

In the early days of the 2001 anthrax attacks, the U.S. Centers for Disease Control and Prevention (CDC), with the limited information available at the time, mistakenly said that anthrax spores could not escape a sealed envelope to threaten postal workers. Later, they corrected that information and admitted their error, adding to the agency's trust for handling future incidents. These examples demonstrate that it is important to tell people what you know and what is not known, and to correct misinformation as soon as it is discovered.

Trust is also affected by who delivers the message. Researchers have found that in emergency risk communication, local officials and emergency responders were more trusted than federal officials (Wray et al. 2006). State or federal government agencies should involve these groups as appropriate during an emergency.

Individual trust still overrides organizational trust. Research has shown that direct personal contact has the most significant effect on a person's willingness to trust and act on health-related information (Covello et al. 2001). Throughout the duration of the Washington, DC, anthrax case, a CDC Epidemiologic Investigation Service officer met repeatedly with the Brentwood postal workers to discuss antibiotics. In a public situation, when the officer gave the recommendation of 30 days of antibiotic therapy in addition to the anthrax vaccine, an activist in the crowd started shouting inflammatory comments. But because the officer had established himself as a credible and trusted source of information, instead of rallying around the activist, the crowd told him to be quiet so they could hear what the officer had to say.

> Individual trust overrides organizational trust.

Community trust can be increased if authorities have displayed competency, fairness, empathy, honesty, and openness before a disaster (Cordasco et al. 2007). For example, public health and emergency response officials charged with planning for disasters should include community representatives, drawn from churches, social clubs, schools, or labor unions, at all levels of disaster planning and response.

Trust is also affected by how an organization responds to a diverse community. In 2005, more than 100,000 New Orleans residents failed to evacuate before Hurricane Katrina's landfall, despite mandatory evacuation orders. Research showed that distrust of authorities played a strong role among those who did not evacuate, especially among poor and minority groups (Cordasco et al. 2007). Some people, in fact, believed that the authorities purposely diverted water into poorer neighborhoods to save the wealthier ones.

During the 1999 West Nile virus outbreak, New York City officials commendably issued brochures and fact sheets in at least 10 languages. However, trust was diminished when communications were neglected for sensitive populations, such as alerting asthmatics about spraying locations and schedules (Covello et al. 2001).

PLANNING FOR THE UNEXPECTED

Planning involves understanding the needs and desires of the community and organizational jurisdictions in an emergency situation, creating and getting approval for a written plan, training staff, educating the public, getting the resources required in the plan, and making sure that the infrastructure is in place to carry it out. Planning should be done with the help of stakeholders and partners, including citizens, who could be affected by or will respond to an emergency, as well as those who will implement the plan. The following subsections provide guidance on making sure that your organization is ready, teaming with other organizations, working with communities in advance, determining appropriate communication methods, and developing an emergency communication plan.

Preparing Your Organization

Many organizations are expected to play a role in responding to emergencies. In the public sector, state and local emergency response units stand ready to save lives and bolster damaged infrastructure. In the private sector, organizations provide needed volunteers and donations of critical goods and services. Although effective response requires the teamwork of many organizations, each one understands its niche in the process. It is the same in risk communication.

Preparing your organization to effectively communicate risk during an emergency takes time and effort. Some of the most critical preparations, however, have to do with attitude and process rather than simply gathering communication tools. What often hinders communication in emergencies is not the lack of infrastructure or skills but the lack of consensus on roles and responsibilities. Also important is making sure that the organization's own employees receive and share information during an emergency.

Check Your Attitudes at the Door

Organizations charged with communicating risk during an emergency must first take inventory of their attitudes and processes. Chapter 4 discusses ways to combat such nonproductive attitudes as malicious compliance with regulations, a pessimistic attitude toward the public, an unwillingness to share power, and management apathy or hostility. That chapter also describes how to overcome ineffective organizational processes such as inappropriate resources, inappropriate review and approval procedures, conflicting organizational requirements, and access to staff and information. Although such difficulties can limit any type of risk communication endeavor, they become magnified when trying to deal with an emergency. It is best to deal with them when calm heads are more likely to prevail.

Organizations also must deal with ethical issues well before an emergency occurs. Chapter 5 deals with such issues as determining those who are potentially at risk, the acceptable magnitude of a risk, representation of those at risk, and the use of persuasion. Each organization will likely respond differently to such issues, but working with teaming agencies to come to consensus on how these issues will be addressed in an emergency is also critical when communicating with the news media and the public during an emergency.

You Want Me to Do What? Determining Staff Roles

Because of the extensive teamwork necessary to respond to an emergency, organizations also need to be clear internally as well as externally about their roles. Are you the first responders, on the scene immediately and gathering information about the depth and breadth of the situation for other team members to communicate, while at the same time working to minimize panic and maximize appropriate responses by the public? Does your organization gather information from all sources, distill it, and provide a cohesive message so that team members can speak with one voice? Are you the spokesperson, interfacing with the news media and public? Do you provide support with staff or equipment and, thus, communicate needs and capabilities to those making decisions?

> All organizations expected to work together on a response need to understand and agree on each other's roles.

All staff likely to be involved in an emergency response need to understand the organization's role in the risk communication process, and all organizations expected to work together on a response need to understand and agree on each other's roles.

Staff within an organization also need to understand their roles within the wider organizational mandate. Who is the contact person for the organization? Who serves as spokesperson? Who analyzes risks and responses? Who ensures communication within the organization? Some government organizations have designated mission-essential functions—those activities that must be maintained should a catastrophic disaster hinder normal functioning. What are your organization's mission-essential functions and which employees will be expected to staff them? How will they communicate with each other, with other members of the organization, with partner organizations, and with the news media?

Table 21-2 shows how typical organizational roles fit within the emergency risk communication process. Note that the roles are patterned after the incident command structure used by many emergency management organizations across the United States. Although titles in some areas may vary, functions should be analogous.

Ensuring Worker Communication

In a wide-scale emergency such as a terrorist attack or disease outbreak, organization staff are as likely to be affected as the public. Even for staff not directly affected or involved in the response effort, the loss of infrastructure such as phone lines may prevent information from reaching them in traditional ways. In addition, emergency communication plans often direct communications externally to the news media and public—and not internally to staff and management. Thus, organizations need to consider how to get information to staff during an emergency.

> Before an emergency, staff members need to know how information is going to reach them and where to go for additional information.

Workers generally have several broad information needs in an emergency. They need to know what is expected of them from a work perspective. If the emergency occurs during off-hours, do they come in as usual, come in on different shifts, report to a different location, or shelter in place? If the emergency happens while they are at work, do they evacuate the building, move to an emergency shelter, or return home?

During an actual emergency, staff members need to know the same things the general public wants to know, including the depth and breadth of the emergency, what is being done to respond, and what they can do to help. Before an emergency, however, staff members need to know how emergency information is going to reach them and where to go for additional information. Some organizations make emergency communication the responsibility of the communications staff, but this can prove problematic if communication staff are also serving as the public information officers in an emergency. The information officer must look outward, to other organizations, to the news media, and to the public. This outward focus and the time commitment associated with it make reaching staff difficult. It is better to make organizational communication the responsibility of another function, such as human resources.

To ensure that staff know their roles in an emergency, they must be trained and their training reinforced. Risk assessment and risk communication researchers who gathered in Israel to consider how to respond to terrorism-related emergencies (Green et al. 2007) suggested that organizations should think through the actions expected for staff. For example, if the staff may be exposed to radiation, how far must they undress to be decontaminated? Will they need something to protect their modesty? If all their belongings, including car keys and cell phones, are confiscated for decontamination, how will they get home?

Emergencies happen too rarely for people to remember how to respond without frequent reminders. Some organizations that use security badges include information on emergency response directly on the badge or on another card worn with the badge. Others issue refrigerator magnets, business-card-size information pieces to be carried in wallets,

Table 21-2. Typical staff roles in the emergency risk communication process

Role in emergency	Role in emergency response risk communication	Example activities
Incident commander—manages the response	Ultimate authority on response, manager of the public information officer	• Holds hourly to daily briefings with response leaders, including public information officer • Shares information with public information officer • Reviews and approves information to be released
Safety officer—ensures that responders have adequate protection	Subject matter expert, providing information on safety (precautions, limitations, etc.)	• Initially briefs public information officer followed by additional briefings as the situation evolves
Public information officer—ensures public and news media have appropriate information	Spokesperson or coach of spokesperson, conduit of information	• Attends briefings with various command staff • Monitors public and media information needs • Suggests data that should be gathered for planning and intelligence functions • Crafts messages to public and media • Serves as or coaches spokesperson • Responds to public and media information requests
Agency liaison—ensures cooperation with appropriate organizations	Subject matter expert, providing information on own agency's support to the effort (number of staff, roles, etc.)	• Initially briefs public information officer followed by additional briefings as the situation evolves
Agency representative—manages agency's portion of response	Subject matter expert, providing information on agency's role, and sharing response information with own agency; may also serve as agency spokesperson	• Responds to questions from public information officer as requested • May serve as agency spokesperson
Planning and intelligence—collects and analyzes information and recommends actions	Data gathering and sharing with response team, including public information officer	• Gathers communication information as well as other emergency information • Briefs public information officer on regular basis
Logistics—provides support, necessary infrastructure, and supplies	Providing necessary equipment for communication internally and externally	• Understands communication needs • Supplies equipment, infrastructure, and other items necessary for communication
Finance/administration—manages costs, payment, and procurement	Collecting and paying costs of communication	• Accrues costs • Pays bills

or small booklets to be kept at home. Other methods employ telephone hotlines and internal websites for staff only.

Teaming with Other Organizations

Once your organization knows its role in an emergency, it is time to find its role with other responding organizations. It takes a number of functions to adequately respond to an emergency, particularly one of regional or national significance. Health care providers, emergency medical technicians, fire, police, and civic organizations respond at a local level. Most have state- and national-level counterparts. Some emergencies involve private organizations as volunteers or critical equipment donors.

> If you have an incident, they will come.
> —Health Officer, Palm Beach County, Florida, commenting on people who responded when anthrax was found there.

Local organizations sometimes worry that they will have to shoulder the burden alone, but experience with emergency events in recent years has shown that a serious event brings help from expected and not-so-expected places. As the Health Officer of Palm Beach County, Florida, eloquently put it after anthrax surfaced there, "If you have an incident, they will come." The question is: what are you going to do with this help?

The U.S. Federal Emergency Management Agency suggests organizing responses around emergency support functions (ESFs). Those most closely related to risk communication are ESF 8, which relates to public health, and ESF 15, which relates to external affairs (public information). Teaming organizations need to determine who takes the lead and who supports each function. How are decisions made among teaming organizations? They also need to determine what support entails, how teaming organizations will be notified, and where they will meet to respond. Such details are typically outlined in an emergency response plan, which all organizations should review and approve.

With organizational roles agreed upon, team members need to train internally and with each other. Such training helps find potential pitfalls in the plan before an emergency occurs and helps staff remember their roles. The news media can be included in training to ensure that they are aware of how information will be shared and with whom (Green et al. 2007). Public health agencies that participated in field exercises with physicians and utility owners (water, electricity) found that the key to effective teaming was to repeatedly validate each other's assumptions. Misunderstandings could then be corrected as soon as they were identified (Parkin et al. 2008).

Note, however, that training is more necessity than luxury. In the aftermath of the New York World Trade Center bombing in 1993 and the Oklahoma City bombing in 1995, a 1998 U.S. Congressional mandate called for practice operations for a terrorist attack. Such exercises are designed to assess the nation's crisis management capabilities under extreme conditions and to identify areas needing additional training and preparation. In reviewing lessons learned from the 2009 H1N1 pandemic, the Government Accountability Office found that relationships built in interagency exercises were valuable during actual responses (GAO 2011).

One of these exercises, a multistate biowarfare attack exercise in 2000 called Operation Topoff, showed participants in Colorado that they had not thought through the consequences of imposing quarantines. In the exercise, the governor had issued a travel restriction order for all of metropolitan Denver, and the CDC had quarantined the entire state of Colorado. The orders created many unforeseen problems, including how to enforce the restrictions, maintain essential community services, and distribute food and medicine (Hoffman 2003).

Topoff 2, a mock nuclear and biological terror attack, was staged in 2003 in Chicago and Seattle by the Department of Homeland Security. Over a week, hundreds of "casualties" filtered through emergency rooms in both cities. The $16-million drill involved more than 85,000 people from more than 100 federal, state, and local agencies, plus several hundred more participants in British Columbia. Participants gained firsthand knowledge about large-scale triaging and isolation to prevent contamination to caregivers and other patients in the facility. In subsequent years, the Department of Homeland Security has continued to conduct large-scale emergency response training exercises under its National Exercise Program. The program provides a framework for prioritizing and coordinating federal, regional, and state exercise activities. Each year, one exercise is designated as the national-level exercise event, with scenarios that involve multiple regions and organizations.

Working with Communities in Advance

Besides identifying organizational and team roles, you need to consider teaming with the public. Emergency services personnel, when focused on carrying out their duties, sometimes think of the public as passive bystanders. At the scene of a traffic accident or crime scene, for example, the public is separated from the response operation by the familiar barrier of yellow tape. But a disaster is an event that generates casualties beyond available resources, shattering the yellow tape phenomenon, argue researchers Glass and Schoch-Spana (2002).

There is a tendency toward adaptability, cooperation, and resourcefulness in times of disaster. In the 2001 terrorist attacks in New York, volunteers and groups converged at ground zero to offer aid and support, despite hazardous conditions and uncertainty about the risks of further attack or collapse of the towers. Volunteers responded rapidly and in large numbers to support search-and-rescue efforts while professional operations were still being put in place.

After the attacks, affected communities organized through local governments, relief groups, and civic organizations, such as churches, neighborhood associations, and labor organizations. During the Persian Gulf War in 1991, Israel effectively used a network of community information centers to dispense medical information, medication instructions, and reports indicating which hospitals, clinics, and pharmacies were open (Sachs et al. 1991).

> A legitimate sense of control can be given to those under threat, especially in advance of an attack, by public education, by public participation in the preparation process, and by providing the public a voice in the decisions that will affect them.
> —Vince Covello et al. (2001, p. 389).

Emergency preparedness programs are increasingly planning ways to capitalize on the work of nonprofessionals, especially in identifying, surveying, and containing a disease outbreak and caring for large numbers of casualties. In responding to a bioterrorism incident, health and biodefense researchers suggest that church groups could distribute antibiotics, convene vaccination meetings, or arrange visits to the homes of people who are ill (Glass and Schoch-Spana 2002). Social groups such as the Kiwanis or Rotary Clubs might activate phone trees to gather case reports, trace contacts, or disseminate instructions on appropriate use of medications. In its SARS communication plan, the CDC recommended that the American Lung Association and other groups could be helpful in disseminating educational messages to community groups about SARS (CDC 2003).

> A response to emergencies should capitalize on people's desire to help others, especially through the use of existing social groups.

Beyond harnessing public volunteers, participatory decision-making processes should be used in developing emergency risk plans. During the West Nile virus outbreak in 1999, risk communication researcher and consultant Vince Covello argues that New York City officials did not do enough to understand stakeholder concern about certain actions the city took. For example, wildlife experts and environmental groups became outraged about the city's decision to use the pesticide Malathion for disease control, using aerial spraying by highly visible helicopters (Covello et al. 2001). He advocates using citizen advisory panels or other community leaders for responses that require a community's ethical judgment. These might include setting priorities for use of scarce medical resources, such as antibiotics and vaccines, or instituting epidemic control measures that compromise civil liberties.

Determining Appropriate Communication Methods

When disaster strikes, people want as many information sources as possible, and technology is increasingly being used to inform people about emergencies. Media such as television and radio are among the most commonly used communication methods in emergencies, although social media is gaining prominence. All three are described in more detail elsewhere in this book. This section describes other forms of communication used in emergencies.

The Internet

Websites that are updated frequently can be good resources in helping the public know what to do in case of emergency and how to respond when one occurs. For example, a chemical weapons depot had a website containing a wide variety of emergency response information. Among the items was a map showing school locations in each surrounding county, with designated emergency plans for each (shelter in place, evacuate, etc.). The site also included instructions for citizens to prepare for an emergency, including sound clips of warning sirens, and what to do during an emergency. Also included was an archive of press releases, updated as new ones are issued, as well as contact information for public information officers.

Hyer and Covello (2005) recommend the following tips for using websites in emergencies:

- Predevelop basic content to be quickly uploaded in the event of an emergency.
- Include appropriate contact information, such as a hotline.
- Ensure that everything on the website has been approved for accuracy.
- Include links to partner websites that also have information on the emergency, and provide information for their websites.
- Update your website as information changes.

For public health emergencies such as epidemics, the Pan American Health Organization suggests that websites include the following information as well:

- Reports from various agencies, including situation reports on health agencies
- Maps and data on damage
- Information on epidemiology and the status of health care facilities
- Photographs associated with the emergency
- Recommendations of actions for the public to take
- Press releases and other information targeted to the news media
- Background information on the location, the disease, and the population affected
- Links to other credible information sources (Barrantes et al. 2009)

Be aware that the Internet, though remarkably resilient in times of emergency, has its limitations. Unprecedented levels of user demand immediately following the 2001 terror attack in New York City severely stressed the server computers of news websites. Web service providers quickly took a number of steps, such as reducing the complexity of web pages, using alternative mechanisms for distributing content, and reallocating computing resources, to respond successfully to demand. In some cases, automatic rerouting at the physical or network levels allowed some Internet traffic to bypass many of the infrastructure's failed parts. More unexpectedly, however, Internet services in parts of Europe and South Africa lost connectivity because of their connections with the rest of the Internet in New York City.

In addition, many people use Internet-assisted technology in emergencies. After the terror attacks, there was much higher usage of e-mailing and instant messaging. One New York City hospital relied on an external Internet link to connect its wireless devices.

The National Research Council (2003) recommended that key businesses and services that must operate in a disaster should:

- Provide redundant network connectivity, from more than one network provider and by way of more than one physical link or conduit
- Perform an end-to-end audit of Internet dependencies
- Establish plans for dealing with greatly increased traffic loads

Voice-Over Internet Protocol and interfaces such as Skype are increasingly being used as alternative methods of telecommunication in emergencies. These methods make it possible to transmit voice through the Internet. Still, even that system can be stressed in times of high usage. An Internet engineering task force made technical recommendations to alleviate these problems (McGregor et al. 2006).

Since the 1990s, citizen volunteers have been using and adapting web-based technologies and satellite-based imagery to respond during and after disasters. Called crowdsourcing, or sometimes volunteer technical communities, this method mobilizes volunteers and technologies to respond rapidly in emergencies and other situations. Such efforts have led to citizen-developed tools such as mapping the areas impacted by crises and finding people who had been forced to evacuate. Crisis Commons is one example of an organized global network of volunteers dedicated to using open technologies to help in times of crisis. As of 2011, the group had claimed more than 2,000 trained volunteers who apply technical, social media, organizational, and other skills to analyze, plan, report, respond, translate, and document. One example was creating a "Craigslist" of self-identified needs and requests by nonprofits assisting in Haiti relief operations after an earthquake.

The United Nations and other organizations whose missions include emergency response are looking at ways to better integrate crowdsourcing in their planning and operations. Risk communicators should determine whether such crowdsourcing efforts may be useful for partnering. Be aware, however, that this is still an emerging phenomenon. Partners should specifically address roles and responsibilities, reliability and accuracy of information created and shared, work scope, products to be created and used, liability, and privacy issues.

Chapters 18 and 19 describe Internet-delivered communication in more detail.

Telecommunications

Although telecommunications systems can be severed during an emergency, many communities rely on them for emergency communication. In the aftermath of the 2001 terrorist attacks in New York City, for example, the public used low-bandwidth e-mail and instant messaging as substitutes for telephone service, especially where telecommunication congestion was high.

Several communities and states have so-called reverse 911 telecommunication systems that phone citizens in targeted geographic areas with a recorded message to notify them about specific threats. Reverse 911 also includes a phone number citizens can call to hear prerecorded information, such as emergency evacuation procedures. In Florida, the Orange County Sheriff's Department used its reverse 911 system in 2002 to warn trailer park residents to evacuate when a hurricane bore down on Orlando. Colorado's system warns homeowners of approaching wildfires.

Electronic communication systems that send information and track responses in real time are becoming more common. Information is communicated instantly to subscribers' mobile devices (cell phones, pagers, and handheld computers) as well as to e-mail systems. Government organizations are increasingly using these wired or wireless emergency communication systems to keep their first responders, employees, and citizens informed.

In the 2008 earthquake in the Sichuan province of China, local officials sent text messages to mobile phones to warn citizens to evacuate when the water level of a local lake began rising dangerously. The U.S. National Hurricane Center offers a mobile device alerting service about approaching hurricanes. Subscribers receive text messages about forecasts and warnings and can view the Hurricane Center's Internet page, including satellite imagery with zoom capability.

In a shooting incident at Virginia Tech in 2007, campus officials were criticized for delaying campus-wide alerts of the first shooting by almost 2 hours, before more people were killed. Since then, many universities have upgraded their campus-wide communication systems. Many are asking students, faculty, and even parents to register or subscribe at a central website to receive alerts so that the universities can send message by phone, cell phone, text message, and e-mail in case of emergency or situations such as weather-related closures. Users must be assured that their information will be kept private and must keep their contact information updated.

Wireless devices also are playing stronger roles in terrorism prevention and response efforts. According to emergency preparedness expert Michael Hopmeier, even when people are prepared ahead of time, terrorism's randomness and uncertainty require authorities to use technologies that reach the greatest number of people in the least amount of time. He also suggested that citizens can also be "digitally deputized" to report suspicious activities by using their cell phones, text messaging, or even by sending photos of suspicious individuals to a central reporting unit. For example, in Iraq, civilians can anonymously pass information to authorities from the safety of their own homes.

Because word travels fast by cell phones and other technology, it is especially important to make sure that messages are accurate and to correct misinformation. In 2008, massive earthquakes occurred in the Sichuan province of China. Students and others quickly disseminated the information via microblogging services such as Twitter and Fanfou, text messaging, and online videos that reached millions. In one case, incorrect predictions of the exact times for aftershock quakes in Beijing and Shanghai were sent via text message, forcing the state-run Xinhua news agency to run a story to forestall widespread panic.

In another case, in 2006, jubilant emergency response workers in West Virginia conveyed the good news that 12 trapped miners were found alive. The word quickly spread via cell phone and was transmitted by national media. Though the emergency plan called for strict vetting and use of an approval chain to release information, people succumbed to the natural desire to celebrate. When the truth came out that only one miner had survived, public outrage ensued.

Creative Alternatives

Lower-tech communication options may be necessary when electricity is out, cell phone towers are inoperable, or computers are slowed by traffic and viruses.

When Hurricane Katrina hit Louisiana in 2005, the CDC found that it could not deliver public health information through its usual electronic channels. In addition, overnight delivery service was curtailed, and even a CDC-chartered truck was turned back because roads were impassable. Instead, the CDC deployed 30 health and communication specialists to local and state health departments in Louisiana, Mississippi, and Texas. At evacuation and emergency response centers, these field workers identified and filled information needs, sometimes even hand-delivering copies of printed health guidance to workers and affected communities. Information included where to find water and ice, avoiding electrical hazards, avoiding injuries during cleanup, and managing stress to prevent violence.

> Lower-tech communication options may be necessary when electricity is out, cell phone towers are inoperable, or computers are slowed by traffic and viruses.

During this time, the CDC restructured its health messages and channels as needed. For example, it used door hangers for door-to-door delivery of carbon monoxide prevention materials, stickers in evacuation centers to remind children about hand hygiene, and later, one-line messages for high-frequency radio broadcasts. CDC also taped public service announcements in Atlanta and hand delivered them to be broadcast in evacuation centers and local hardware stores.

In 2003, a soy-based baby formula lacking vitamin B1 was found to have caused several infant deaths in Israel. The milk substitute had been widely used in orthodox Jewish communities because of its kosher certification. When the problem was discovered, the Jewish Sabbath, which forbids electricity use, had already begun. Lacking access to electricity-powered mass media to reach the target audience, Israeli health authorities sent trucks equipped with loudspeakers into Orthodox Jewish neighborhoods, warning parents not to use the formula.

The public warning system for a chemical emergency at a munitions depot in Oregon included outdoor sirens and electronic highway billboards. Staff distributed battery-powered tone alert radios to thousands of people, to provide warnings and emergency instructions.

Developing an Emergency Risk Communication Plan

Chapter 12 includes most of what is needed for emergency risk communication planning. However, because of the differences alluded to earlier, emergency risk communication plans include additional elements or cover some elements in greater detail. For example, a group of researchers on risk assessment and risk communication who gathered in Israel to determine best practices during emergencies recommended that pre-event risk communication plans should include actions that would share clear precautions, reassure the public, reduce unnecessary stress, and limit demands on health care (Green et al. 2007).

Many of the following suggestions are adapted from guidance by the CDC, which has done extensive public health emergency planning and training. See the "Additional Resources" section of this chapter for other organizations with emergency communication plans. You can also request the emergency communication plan for your state and other public jurisdictions. By law, these plans must be created and maintained.

In addition to the elements in a typical risk communication plan, an emergency risk communication plan needs to pay particular attention to the following areas:

- **Identification of organizations and individuals who are responsible for various activities.** These groups include the public information team, public health officials, emergency responders, law enforcement agencies, elected officials, and community organizations. A key role is played by the spokesperson for various issues. According to Catherine DesRoches of the Harvard School of Public Health, the most credible spokespersons in an emergency such as a disease outbreak or bioterrorism are a person's doctor, the director of the local fire department, director of the local hospital, director of state or local police, and director of the state or local health department (DesRoches 2003).
- **Identification of organizations and stakeholders who need to receive and, in some cases, convey information during the emergency.** Emergency risk communication plans should describe stakeholders' likely concerns and how those will be addressed. Table 21-3 shows some examples of concerns associated with various groups.
- **The process for information verification and approval.** Especially when many organizations are required to team to respond to an emergency, the process for determining which information is accurate and who can approve its issuance is vital to successful communication. The simpler the process, the better, for time is always a scarce commodity in an emergency.
- **Procedures to get the needed resources.** In an emergency, public information offices will need space, equipment, staff, and supplies, around the clock. The Texas Division of Emergency Management, for example, suggests in its emergency management plan that the public information office be given separate space adjoining the emergency command center, with a dedicated phone line, computer, and direct links to key personnel (Texas DEM 2003).
- **Identification of communication methods.** Include how information will be disseminated and gathered (such as about disease spread) and how questions will be answered (media, hotlines, websites, e-mail lists or listservs, phone banks, town hall meetings, broadcast fax, conference calls, telebriefings, door-to-door canvassing, stakeholders, partners, etc.). Include alternative methods for times when regular channels of communication are disrupted.

Table 21-3. Examples of stakeholders and their concerns in an emergency*

Stakeholder	Likely concerns
Public in the disaster, for whom action messages are intended	Personal, family, and pet safety; stigmatization; and property protection
Public immediately outside the disaster, for whom action messages are not intended	Personal, family, and pet safety; interruption of normal-life activities
First responders	Resources to accomplish response and recovery; personal, family, and pet safety
Public health and medical professional responders	Personal safety and resources adequate to respond
Family members of victims and first responders	Personal safety, safety of victims and response workers
Health care professionals outside response	Vicarious rehearsal of treatment recommendations, ability to respond to patients with appropriate information, and access to treatment supplies
Civic leaders: local, state, and national	Leadership, response and recovery resources, quality of response and recovery planning and implementation, expressions of concern, liability; international relations
Congress	Informing constituents, review of statutes and laws for adequacy and adjustment needs, and expressions of concern
Trade and industry	Business issues (protection of employees, loss of revenue, liability, and business interruption)
National community	Vicarious rehearsal and readiness efforts
International neighbors	Vicarious rehearsal and readiness efforts
International community	Vicarious rehearsal and exploration of readiness
Stakeholders and partners	Included in decision making and access to information specific to the emergency
Media	Personal safety, access to information and spokespersons, and deadlines

*Adapted from "Immediate Response Communication Plan," from CDC et al. (2003).

- **How vulnerable populations will be informed.** These populations could include, for example, the elderly, unvaccinated, non-English speakers, and people with chronic respiratory illness. (See the section "Vulnerable Populations" later in this chapter.)
- **Methods for analyzing media content and public information calls.** Emergency risk communication, like other forms of risk communication, needs to be two way. Determine how you will gather information from stakeholders in real time during an emergency. Use stakeholder input to make sure that accurate information is conveyed and to understand what still needs to be addressed.

The emergency risk communication should also include a contact list for local and regional media, with information on after-hours news desks. Also useful is a list of subject matter experts outside your organization who can speak to other aspects of the emergency and can augment your recommendations during an emergency. Researchers studying emergency risk communication around the globe suggested that this information should be in both printed copy and electronic version, and include, in addition to the contacts listed earlier in this chapter, contacts for external crisis communication consultants (George 2012).

Rene Henry, former head of communications for the Atlantic States region for the U.S. Environmental Protection Agency, suggests some additional things to include in an emergency risk communication plan (Henry 2000):

- A list of all communications team members and work and home contact information
- The phone tree showing how to contact other organization staff
- A policy statement regarding who can speak for the organization and under what conditions
- Guidelines for communicating with the news media, including how information can be released, the locations where they will be briefed, how their credentials will be validated, where their television and satellite uplink trucks can be parked, and alternative locations and plans
- Procedures to follow in the event of loss of electricity or phone service or if the primary communication location is unavailable
- Contact information on all stakeholders who will be notified in an emergency and the priority in which they will be notified.

Additional Considerations in Emergency Risk Communication Planning

Besides completing the emergency risk communication plan, organizations need to make sure that the following equipment will be readily available during an emergency:

- Computers on a local area network with e-mail listservs designated for partners and media
- Website capability around the clock, so new information can be posted as quickly as possible
- Printers, copiers, paper, audiovisual equipment, and office supplies
- Cell phones, pagers, and handheld computers
- Visible calendars, flow charts, bulletin boards, message boards, and easels
- Portable microphones, podium, and TV with cable hookup
- Small refrigerator and microwave oven for staff working around the clock

In addition, the CDC recommends portable "go kits" for public information specialists who may have to abandon their normal place of operation. These kits should include a computer capable of linking to the Internet and receiving e-mail, computer disks or

"Go kits" include a computer, the risk communication plan, a cell or satellite phone, and other pieces of equipment and information communicators need if they must abandon their normal place of operation.

thumb drives containing the elements of the emergency risk communication plan as well as contact information, a cell or satellite phone and/or pager, a credit card or other funding mechanism to purchase operational resources, and background information to provide to the public and media.

Certain pieces of information can also be prepared ahead of time. Rene Henry, who lived through a number of emergencies, recommends developing the following materials and keeping them current for use in emergencies:

- Fact sheets on the organization. For industry, this information should include products and services; for government agencies, this information should include their role in an emergency.
- Biographical sketches of key managers and spokespeople
- Maps, diagrams, and other basic information about facilities
- General news releases about the organization (Henry 2000)

Vulnerable Populations

Planning for communication before and during an emergency is especially important for vulnerable or at-risk populations. The Pandemic and All-Hazards Preparedness Acts defines at-risk populations as those whose needs are not fully addressed by traditional service providers or who feel that they cannot comfortably or safely use the standard resources offered during preparedness, response, and recovery efforts. These groups include people who are physically or mentally disabled, people with limited English language skills, geographically or culturally isolated people, homeless people, senior citizens, and children.

Why is it important to account for at-risk populations in emergency communication? Federal requirements set some ground rules. Limited-English populations qualify for the same antidiscrimination protection as that for race, color, or national origin under Title VI of the Civil Rights Act. The Post-Katrina Emergency Management Reform Act of 2006 includes provisions to account for limited-English groups in disaster-planning processes. The 2006 Pandemic and All-Hazards Preparedness Act required the U.S. Department of Health and Human Services to integrate the needs of at-risk individuals into emergency planning.

Research reinforces the need for special outreach to vulnerable groups. Racial and ethnic communities have been found to be more vulnerable to natural disasters, as a result of such factors as language, housing patterns, building construction, community isolation, and cultural insensitivities (Fothergill et al. 1999). Some have settled where land is less expensive, such as low-lying flood plains, and, consequentially, is disproportionately vulnerable to flooding following tropical storms. In addition, many minorities may find it more difficult to recover from a disaster because of lower incomes, fewer savings, greater unemployment, less insurance, and less access to communication channels and information. The existence of other health disparities, such as heart disease and diabetes, compounds the negative impact of a natural disaster among these groups.

> I have juvenile rheumatoid arthritis and use a wheel chair. We had a bomb threat at work, which was very scary. Everyone evacuated, but I was still left on the 3rd floor by the stairwell for the firefighter to come get me. But, no one came. It was scary just to realize that there are not really any procedures in place to help someone like me in an emergency.
> —Anonymous survey respondent, Nobody Left Behind Research and Training Center on Independent Living, University of Kansas.

Another challenge is access to communication sources. Studies have shown that that low-income African Americans have been less likely to use the Internet to seek information because of a lack of access to computers (James et al. 2007). Before Hurricane Katrina, low-income African Americans and Latinos used friends and relatives, radio, and television as primary information sources (Hilfinger Messias and Lacy 2007; James et al. 2007).

Those who do not plan in advance to reach at-risk groups have suffered the consequences. After the hurricane and Fukushima nuclear reactor accident in Japan in 2011, the government was criticized for not clearly explaining the variations in the risk of radiation exposure regarding groups that may have been more susceptible, such as infants, youths, and expectant mothers (The National Diet of Japan 2012). The Government Accountability Office criticized the CDC for failing to reach non-English-speaking groups during the H1N1 pandemic of 2009 (GAO 2011).

During Hurricane Katrina in 2005, Latinos with limited English proficiency had difficulties understanding warnings and instructions in English. Immigrants accustomed to the metric system had difficulty interpreting weather reports that referred to the storm's strength and direction expressed in miles per hour. Some of the newer Latino immigrants from Mexico had little or no prior experience with hurricanes and did not anticipate the storm's severity. The importance of the extended family, which is often more cohesive in minority communities, also played a part. Obligations to the extended family, especially elderly members who resisted evaluation, delayed Latino responses during hurricane warnings (Eisenman et al. 2007).

The good news is that there are a variety of resources to help (Andrulis et al. 2007). At the time of this book's publication, an excellent starting point was the website of the National Resource Center on Advancing Emergency Preparedness for Culturally Diverse Communities (http://www.diversitypreparedness.org). This center, a project of Drexel University School of Public Health with federal support, is a clearinghouse of current information and resources to plan, serve, and engage culturally diverse communities in emergency preparedness, response, and recovery. Resources include links to data sets, policy, research results, guidelines and toolkits, training and education, and materials about disaster preparation and survival. Users can search for resources that apply to their own states, populations of interest, language, organization, and target audience.

The CDC has an online workbook providing guidance to state and local planners to reach at-risk populations, including racial and ethnic minorities, in disasters (Office of Public Health Preparedness and Response, Centers for Disease Control and Prevention, Department of Health and Human Services, No Date). The workbook is a treasure trove of practical and useful tools, templates, and checklists for defining, locating, and reaching special populations. It even describes how to use free online software to generate digital maps that show where the vulnerable populations and outreach resources are located in your community.

Researchers and experienced practitioners generally recommend the following elements for risk communication that serves vulnerable populations:

- **Identify applicable policies, existing guidelines, and roles of existing** organizations related to emergency and public health communication in the geographic region of interest.
- **Develop partnerships** with organizations that are already linked with the populations of interest, such as human service agencies and faith-based organizations. They already work with the people you are trying to reach and are viewed as a trusted source of information by their members. For example, the Minnesota Department of Health and that state's Emergency and Community Health Outreach (ECHO)

used its network of community resources, including trained translators, to reach Somalis during a measles outbreak and ethnic populations during a fish contamination incident. After reports of explosives-packed printer cartridges directed at Jewish entities in 2010, the Greater Boston Jewish Emergency Management System quickly communicated information from U.S. government agencies to 200 Jewish agencies in the network.

- **Define and locate the at-risk populations and their spokespersons/gatekeepers.** This many include establishing criteria for special populations to be served, conducting needs assessments, and collecting data. Geographic mapping is useful for understanding where major groups reside; see examples done in the Seattle regional area for linguistically isolated and sensory disabled populations at http://www.apctoolkits.com/vulnerablepopulation/knowing/geographic_mapping/.

- **Understand each group's knowledge, attitudes, and practices related to emergency preparedness**, including whom they trust and how they receive information. Some populations, for example, may have a distrust or fear of government and health institutions and may be more likely to trust local officials and agencies than federal ones (Wray et al. 2006). New American Media, as association of ethnic media organizations, is a good source for determining which mass media outlets reach certain populations.

- **Involve representatives of the populations when developing emergency communication plans and information materials.** Culturally tailored and linguistically appropriate emergency risk communication goes beyond literal translation.

- **With community partners, conduct training and, if appropriate, drills or exercises focused on the vulnerable populations, then modify plans accordingly.** In San Francisco, for example, the NICOS Chinese Health Coalition established a Chinatown disaster response program, including an annual large-scale disaster drill.

COMMUNICATING DURING AN EMERGENCY

Communicating ineffectively during an emergency can have negative or even disastrous consequences. In many ways, the tsunami and subsequent nuclear reactor accident at Fukushima, Japan, in 2011 is a prime example of what not to do. According to an official report (The National Diet of Japan 2012), the key risk communication mistakes and their consequences during the emergency were:

- **Communication delays.** Plant operator Tokyo Electric Power Company (Tepco) was slow to relay information to the government. The prime minister's office waited too long to declare a state of emergency. In communicating with the public near the plant, officials emphasized detailed accuracy at the expense of quickly getting the information to those who needed it for informed decisions.

- **Confusion about roles, lack of trust.** The prime minster traveled to the plant during the crisis to direct the workers who were dealing with the damaged core. This "unprecedented direct intervention . . . diverted the attention and time of the on-site operational staff and confused the line of command," the report said. "Had the head office of Tepco actively communicated the on-site situation from the start, and explained the severity of the situation to the other parties," the report said, "there is a possibility that the distrust—and the confusion in the chain of command that followed—could have been prevented."

- **Lack of preparation.** Regulators' negative attitudes toward revising and improving existing emergency plans resulted in a lag in upgrading preparedness and disaster countermeasures. The lack of updated emergency response plans contributed to delayed and confusing evacuation orders and public communication about radiation risks.
- **Inaccurate and confusing instructions to the public.** The cabinet secretary repeatedly stated that there were no immediate health effects from the release of radiation, giving the public a false sense of security. Residents therefore did not understand why an evacuation was necessary or urgent. Even as damage from the accident began to escalate, evacuation orders were chaotic. Many residents were unaware of the accident, its severity, or the radiation release. Evacuation orders were repeatedly revised as the evacuation zones expanded from the original 3-km radius to 10 km and later, 20 km, all in one day. Each time the evacuation zone expanded, the residents were required to relocate. Some evacuees were unaware that they had been relocated to sites with high levels of radiation. Dozens of patients in hospitals and nursing homes died from evacuation-related complications.

These unfortunate mistakes underscore how communication in the midst of an emergency can quickly spiral out of control. Understanding the principles described here, and conducting emergency response exercises with all stakeholders to surface issues in advance, will go a long way to communicating effectively in a disaster.

> Vicarious rehearsal occurs when those not susceptible to the risk believe themselves to be at risk and act accordingly. Those "worried well" may flood hospital emergency rooms or otherwise overload public health resources. They may also divert resources away from more urgent use elsewhere.

Studies show that the earliest information needs to address people's basic needs, including food and water, safety, and shelter (Green et al. 2007). As one expert said in retrospect about the 2011 Fukushima nuclear power plant accident, "Forget the educational messages that we prepared. The public wants to know if it is safe for themselves and for their kids. And, if not, what do they do about it, period. They don't care what a Sievert is" (Center for Biosecurity of UPMC 2012).

The first 48 hours of an emergency are likely to be the most challenging. Table 21-4 shows communication actions that the CDC recommends in this critical time period.

The CDC website has research-based resources to help health officials as they communicate with the public in the first hours of an emergency. Resources on the website at the time of this book's publication included customizable message templates, radio scripts, broadcast media resources in English and Spanish, and fact sheets (called "creative briefs") that can be used by communicators in various emergency scenarios.

The CDC commissioned university research on best practices in communicating about terror threats and other emergencies in the Pre-Event Message Development Project. The universities made recommendations about communication during scenarios such as botulism, pneumonic plague, sarin-type chemical threats, and dirty bombs (Vanderford 2004). Findings that crossed all the areas that messaging should address the following factors:

- The nature of the threat
- How to detect exposure and symptoms
- How to protect themselves by avoiding or reducing exposure
- When to seek medical attention

Table 21-4. Communication actions in the first 48 hours of an emergency

Action	Details
Verify the situation	Determine the type, scope, and severity of the event. To get as accurate information as possible, verify each piece of information with more than one source if possible.
Conduct notifications	Notify the appropriate organizations and individuals, using the call list in your communication plan. This should include your organization, stakeholders (including elected officials), and partners. Tell them about the emergency and what actions are being taken. Determine how often the decision-making team will get back together during the day to update each other.
Assess the level of crisis	Identify the severity and character of the crisis to help make decisions about hours of operation for communication team, jurisdictions, and other factors.
Organize and delegate assignments	Activate the teams identified in the communication plan, including spokespeople. Communication team members may be conducting activities like media interaction, staffing the hotline, updating websites, developing information materials, and clearing information for release. Science or medical team members may be defining medical issues and treatments and communicating with health professionals. Various team members may be interacting with emergency responders and law enforcement.
Prepare information and obtain approvals	Include information that addresses questions and concerns of stakeholders and information they need to know including protective actions, while expressing empathy and caring.
Release information	Try to get each new wave of information out to all audiences, all venues, at the same time in a coordinated way. Methods and audiences can include media, websites, hotlines, employees, partners, legislators and special interest groups, and community members.

- How to treat symptoms that have already appeared
- The potential for long-term health effects
- Progress in apprehending the perpetrators

Researchers made these recommendations:

- Ensure that the media and local authorities are continually informed and prepared to speak.
- Use well-known, well-respected public figures and subject matter experts to address issues on protective actions and health.
- Develop message materials that answer the key questions stated earlier, provide clear action steps, are clear and easily understood, include sources for credibility, and reflect full government disclosure.
- Use an information dissemination plan to ensure that critical information will be available when people need it and where they look.

In addition, a nationwide study to develop a model for how people decide to take emergency preparedness actions found that people were more likely to act when given "dense information." Dense information was defined as consistent information from all communicators (federal, state, and local government; nongovernmental organizations and nonprofit organizations) that comes through many different communication methods and repeats over time. Repetition was the key to getting people to take action (Wood et al. 2012).

Remember that during an event, it is your responsibility to communicate. If the public or news media misunderstands, you bear the responsibility to adapt methods or messages to get critical points across. Lessons learned from emergencies in recent years also suggest the following:

- **Be prepared to counter the desire to downplay the risk.** Research and actual disasters have demonstrated the tendency for people to feel safer than warranted, even when told to evacuate their homes. This was shown to be true in earthquake-prone Bucharest, Romania, as well as after a major flood in Switzerland (Armas 2006; Siegrist and Gutscher 2008). It has also happened with several hurricanes in the United States in recent years.
- **Adapt communication messages and materials as needed.** After Hurricane Katrina in 2005, the CDC created special communications for people in evacuation centers, including topics such as handling stress to prevent bullying, shaken baby syndrome, suicide, and sexual abuse. The CDC also created a set of "playing cards" in various languages containing simple, illustrated prevention messages on topics such as stress and relationships, parenting under stress, preventing violence, and rape prevention.
- **Realize that one of the biggest issues is control.** Nobody wants to feel like a victim, even vicariously. To help people regain some sense of control, give them something to do. This "something" needs to be positive ("do this," rather than "don't do that"), actionable (they know what to do and when they have accomplished it), and real (no placebos). One of the frustrating things for Americans following the September 11 attacks was the call from the nation's leaders to "be vigilant." While positive, it was difficult to implement and thus not satisfying as a personal response.
- **Remember that transparency and process still matter.** In stakeholder involvement, the process of communicating risk has often been just as important to the success of the effort as the actual risk communication product. This fact is proving even more so for emergency situations. Jennifer Leaning of the Harvard School of Public Health looked at ethical issues for public-health-related terrorist events and found that the integrity of the search for answers was just as important as the answers themselves (Leaning 2003).
- **Include the individual as well as the group.** Leaning also found that although public health practitioners are taught to seek the greatest good for the greatest number, the needs of individuals in emergencies cannot be ignored. Leaning also stresses attention to psychological distress (Leaning 2003).

The following sections provide additional advice on communicating during an emergency, including communicating from an emergency operations center, working with the news media, answering public questions, and supporting a family assistance center.

Emergency Operation Centers

An emergency operation center, which may include or be called a joint information center, is activated in an emergency to distribute consistent and accurate information. More and more, these centers are run according to the incident command structure mentioned earlier. In a public health emergency such as SARS, such centers have the following communication responsibilities (CDC et al. 2003):

- Issue local public health announcements and updated information on the outbreak and the response.
- Disseminate information about the crisis, its management, and the possible need for travel restrictions, isolation, and quarantine.
- Establish a news desk operation to coordinate and manage media relations activities.
- Provide a location for state, local, and federal communication and emergency response personnel to meet and work side by side in developing key messages, handling media inquiries, and writing media advisories and briefing documents.
- Respond to frequently occurring questions by developing fact sheets, talking points (key messages), and question-and-answer documents.
- Coordinate requests for spokespersons and subject matter experts.
- Issue media credentials.
- Address other local/regional information requests related to the outbreak that require distribution to the media and the public.
- Develop, coordinate, and manage local websites, as required.

Such centers may also house the public information hotline and operators (see later in this chapter).

A chemical weapons depot in Oregon created a "Smart Book" that guided operations in its joint information center in case of an emergency. Topics ranged from tips for answering calls to information about agricultural and livestock exposure to steps for sheltering in place.

Those who are charged with communicating risk must understand their role in the structure of an emergency operations center (public information officer, subject-matter expert, gathering and analyzing information, etc.), and provide around-the-clock staff trained in that role. Another role that risk communicators may have to play is in the development of Situation Reports or SITREPs. These concise reports written in nontechnical language provide information on the unfolding situation of the emergency and serve to keep all members of the emergency operations center up to date on activities. Although different situations and incident commanders may require different content, some general topics SITREPs could cover include the following:

- Condition of the affected area
- Characterization of the affected population, including identification of the most seriously affected or most vulnerable groups
- Health conditions, both for people and in the environment
- Main needs, including which are currently covered
- Accomplishments since the last SITREP
- Actions in progress
- Condition of response, whether additional help is needed, and what type of assistance to request (Barrantes et al. 2009)

Working with the Media in an Emergency

TV and radio are important information sources for most people in an emergency (Hasson and Holmes 2003). Newspapers remain important, especially for describing the final result

> What can you do to make media interactions as productive as possible? Start by finding a method to coordinate public information personnel from a range of federal, state, and local agencies, working together to respond to media inquiries, writing releases, and providing information on their agencies.

of an event and for putting it in context. What can you do to make media interactions as productive as possible? Start by finding a method to coordinate public information personnel from a range of federal, state, and local agencies, working together to respond to media inquiries, writing releases, and providing information on their agencies. If a joint information center is not activated for this, the participants should establish a daily briefing among themselves to coordinate and communicate on media briefings and materials.

The Emergency Management Laboratory (2001) and Hyer and Covello (2005) have several suggestions to accommodate reporters on the scene of an accident or disaster. First, try to make sure that they have access to the resources necessary to do their jobs:

- Ample electrical power
- Sufficient light for auditorium and meeting space
- Sufficient space for speakers, cameras, lights, and microphones
- Internet access for filing stories electronically
- Use of a multiplex remote sound box
- Access to high-quality graphics
- Access to parking near the scene
- Arrangements with local police/city agencies to block off congested streets or areas

Consider the timing of news conferences. Because of deadlines, the best time to hold a media event may be around 10:00–11:00 am on a weekday morning or 3:00–4:00 pm on a weekday afternoon, though this may vary by country or locality. If possible, plan around competing events that may prevent journalists from attending. For fast-breaking emergencies, consider holding at least two news conferences per day, thereby allowing the spokesperson to gather and share more information.

Recognize that the media will seek certain information and behave in a certain way during an emergency. They tend to:

- Search for background information
- Dispatch reporters/resources to the scene (may include local and national coverage)
- Get access to the site or spokesperson
- Dramatize the situation (which includes looking for the most dramatic video or photo possible)
- Expect an instant briefing, complete with written information
- Find immediate victims and other affected people
- Find filler for stories if credible information is not available, using sources such as nearby residents and volunteer rescue workers

TV, radio, and web-based media usually have faster deadlines than print media. They can go with very brief information at the beginning and do not need to wait to have a more thorough story before distributing it. These media outlets might need just the basics within 30 minutes, and more later. Thus, in an emergency, organizations need to be thinking about "what's good enough for now" instead of waiting 2 hours to respond with a full description of the incident.

> When you release your own bad news, you decrease the likelihood of rumor, supposition, half-truths, and misinformation.
> —Kathleen Fearn-Banks (1996, p. 65).

Be prepared to answer the following questions:

- What happened and where?
- Who was affected and how?
- What is the extent of damage?
- What caused the problem?
- Who is to blame?
- Has anyone broken the law?
- Has this ever happened before?
- Is there danger now?
- What else can go wrong?
- What are you doing about it?
- How can we find out more?

For emergencies involving disease outbreaks, Hyer and Covello (2005) list dozens of likely media questions that can be prepared for in advance, such as:

- How contagious is it, and how is it spread?
- Can people be vaccinated or otherwise be treated, and how effective are those treatments?
- How will medicines be distributed, and is there an adequate supply?
- What are the symptoms of the disease, and what should people do if they think they've been exposed?
- Who is in charge, and how are you coordinating with other organizations?
- Is this caused by terrorism?
- What is being done to stop the disease?
- What kinds of medical care are available, and what happens if facilities are overcome by demand?
- How many people could get sick or die?
- What are you recommending for your own family?
- Is any quarantine or isolation necessary, and if so, how will that work?

The CDC makes the following recommendations about working with media during an emergency:

- Have staff assigned to answer calls before you release information.
- Put media info out via blast fax, newswire, phone, briefing, website, and appropriate social media channels.
- Set up a media command post where the media can consolidate information to deliver to their audiences.
- Have areas for TV media to do their "stand-ups." If this is at your site, it will usually be where the building displays your organization's logo.

- Let the media know when updates will be given. Give them as promised, even if there is nothing new to say.
- Distribute a copy of any official statements and a fact sheet on the situation and the organization.

When conducting a news conference, personnel involved in handling the emergency, spokespersons, and technical advisors should agree on the following before the news conference:

- What information is most important?
- Who will speak for what specific issues?
- What are the key messages?
- What questions are likely to arise?
- What visuals could be used?
- Who will take notes about any information or resources that need to be followed up?

In addition, consider selecting a press room with more than one exit/entry so spokespeople can return to handling the emergency while others speak with the news media (Henry 2000).

During the news conference, each person who speaks should give his or her name, role, and organization represented. With many organizations involved, it helps to have the news conference manager moderate the question-and-answer session by referring the questions to the appropriate person. Tell reporters when the next news conference will occur, if known. Tell them how they can get questions answered in the meantime. After the news conference, continue to monitor media coverage. Debrief with the team to see if any misinformation needs to be corrected later or other changes made for the next interaction.

> After the first 48 hours, the public and media will begin to focus harder on why this event happened, what lessons were learned, and what is being done to keep it from happening again.

After the first 48 hours, the public and media will begin to focus harder on why this event happened, what lessons were learned, and what is being done to keep it from happening again. Media competition may intensify to keep the story going with new angles. Monitor the event for new information, monitor media coverage, and continue implementing the plan, making adjustments as necessary. Determine whether the emergency is changing in any way, and address any rumors or points of conflict. Add any new resources needed; relieve staff or return them to normal duties.

Hotlines

One of the most frequently used methods in emergency risk communication is the hotline. This single, publicized phone number should be available toll-free to anyone looking for more information. Operators must be available 24 hours a day, although some organizations have found that calls can be serviced by starting with a short, prerecorded message (1 minute or less) that outlines answers to most frequently asked questions.

When planning for an emergency hotline, plan big. During the 1999 West Nile virus outbreak in New York City, hotline staff, 27–75 per shift, answered calls around the clock, fielding a total of more than 150,000 inquiries in 7 weeks (Glass and Schoch-Spana 2002). The 2002 anthrax incident in Palm Beach County, Florida, saw such an influx of questions to the hotline, by callers who were remarkably well informed, that operators were quickly overwhelmed.

> During the 1999 West Nile virus outbreak in New York City, hotline staff, 27–75 per shift, answered calls around the clock, fielding a total of more than 150,000 inquiries in 7 weeks.

Giving hotline employees the answers to frequently asked questions, prepared in advance, is one way to help them field calls more efficiently. People typically ask the following basic questions:

- What happened?
- Where is it?
- Who is/will be affected, and how?
- What is being done?
- What can I do?
- How long will it last?
- Will it happen again?
- Who was at fault?

In the case of communicable diseases, people also want to know whether it is contagious and how, who is at risk, whether there is treatment or a cure, whether it is spreading, and when it will end (Green et al. 2007). People may also want to know how to prevent getting the disease.

The following categories suggested by the CDC may be helpful for organizing responses to questions, especially where public health is involved:

- Information about the event or threat
- Tip line, with actions people can take to protect themselves and others
- Reassurance/counseling
- Referral information for health care workers
- Referral information for epidemiologists or others to report cases
- Lab/treatment protocols

In developing responses in advance, consider people's mental models. Understanding what people do not know and what misperceptions they have helps you provide accurate information and counter rumors. When the Allegheny County Department of Health in Pennsylvania was dealing with the flood of anthrax-related calls in 2001, they realized that people did not know they had to be exposed to be at risk. Risk communication researcher Baruch Fischhoff helped them develop a model of the probability of anthrax exposure that could be used to design risk communications. The model included concepts such as exposure route, dose, anthrax strain, vaccination status, and health status of recipient (CDC et al. 2003).

Monitor your hotline for midcourse corrections. A simple response form, such as a shared database, can be used to record questions asked and responses given. Reviews of these responses will reveal whether correct and adequate information is given, new information is needed, and any patterns exist to the questions. Necessary corrections can be shared with hotline staff and in subsequent media interactions.

COMMUNICATING AFTER AN EMERGENCY

Risk communication does not stop when the emergency does. One example of the importance of effective postemergency response was described by the commission that investigated the 2011 Fukushima nuclear plant disaster. Even a year after the disaster, residents in the affected area were still struggling from the effects of the accident (The National Diet of Japan 2012). They continued to face grave concerns, including the health effects of radiation exposure, displacement, the dissolution of families, disruption of their lives and lifestyles, and the contamination of vast areas of the environment. "The government has yet to address the impact of radiation on the health of residents," the report said. "The government has not seriously undertaken programs to help people understand the situation well enough to make their own behavioral judgments. Although [radiation exposure] standards have been categorized in detail, it is more important that the government communicates in ways that are clearly helpful to the public: identifying what is edible, what is the tolerable intake level, which foods continue to be safe, and whether tests are reliable."

> After an emergency, communicators should help move stakeholders to resolution and recovery, improve future public response, and learn from experience.

Information needs vary depending on how much time has elapsed since the emergency. The CDC recommends a phased message approach for communicating after large storms, for example (Table 21-5).

When the emergency situation stabilizes, the work of the risk communicator continues. Information is needed to help stakeholders (affected people, the broader public, and the media) move from the emergency situation to resolution and recovery, improve public response to future similar emergencies, and learn from experience. Information is also needed to convey relief and thanks to the response team, evaluate the response effort, and conduct public education.

In the case of mass casualties, one of the more difficult tasks of those communicating risk may be to support the family assistance centers or medical examiner's/coroner's offices (Office for Victims of Crime 2001). In its November 2002 OVC Bulletin, the Office for Victims of Crime in the U.S. Department of Justice issued an outstanding template for communities and agencies considering crisis response plans (Blakeney 2002). The report was written by Ray L. Blakeney, Director of Operations for the Office of the Chief Medical Examiner of the State of Oklahoma; Blakeney was a key responder to the Oklahoma City bombing and the Oklahoma City tornado of 1999. The bulletin describes lessons learned from mass fatalities on how to provide relief to victims' families. Overarching recommendations include the need for all communities to have an effective crisis response plan and the need for all agency personnel who will interact with families to be trained in communicating effectively, compassionately, and sensitively.

Of particular interest to risk communicators is the list of questions most often asked by families and how to respond. For example:

Table 21-5. Phased message dissemination for hurricanes and floods*

Period of dissemination	Topics
Immediately preceding landfall through first 24 hours after the storm	Hurricane readiness, preparation for power outages, preparation related to prescription medications, evacuating the area of a hurricane, staying safe in your home during a hurricane, worker safety in a power outage, carbon monoxide poisoning prevention, flood readiness, electrical safety, prevention of heat-related illnesses, hand hygiene in emergency situations, copying with a traumatic event, emergency wound care, protecting your pets, animals in public evacuation centers
1–3 days after the storm	Reentering your flooded home, how to clean a flooded home safely, worker safety after a flood, preventing chain-saw injuries during tree removal, preventing injuries from falls (ladders/roofs), personal protective equipment and clothing for flood response, managing acute diarrhea after a natural disaster, cleaning and sanitation after an emergency, keeping food and water safe after a natural disaster or power outage
3–7 days after the storm	Protection from animal- and insect-related hazards, electrical safety and generators, infection control and prevention in evacuation centers, impact of power outages on vaccine storage and other medicines, preventing violence after a natural disaster, animal disposal after a disaster
2–4 weeks after the storm	Rodent control after hurricanes and floods, trench foot or immersion foot, environmental health needs and habitability assessments, protection from chemicals released during a natural disaster, respiratory protection for residents reentering previously flooded areas and homes
1 month and after the storm (emphasis is on long-term health consequences)	Suicide prevention, issues surrounding school-age hurricane evacuees attending new schools, mold removal from flooded homes, mold allergies related to flood cleanup

*Source: Vanderford et al. (2007).

1. **How will families be notified if their loved ones are recovered and identified?** Responders must identify who will pass on this information and how. While a central point of contact is critical, some organizations such as the police, fire department, and military have their own systems. Families must know who will provide them with accurate information.

2. **What is the condition of the body?** Describing the horrific condition of bodies after an airplane crash or bombing requires compassion, honesty, and tact. Although risk communication advice generally suggests specific language, Blakeney instead suggests more general words like "severe," "significant," and "trauma." Responders should listen to the family and give only the information wanted. Anything more may overwhelm.

3. **How do families know that the information they receive is accurate?** Responders must identify authorized sources of information and make arrangements such as conference calls to relay information to families who would otherwise have to travel to the site. Responders should also provide written information to augment

verbal communications, as people under stress have a hard time recalling information.

Those who remain also need risk information. Research has shown that a community is most responsive to risk avoidance and mitigation education directly after a disaster because community members have been sensitized. People want to hear about lessons learned and steps taken to prevent the situation from recurring. People want to be reassured of their safety and attain closure. Particularly in emergencies involving violence such as terrorism or bioterrorism, people need help to deal with the issues.

> A community is most responsive to risk avoidance and mitigation education directly after a disaster.

Example reactions include shock and numbness, intense emotion, fear, guilt, anger and resentment, depression and loneliness, isolation, and panic. Those who live through such attacks may also exhibit physical symptoms like headaches, fatigue, nausea, sleeplessness, loss of sexual feelings, and weight gain or loss. Posttraumatic stress has been recorded even in people in Italy and India who had only viewed television coverage of the 2001 terrorist attacks in the United States (Green et al. 2007). Many people in such situations find it difficult to resume normal activity.

On the other hand, volunteers often come forward looking for ways to help. In Japan after the Fukushima accident, volunteers were decontaminating homes, streets, and schools. In the United States, gulf state residents volunteered to clean up following the Deepwater Horizon oil spill, only to have their services refused because of liability concerns. These lessons and others like them show that it is important to communicate how volunteers can safely and effectively be incorporated into disaster recovery efforts.

Some ways that risk communicators can help include the following:

- Make counselors, clergy members, and other survivors available to talk to those having difficulty coping.
- Be ready to answer questions about types of assistance, payment for travel and other expenses, and how to query insurance companies.
- Remember children, who are often overlooked in times of crisis.
- Give volunteers opportunities to help others.

CHECKLIST FOR EMERGENCY RISK COMMUNICATION

Before an emergency:

☐ Organizations and individuals responsible for responding, reporting, and other activities have been identified.
☐ Those who will need to receive information have been identified.
☐ Vulnerable populations and their needs have been identified.
☐ An emergency communication plan has been prepared and shared with stakeholders.
☐ Exercises and/or drills have been implemented and plans adjusted accordingly.

□ Background information has been prepared (for example, maps, description of organizational roles in an emergency, and website content including contact information).

During an emergency:

□ Emergency operations centers and/or hotlines are activated, if appropriate.
□ The risk communication plan is implemented, with messages and materials adapted as necessary as the emergency unfolds.

After an emergency:

□ Victims' families are communicated with, in partnership with other organizations as appropriate.
□ Phased messages are conveyed that align with evolving risks and concerns as time passes.
□ Lessons learned and ways to prevent or mitigate similar emergencies in the future have been conveyed.
□ The emergency risk communication plan has been updated to reflect lessons learned and new information.

REFERENCES

Andrulis, D. P., N. J. Siddiqui, and J. L. Gantner. 2007. "Preparing Racially and Ethnically Diverse Communities for Public Health Emergencies." *Health Affairs*, 26(5):1269–1279.

Armas, I. 2006. "Earthquake Risk Perception in Bucharest, Romania." *Risk Analysis*, 26(5): 1223–1234.

Barrantes, S. A., M. Rodriguez, and R. Perez. 2009. "Information Management and Communication in Emergencies and Disasters." Pan American Health Organization, Regional Office of the World Health Organization, Washington, DC.

Blakeney, R. L. 2002. "Providing Relief to Families after a Mass Fatality: Roles of the Medical Examiner's Office and the Family Assistance Center." *OVC Bulletin*, November 2002. http://www.ojp.usdoj.gov/ovc/publications/bulletins/prfmfl1_2001/188912.pdf (accessed January 22, 2013).

CDC (Centers for Disease Control and Prevention). 2003. "Public Health Guidance for Community-Level Preparedness and Response to Severe Acute Respiratory Syndrome (SARS), Draft." October 2003. http://www.cdc.gov/sars/index.html (accessed January 29, 2013).

CDC (Centers for Disease Control and Prevention), Agency for Toxic Substances and Disease Registry, Oak Ridge Institute for Science and Education, and the Prospect Center of the American Institutes of Research. 2003. Emergency Risk Communication CDCynergy (CD-ROM, February 2003). http://www.orau.gov/cdcynergy/erc/ (accessed January 22, 2013).

Center for Biosecurity of UPMC. 2012. "After Fukushima: Managing the Consequences of a Radiological Release." Baltimore, Maryland.

Chess, C., K. L. Salomone, B. J. Hance, and A. Saville. 1995. "Results of a National Symposium on Risk Communication: Next Steps for Government Agencies." *Risk Analysis*, 15(2): 115–125.

Cordasco, K. M., D. P. Eisenman, D. C. Glik, J. F. Golden, and S. M. Asch. 2007. "'They Blew the Levee': Distrust of Authorities among Hurricane Katrina Evacuees." *Journal of Health Care for the Poor and Underserved*, 18:277–282.

Covello, V. T., D. B. McCallum, and M. T. Pavlova. 1989. "Principles and Guidelines for Improving Risk Communication." V. T. Covello, D. B. McCallum, and M. T. Pavlova, eds., *Effective Risk Communication: The Role and Responsibility of Government and Nongovernment Organizations*. Plenum Press, New York, pp. 3–16.

Covello, V. T., R. G. Peters, J. G. Wojtecki, and R. C. Hyde. 2001. "Risk Communication, the West Nile Virus Epidemic, and Bioterrorism: Responding to the Communication Challenges Posed by the Intentional or Unintentional Release of a Pathogen in an Urban Setting." *Journal of Urban Health: Bulletin of the New York Academy of Medicine*, 78(2):382–391.

DesRoches, C. M. 2003. "Opinion Surveys and Risk Communication," quoting the Harvard School of Public Health/Robert Wood Johnson Foundation Survey Project on Americans' Response to Biological Terrorism, October 24–28, 2001. Harvard School of Public Health presentation to the Maine Institute for Public Health.

Eisenman, D. P., K. M. Cordasco, S. Asch, J. F. Golden, and D. Glik. 2007. "Disaster Planning and Risk Communication with Vulnerable Communities: Lessons from Hurricane Katrina." *American Journal of Public Health*, 97(S1):S109–S115.

Emergency Management Laboratory. 2001. "Emergency Public Information Pocket Guide." Oak Ridge Institute for Science and Education, Oak Ridge, Tennessee. http://orise.orau.gov/emi/epi/files/epi-booklet.pdf (accessed February 7, 2013).

Fearn-Banks, K. 1996. *Crisis Communications: A Casebook Approach*. Lawrence Erlbaum Associates, Mahwah, New Jersey.

Fischhoff, B. 2002. Remarks delivered at the 27th Annual AAAS Colloquium on Science and Technology Policy on April 11–12, 2002, Washington, DC.

Fothergill, A., E. G. Maestas, and J. D. Darlington. 1999. "Race, Ethnicity and Disasters in the United States: A Review of the Literature." *Disasters*, 23(2):156–173.

GAO (Government Accountability Office). 2011. "Influenza Pandemic: Lessons from the H1N1 Pandemic Should Be Included into Future Planning." GAO-11-632, Government Accountability Office, Washington, DC.

George, A. M. 2012. "The Phases of Crisis Communications." A. M. George and C. B. Pratt, eds., *Case Studies in Crisis Communication: International Perspectives on Hits and Misses*. Routledge, New York.

Glass, T. A. and M. Schoch-Spana. 2002. "Bioterrorism and the People: How to Vaccinate a City against Panic." *Chemical Infectious Diseases*, 34:217–223.

Green, M., J. Zenilman, D. Cohen, I. Wiser, and R. Balicer. 2007. *Risk Assessment and Risk Communication Strategies in Bioterrorism Preparedness*, NATO Security through Science Series-A: Chemistry and Biology. Springer, Dordrecht, Netherlands.

Hasson, J. and A. Holmes. 2003. "Who We Believe." *Federal Computer Week*, 17(30):18–25.

Henry, R. 2000. *You'd Better Have a Hose if You Want to Put Out the Fire: The Complete Guide to Crisis and Risk Communications*. Gollywobbler Productions, Windsor, California.

Hilfinger Messias, D. K. and E. Lacy. 2007. "Katrina-Related Health Concerns of Latino Survivors and Evacuees." *Journal of Health Care for the Poor and Underserved*, 18:443–464.

Hoffman, R. E. 2003. "Preparing for a Bioterrorist Attack: Legal and Administrative Strategies." *Emerging Infectious Diseases*, 9(2):1–11. http://www.cdc.gov/ncidod/EID/vol9no2/020538.htm (accessed January 22, 2013).

Hyer, R. N. and V. T. Covello. 2005. "*Effective Media Communication During Public Health Emergencies: A WHO Handbook.*" WHO/DCS/2005.31. World Health Organization, Geneva, Switzerland. http://www.paho.org/cdmedia/riskcommguide/Effective%20Media%20Communication%20Handbook.pdf (accessed January 22, 2013).

James, X., A. Hawkins, and R. Rowel. 2007. "An Assessment of the Cultural Appropriateness of Emergency Preparedness Communication for Low Income Minorities." *Journal of Homeland Security and Emergency Management*, 4(3):Article 13.

Leaning, J. 2003. "Bioterrorism and Public Health: The Ethics of Public Health Practice in Crisis Settings." Harvard School of Public Health, Harvard Medical School, presentation to the Maine Institute for Public Health.

McGregor, P., R. Kaczmarek, V. Mosley, D. Dease, and P. Adams. 2006. "National Security/ Emergency Preparedness and the Next-Generation Network." *IEEE Communications Magazine*, May:133–143.

National Research Council. 2003. *The Internet under Crisis Conditions: Learning from September 11*. The National Academy Press, Washington, DC.

Office for Victims of Crime. 2001. OVC Handbook for Coping after Terrorism, September 2001. NCJ 190249, Office for Victims of Crime, Office of Justice Programs, U.S. Department of Justice, Washington, DC.

Office of Public Health Preparedness and Response, Centers for Disease Control and Prevention, Department of Health and Human Services. No Date. "Public Health Workbook to Define, Locate, and Reach Special, Vulnerable, and At-risk Populations in an Emergency." CS211575A. http://www.bt.cdc.gov/workbook/pdf/ph_workbookFINAL.pdf (accessed January 22, 2013).

Parkin, R., L. Ragain, R. Bruhl, H. Deutsch, and P. Wilborne-Davis. 2008. "Advancing Collaborations for Water-Related Health Risk Communication." Jointly published by the American Water Works Association Research Foundation, American Water Works Association, and IWA Publishing, Denver, Colorado.

Revkin, A. C. August 26, 2011. "As Irene Approaches, so Does Challenge of Heeding Warnings about Rare Threats." DotEarth, The New York Times blog. http://dotearth.blogs.nytimes.com/2011/08/26/as-irene-approaches-so-does-challenge-of-heeding-warnings-about-rare-threats/ (accessed January 22, 2013).

Sachs, Z., Y. L. Danon, R. Dycian, and Y. Shapiro. 1991. "Community Coordination and Information Centers during the Persian Gulf War." *Israel Journal of Medical Sciences*, 27:696–700.

Siegrist, M. and H. Gutscher. 2008. "Natural Hazards and Motivation for Mitigation Behavior: People Cannot Predict the Affect Evoked by a Severe Flood." *Risk Analysis*, 28(3): 771–778.

Slovic, P. 1999. "Trust, Emotion, Sex, Politics, and Science: Surveying the Risk-Assessment Battlefield." *Risk Analysis*, 19(4):689–701.

Texas DEM (Division of Emergency Management). 2003. "Media and Public Information Office (PIO) Observations." Texas DEM Web Site for WMD/Terrorism Domestic Preparedness.

The National Diet of Japan. 2012. The Official Report of the Fukushima Nuclear Accident Independent Investigation Commission. http://www.nirs.org/fukushima/naiic_report.pdf (accessed February 7, 2013).

Thomas, C. W. 1998. "Maintaining and Restoring Public Trust in Government Agencies and Their Employees." *Administration and Society*, 30(2):166–193.

Vanderford, M. L. 2004. "Breaking New Ground in WMD Risk Communication: The Pre-Event Message Development Project." *Biosecurity and Bioterrorism: Biodefense Strategy, Practice, and Science*, 2(3):193–194. U.S. Centers for Disease Control and Prevention, Atlanta, Georgia.

Vanderford, M. L., T. Nastoff, J. L. Telfer, and S. E. Bonzo. 2007. "Emergency Communication Challenges in Response to Hurricane Katrina: Lessons from the Centers for Disease Control and Prevention." *Journal of Applied Communication Research*, 35(1):9–25.

Wood, M. M., D. S. Mileti, M. Kano, M. M. Kelley, R. Regan, and L. B. Bourque. 2012. "Communicating Actionable Risk for Terrorism and Other Hazards." *Risk Analysis*, 32(4): 601–615.

Wray, R., J. Rivers, A. Whitworth, K. Jupka, and B. Clements. 2006. "Public Perceptions about Trust in Emergency Risk Communication: Qualitative Research Findings." *International Journal of Mass Emergencies and Disasters*, 24(1):45–75.

ADDITIONAL RESOURCES

Agency for Toxic Substances and Disease Registry. http://www.atsdr.cdc.gov (accessed January 22, 2013).

Association for Professionals in Infection Control and Epidemiology. http://www.apic.org (accessed January 22, 2013)

Association of State and Territorial Health Officials. 2002. "Communication in Risk Situations: Responding to the Communication Challenges Posed by Bio-Terrorism and Emerging Infectious Diseases." Washington, DC.

California Governor's Office of Emergency Services. 2001. "Risk Communication Guide for State and Local Agencies." Office of Emergency Services, Sacramento, California.

Canadian Centre for Emergency Preparedness. http://www.ccep.ca (accessed January 22, 2013).

CDC (Centers for Disease Control and Prevention). 2003. "A National Public Health Strategy for Terrorism Preparedness and Response 2003–2008." Atlanta, Georgia.

Federal Emergency Management Agency. http://www.fema.gov (accessed January 22, 2013).

New York State Health Department. "How to Create Effective Health Message for People with Disabilities." http://www.nirs.org/fukushima/naiic_report.pdf (accessed February 7, 2013).

Stuver, P. 2006. "Maximizing Emergency Communication." *Risk Management*, 53(5):30–34.

The Commons Lab. "Wilson Center's Science and Technology Innovation Program." Woodrow Wilson International Center for Scholars. Washington, DC. http://wilsoncommonslab.org/ (accessed January 22, 2013).

U.S. Department of Health and Human Services. 2002. "Communicating in a Crisis: Risk Communication Guidelines for Public Officials." Substance Abuse and Mental Health Services Administration, Rockville, Maryland. http://www.hhs.gov/od/documents/RiskCommunication.pdf (accessed February 7, 2013).

U.S. Department of Homeland Security, Ready.gov. http://www.ready.gov (accessed January 22, 2013).

22

INTERNATIONAL RISK COMMUNICATION

In writing this book, we focused heavily on the United States because it is our home country and, thus, our area of expertise. But we also know that an increasing number of our readers are communicating risks in other countries, or advising those who do. In fact, people from more than 20 countries are using previous editions of this book.

Why the rise in international risk communication? Some risks, such as foodborne illnesses, start in one country but travel rapidly around the world. And risk communication travels just as quickly. People use social media to report on earthquakes and other disasters literally while they are happening.

In our shrinking world, countries are looking to each other for communication strategies that have proven useful. For example, nine countries in Southwest and Central Asia, in one of the most seismically active areas of the world, joined in 2003 to share risk and knowledge-management strategies for emergency preparedness and recovery. U.S. officials sought flood control advice from the Netherlands, a country mostly under sea level, after Hurricane Katrina struck the United States in 2005.

> Some risks, such as foodborne illnesses, start in one country but travel rapidly around the world. And risk communication travels just as quickly.

Specific strategies for international risk communication appear throughout this book, in the appropriate sections. In addition, however, we offer these broad principles for communicating risks in counties outside the United States. Most are drawn from published research and case studies, or highly publicized events. One important caution: International risk communication is complicated by many factors, and this section merely touches

Risk Communication: A Handbook for Communicating Environmental, Safety, and Health Risks,
Fifth Edition. Regina E. Lundgren and Andrea H. McMakin.
© 2013 The Institute of Electrical and Electronics Engineers, Inc., and Regina E. Lundgren and
Andrea H. McMakin. Published 2013 by John Wiley & Sons, Inc.

the surface. To plan risk communication strategies for individual countries, be sure to consult experts from those locations and review country-specific research.

RECOGNIZE THE SIMILARITIES

Do not immediately assume that risk communication is wildly different worldwide. Many of the constraints, ethical issues, general principles, planning steps, and actions described in this book hold true regardless of the geographic area or whether you are conducting care, consensus, or crisis communication. For example, people everywhere may become angry when a risk is mishandled or not communicated properly. Parents of children who died in poorly built schools in the 2008 earthquake in Sichuan, China, expressed the same kind of outrage as Louisiana residents who had no way to evacuate after Hurricane Katrina struck.

The same is true for planning. Residents who live in high-danger coastal areas all need to understand what to do in case of a violent storm, whether their country calls it a hurricane, typhoon, or cyclone. And the responsible stakeholders, whether government officials, private aid groups, or media, need to know exactly how to respond in the emergency.

Start with key principles, but investigate further to ensure that they will work for your audience. Customize your activities accordingly. Look for multicountry guidance on particular risk situations as well. For example, the World Health Organization has published a guide to effective media communication during public health emergencies (WHO 2005). The International Atomic Energy Agency can provide good background materials on nuclear risks.

ACCOUNT FOR CULTURAL DIFFERENCES

Though some principles are universal, others are not. Each country and population group may have its own characteristics that affect how people perceive and communicate risks. These characteristics may include religious beliefs, health and environmental regulations, and community traditions, all of which can affect how people perceive and respond to risk information.

That culture affects risk communication is not a new concept. Cvetkovich and Earle (1990) described the concept of cultural relativism in risk communication, saying "Culture can provide powerful lenses for seeing through the fog of uncertainty." For example, the commission that investigated the 2011 Fukushima nuclear plant accident blamed the poor disaster response in part to cultural conditions. As the report stated, "What must be admitted—very painfully—is that this was a disaster 'Made in Japan'. Its fundamental causes are to be found in the ingrained conventions of Japanese culture: our reflexive obedience; our reluctance to question authority; our devotion to 'sticking with the program'; our groupism; and our insularity" (The National Diet of Japan 2012).

> Each country and population group may have its own characteristics that affect how people perceive and communicate risks. These characteristics may include religious beliefs, health and environmental regulations, and community traditions.

Researchers have suggested that risk perceptions in China draw from the Confucian heritage. Researchers have shown that people in Western countries are more likely to judge unknown or less controllable threats as more threatening than others. But Hong Kong Chinese in general judged less familiar threats as less troubling and something that can be thought about with reasonable calmness (Lai and Tao 2003). In that study, the authors explained the difference in part by the Confucian teachings about focusing on this life, not on things that cannot be controlled or understood through personal experience.

In another study, Chinese citizens clearly believed much more in the controllability of risks in nature, society, and technology than did those from Austria (Schmidt and Wei 2006). The researchers explained this discrepancy by the theory of reflexive modernization—the more modernized a country becomes, the less willing its citizens are to believe that scientific innovations are generally positive and can be mastered at will.

Perhaps nowhere is the difference in culture among countries shown than in the roles of women. The Gender and Disaster Network, a virtual international group of researchers and consultants interested in gender relations in disaster contexts, found that risk communication messages were often too general, failing to take into account cultural, age, and gender factors of the community. Risk communicators in other nations also frequently failed to include women and their networks when developing, conducting, and evaluating risk communication efforts (Gender and Disaster Network 2009).

Cultural differences also can mean how governments handle risk communication, and these differences can evolve over time. For example, during the Cold War, Russian scientists visiting a U.S. nuclear site expressed astonishment over the amount of environmental and health monitoring data collected and shared with the local community. "You will only make them afraid!" they warned.

Another example comes from Holland. The Dutch have had a past tradition of being a somewhat paternalistic society. Many local governments argued that it was better to not actively inform citizens about risks in their direct environment because this could lead to irrational fear in their communities. The general opinion was that governments should take appropriate measures and develop plans to protect citizens against risks, but citizens did not have to know about these risks (Meijer 2005). But a deadly 2000 fireworks factory explosion in an urban area changed that. Many citizens were outraged because they had no idea that such a factory existed near them. Dutch government organizations began paying more attention to risk management and communication of risks. In 2004, the government required risk maps for all its provinces—Internet-based maps showing the location of certain hazards in geographic areas. Though some people question whether these kinds of maps could inadvertently help terrorists, the maps caused government officials and companies to be much more transparent about industrial permits that involve hazards.

Some problems occur in more than one country but require very different communication approaches because of cultural and social differences and communication infrastructures. Hundreds of studies have been done on HIV/AIDS communication and prevention worldwide. Strategies that work in Thailand and India, for example, do not necessarily work in Brazil, the United States, or Africa, but some have proven adaptable for a variety of cultures. Stories, for example, have proven to be a universal way of presenting risk information (Hillier 2006). Noted health educators Rogers and Singhal (2003) describe how "culturally shareable" education/entertainment programs can reduce stigma and promote behaviors that reduce HIV/AIDS transmission.

Our advice is simple but essential. Know your audience! Understand the cultural attributes of the areas in which you are communicating and form your strategies around them.

LOOK FOR "YOUR" RISK IN OTHER COUNTRIES

When investigating international communication strategies, it is important to seek countries with social structures that have the most in common with the one you are targeting. In China, where more than two-thirds of men smoke, pregnant women were given skills to ask their men to stop smoking in ways that did not compromise family harmony (Lee 2008). In the United States, researchers designed a more overt approach. The "Not in Mama's Kitchen" campaign targeted African American females, many of whom head households, to reduce second-hand smoke by prohibiting smoking in their homes and cars (Bankston-Lee 2005). And in Russia and Finland, smoking cessation campaigns were based on social competition and positive role models (McAlister et al. 2000). The lesson: the same risk may require different strategies in different countries.

Of course, geographic similarity is no guarantee of similar reactions to risk. One study showed that people living in different parts of an urban neighborhood of Beirut, Lebanon, ranked environmental priorities, ranging from air quality to mismanagement of waste containers, differently (Abbas et al. 2006). The researchers attribute this difference to variations in socioeconomic status, locality, health, behavior, and environmental beliefs. They contend that identifying these "divide lines" among priorities can help bring about more inclusive and effective participatory environmental management.

> In China, where more than two-thirds of men smoke, pregnant women were given skills to ask their men to stop smoking in the home in ways that did not compromise family harmony.

When seeking to communicate risks in one country, it may be helpful to consider other countries that have experienced the same risks. Food quality and safety concerns have long been prominent in certain European countries and Canada, likely because of mad cow disease, the controversy over genetically modified foods, and various food scares in the past decade (De Jonge et al. 2004; French et al. 2005; Leiss and Powell 2005; Lofstedt 2006; Verbeke et al. 2007). Among the risk communication strategies used in Europe to boost consumer confidence are food traceability, food source labeling, and use of the precautionary principle, which states that when there is reasonable suspicion of harm, the lack of scientific certainty or consensus must not be used to postpone preventative action (Aslaksen et al. 2006; van Rijswijk et al. 2008).

Not surprisingly, countries that are prone to certain disasters typically have strong communication and preparedness in those areas. For example, a robust body of research exists on volcanic crises in Indonesia, the Caribbean, Mexico, and other areas of volcanic activity (for example, De la Cruz-Reyna and Tilling 2008; Haynes et al. 2008; Lavigne et al. 2008). Many researchers have documented the role of

> Among the risk communication strategies used in Europe to boost consumer confidence are food traceability, food source labeling, and use of the precautionary principle, which states that when there is reasonable suspicion of harm, the lack of scientific certainty or consensus must not be used to postpone preventative action.

cultural, religious, and socioeconomic factors in the willingness, or lack thereof, to evacuate from volcanic danger zones.

Seek out and learn from the risk communication successes and mistakes of other countries. As many countries have tragically learned, "We know best" and "It can't happen here" attitudes may engender patriotism, but they can make risk consequences worse. More positively, you can often improve and streamline your communication program by learning what others have done, then adapting it for your situation. For best results, pretest approaches, messages, and materials with your target audiences.

PLAN FOR CROSS-COUNTRY COMMUNICATION

Risk communication becomes more complicated when it crosses borders. Not only have you added cultures and languages, but you also may face competing regulations, political climates, and the need to coordinate organizations that are not necessarily accustomed to working with each other.

Laws about what constitutes a risk, and how to communicate them, vary from country to country. One prominent example is warning labels on tobacco packages. The United States simply requires the use of text-based health warnings, and, as of 2012, U.S. courts have turned down the Food and Drug Administration's request for cigarette companies to include large, graphic warnings on their packaging. But countries that ratified the 2005 World Health Organization tobacco control treaty, known as the Framework Convention on Tobacco Control, go further. They mandate large, graphic health warnings that cover at least 50% of the front and back of the tobacco package. More than 10 countries in North and South America, Asia, the South Pacific, Europe, and the Middle East require picture-based health warnings such as photos of diseased lungs (see examples in Chapter 14).

> Laws about what constitutes a risk, and how to communicate them, vary from country to country.

Another challenge is when governments coexist in a single country. U.S. military bases overseas and Native American lands in the United States are examples of autonomous governments within a country with a different culture. Make sure that you know which regulations apply to your situation and what your audience expects from the risk interaction, particularly in consensus communication efforts.

Sometimes regulations that vary across countries can cause confusion. For example, a 2004 U.S. study found chemicals in farmed salmon. In response, the United States recommended limiting consumption, whereas Canada told its citizens that salmon was perfectly safe. Despite the opposing recommendations, salmon sales dropped worldwide. The reason, two scholars say, is because of the risk communication message (Leiss and Nicol 2006). The U.S. study was a clear and prescriptive narrative that told consumers how much and which type of fish they could safely consume, whereas the Canadian message was much more vague, promising that salmon "does not pose a health risk to consumers, based on Canadian regulations." The authors contend that Health Canada has not done a good job of providing risk information to the public or understanding their perceptions, and that Canadians are not used to trusting the government for food risk issues. Though Health Canada added a new risk communication tool to its website in 2006, which provides warnings about potential health risks, the authors argued that this was not transparent or participatory enough.

Recognizing the cross-border nature of public health threats, more countries are formalizing agreements and coalitions to share information and cooperate on response activities. Risk communicators should inform themselves of any such agreements that may affect their activities.

For example, all member nations of the World Health Organization participate in an international agreement known as International Health Regulations. The agreement sets forth guidelines to monitor, report on, and respond to any events that could pose a threat to international public health, including certain diseases, as well as biological, chemical, radiological, and other threats. Each participating country designates a National Focal Point who is responsible for communications with the World Health Organization. Requirements include providing information-sharing links with health care facilities, national entry points, and other key operational areas, as well as maintaining a national public health emergency response plan. There are time requirements for reporting events and for responding to World Health Organization requests for information.

Building on the North American response to the H1N1 flu pandemic in 2009, Canada, the United States, and Mexico released the "North American Plan for Animal and Pandemic Influenza" in 2012 (U.S. Department of Health and Human Services 2012). The three countries established the policy framework to address not only communication among the relevant authorities but also a faster and more coordinated response to future outbreaks. The agreement describes actions to develop current contact lists, as well as sharing communication strategies and plans, public messaging, best practice strategies, and post-event evaluations. At a more tactical level, the Pan American Health Organization/World Health Organization (2009) created a communication strategy for pandemic influenza.

At least five countries formed the International Center of Excellence in Food Risk Communication in 2011. The organization of global food and health organizations, government agencies, academic institutions, and expert nonprofits was founded to have an international resource of food-specific risk communication materials enabling informed decision making to promote global health. The organization makes online resources available including best practices, guidance, learning modules, and research results. Other types of disasters also engender international cooperation. For example, nations abiding by the World Health Organization's tenants have agreed that all disaster management plans will be consistent.

European Union countries are bound by regulation to inform the general public about the risks of chemicals. To assist its members in this activity, the European Chemicals Agency, a regulatory authority, launched the Risk Communication Network in 2008, which brings together country representatives to communicate with the general public on the safe use of chemicals and the risks of substances. The agency published guidance on communicating information on the risks and safe use of chemicals (European Chemicals Agency 2010). Regulations also require that chemical hazards are clearly communicated to workers and consumers in the European Union through classification and labeling of chemicals.

Multicountry communication, done properly, can be very effective. In 2003, the European Food Safety Authority worked with its partners to communicate findings and recommendations regarding a sensitive issue: a potential carcinogen in baby food (Gassin and Van Geest 2006). The Authority coordinated the communications of its risk assessment findings with the European Commission, national food safety authorities, and other stakeholders to provide European consumers with an accurate and meaningful message. Though the resulting media coverage appeared a bit more alarmist in Great Britain than in other countries, it did not create a controversy or food scare among the public. The authors attribute this to the fact that the effort addressed public concerns and established the European Food Safety Authority as a credible information source.

The European Agency for Safety and Health at Work has run an annual European campaign since 2000, emphasizing a specific workplace issue each year (http://osha. europa.eu/en/campaigns). Topics have included, for example, musculoskeletal disorders, noise, dangerous substances, accident prevention, asbestos, and psychosocial risks. The Agency designs and develops resources for micro-, small- and mid-sized enterprises to help them assess their workplace risks, share knowledge, and apply good practices on safety and health. More than 20 countries participate, receiving campaign toolkits of information materials and other resources in more than 20 languages. Admirably, the agency conducts a wide variety of evaluations to gauge the effectiveness of the campaigns and publishes the results and recommendations on its website.

A situation in 2008 was not handled well, demonstrating the downside of inadequate multicountry risk communication. The chemical melamine was found in milk products in China, sickening thousands of babies and killing some. Once the crisis surfaced, countries began taking actions on their own, testing and yanking Chinese milk products from their shelves. Though the Chinese government launched an investigation, milk producers and government officials came under fire for not communicating about the hazard sooner.

> When a risk emerges in another country, and your organization is responsible for a similar risk in your location, activate your plan to explain what you are doing to prevent the risk and to mitigate it if it does occur.

When communicating across countries, research all applicable laws and regulations that affect the situation. Understand and gain agreement on the roles of all stakeholder organizations. When a risk emerges in another country, and your organization is responsible for a similar risk in your location, activate your plan to explain what you are doing to prevent the risk and to mitigate it if it does occur.

CHECKLIST FOR INTERNATIONAL RISK COMMUNICATION

In communicating risk outside the United States, or across countries:

- ☐ Cultural attributes, including risk perceptions, are identified and accounted for.
- ☐ Successful and less successful risk communication cases have been identified in other countries that have the most in common with those being targeted.
- ☐ Laws and regulations related to risk communication in the targeted countries have been identified.
- ☐ Existing multicountry agreements or coalitions have been integrated into planning.

REFERENCES

Abbas, E., R. Nasrallah, I. Nuwayhid, L. Kai, and J. Makhoul. 2006. "Why Do Neighbors Have Different Environmental Priorities? Analysis of Environmental Risk Perception in a Beirut Neighborhood." *Risk Analysis*, 26(2):423–435.

Aslaksen, I., B. Natvig, and I. Nordal. 2006. "Environmental Risk and the Precautionary Principle: Late Lessons from Early Warnings Applied to Genetically Modified Plants." *Journal of Risk Research*, 9(3):205–224. http://dx.doi.org/10.1080/13669870500419586 (accessed January 22, 2013).

Bankston-Lee, K. 2005. "Evolution of Not in Mama's Kitchen Secondhand Smoke Campaign: The California Experience." Presentation at the National Conference on Tobacco or Health, May 4–6, 2005, Chicago, Illinois.

Cvetkovich, C. T. and T. C. Earle. 1990. "Risk, Culture, and Psychology." *Cross-Cultural Psychology Bulletin*, 24:3–10.

De Jonge, J., L. Frewer, H. van Trijp, R. J. Renes, W. de Wit, and J. Timmers. 2004. "The Development of a Monitor for Consumer Confidence in Food Safety: Results of an Exploratory Study." *British Food Journal*, 106:837–849.

De la Cruz-Reyna, S. and R. I. Tilling. 2008. "Scientific and Public Responses to the Ongoing Volcanic Crisis at Popocatépetl Volcano, Mexico: Importance of an Effective Hazards-Warning System." *Journal of Volcanology and Geothermal Research*, 170:121–134.

European Chemicals Agency. 2010. "Guidance on the Communication of Information on the Risks and Safe Use of Chemicals." Version 1. ECHA-2010_G-21-EN, Helsinki, Finland. http://echa.europa.eu/documents/10162/13639/risk_communications_en.pdf (accessed January 22, 2013).

French, S., A. J. Maule, and G. Mythen. 2005. "Soft Modelling in Risk Communication and Management: Examples in Handling Food Risk." *Journal of the Operational Research Society*, 56:879–888.

Gassin, A. and I. Van Geest. 2006. "Communication in Europe on Semicarbazide and Baby Food." *Journal of Risk Research*, 9(8):823–832.

Gender and Disaster Network. 2009. *Gender Note #5: Women, Gender, and Disaster Risk Communication*. Gender and Disaster Network at http://www.gdnonline.org/ (accessed January 22, 2013).

Haynes, H., J. Barclay, and N. Pidgeon. 2008. "Whose Reality Counts? Factors Affecting the Perception of Volcanic Risk." *Journal of Volcanology and Geothermal Research*, 172:259–272.

Hillier, D. 2006. *Communicating Health Risks to the Public: A Global Perspective*. Gower Publishing, Aldershot, Hampshire, England.

Lai, J. C. and J. Tao. 2003. "Perception of Environmental Hazards in Hong Kong Chinese." *Risk Analysis*, 23(4):669–684.

Lavigne, F., B. De Coster, N. Juvin, F. Flohic, J. Gaillard, P. Texier, J. Morin, and J. Sartohadi. 2008. "People's Behaviour in the Face of Volcanic Hazards: Perspectives from Javanese Communities, Indonesia." *Journal of Volcanology and Geothermal Research*, 172:273–287.

Lee, A. 2008. "A Pilot Intervention for Pregnant Women in Sichuan, China on Passive Smoking." *Patient Education and Counseling*, 71:396–401.

Leiss, W. and A. Nicol. 2006. "A Tale of Two Food Risks: BSE and Farmed Salmon in Canada." *Journal of Risk Research*, 9(8):891–910. http://dx.doi.org/10.1080/13669870600924584 (accessed January 22, 2013).

Leiss, W. and D. Powell. 2005. *Mad Cows and Mother's Milk: The Perils of Poor Risk Communication*, 2nd ed. McGill-Queen's University Press, Montreal, Quebec, Canada.

Lofstedt, R. E. 2006. "How Can We Make Food Risk Communication Better: Where Are We and Where Are We Going?" *Journal of Risk Research*, 9(8):869–890.

McAlister, A. L., T. Gumina, E. Urjanheimo, T. Laatikainen, M. Uhanov, R. Oganov, and P. Pekka. 2000. "Promoting Smoking Cessation in Russian Karelia: A 1-year Community-Based Program with Quasi-Experimental Evaluation." *Health Promotion International*, 15(2):109–112.

Meijer, A. J. 2005. "Risk Maps on the Internet: Transparency and the Management of Risks." *Information Policy*, 10:105–113.

Pan American Health Organization/World Health Organization. 2009. "Creating a Communication Strategy for Pandemic Influenza." Washington, DC. http://www.paho.org/English/AD/PAHO_CommStrategy_Eng.pdf (accessed January 22, 2013).

Rogers, E. M. and A. Singhal. 2003. *Combating AIDS: Communication Strategies in Action*. Sage Publications, New Delhi, India.

Schmidt, M. R. and W. Wei. 2006. "Loss of Agro-Biodiversity, Uncertainty, and Perceived Control: A Comparative Risk Perception Study in Austria and China." *Risk Analysis*, 26(2):455–470.

The National Diet of Japan. 2012. *The Official Report of the Fukushima Nuclear Accident Independent Investigation Commission*. http://www.nirs.org/fukushima/naiic_report.pdf (accessed February 7, 2013).

U.S. Department of Health and Human Services. 2012. "North American Plan for Animal and Pandemic Influenza." Office of the Assistant Secretary for Preparedness and Response, Washington, DC. http://www.phe.gov/Preparedness/international/Documents/napapi.pdf (accessed January 22, 2013).

van Rijswijk, W., L. J. Frewer, D. Menozzi, and G. Faioli. 2008. *Food Quality and Preference*, 19:452–464.

Verbeke, W., L. J. Frewer, J. Scholderer, and H. F. De Brabander. 2007. "Why Consumers Behave as They Do with Respect to Food Safety and Risk Information." *Analytica Chimica Acta*, 586:2–7.

WHO (World Health Organization). 2005. *Effective Media Communication during Public Health Emergencies*. http://www.who.int/csr/resources/publications/WHO_CDS_2005_31/en/ (accessed January 22, 2013).

ADDITIONAL RESOURCES

Barrantes, S. A., M. Rodriguez, and R. Perez. 2009. *Information Management and Communication in Emergencies and Disasters*. Pan American Health Organization, Regional Office of the World Health Organization, Washington, DC.

Campaign for Tobacco-Free Kids. http://www.tobaccofreekids.org/index.php (accessed January 22, 2013).

Centers for Disease Control and Prevention Global Health Marketing. http://www.cdc.gov/healthmarketing/ihm.htm (accessed January 22, 2013).

European Chemicals Agency. 2012. "Communication on the Safe Use of Chemicals: Study on the Communication of Information to the General Public." ECHA-12-A-01-EN, Helsinki, Finland. http://echa.europa.eu/documents/10162/13559/clp_study_en.pdf (accessed January 22, 2013).

George, A. M. and C. B. Pratt, eds. 2012. *Case Studies in Crisis Communication: International Perspectives on Hits and Misses*. Routledge, New York.

Ungar, S. 2008. "Global Bird Flu Communication: Hot Crisis and Media Reassurance." *Science Communication*, 29(4):472–497. http://scx.sagepub.com/cgi/content/abstract/29/4/472 (accessed January 22, 2013).

23

PUBLIC HEALTH CAMPAIGNS

Public health campaigns are a specific type of care communication that are almost ubiquitous. They are designed to prompt long-term changes in knowledge, attitudes, behaviors, and, sometimes, public policy. For example, campaigns may encourage people to get their children vaccinated, reduce pesticide use, drive without texting, increase regular exercise, reduce yard-waste burning on high-smog days, know what to do in case of a fire, or handle raw food safely.

Social marketing is another method that is closely related to public health campaigns. Social marketing borrows concepts and techniques from commercial sector marketing, such as customer segmentation and advertising, to promote increases in knowledge, behavior change, and social change. How the "customer" thinks and acts continuously shapes the marketing process.

A large body of research and case studies exists on the design, implementation, and measurement of public health and social marketing campaigns. A sample of these published studies is listed in Additional Resources at the end of this chapter. The wise campaign planner will draw on the lessons of others to make the best use of the significant time and funding necessary for an effective campaign. Here, we present some guidelines based on research and best practices. Though many of the principles described in this book also apply to public health campaigns, we highlight particular guidelines here that are worth special mention.

Risk Communication: A Handbook for Communicating Environmental, Safety, and Health Risks,
Fifth Edition. Regina E. Lundgren and Andrea H. McMakin.
© 2013 The Institute of Electrical and Electronics Engineers, Inc., and Regina E. Lundgren and
Andrea H. McMakin. Published 2013 by John Wiley & Sons, Inc.

UNDERSTAND YOUR GOALS

Understanding the goal of the campaign drives the communication plan, including measurements of effectiveness. Design your campaign to achieve one or more goals such as:

- Influencing knowledge and attitudes about a behavior and its consequences
- Changing behavior
- Increasing the visibility of an issue and its importance
- Affecting perceptions of social issues and who is seen as responsible
- Increasing knowledge about solutions
- Affecting criteria used to judge policies and policy makers
- Engaging and mobilizing constituencies to action to develop or change policy

You may wish to set quantifiable objectives. Examples from some campaigns include the following:

- Increase vaccinations by X percent in a certain region
- Increase the percent of a specific population obtaining cancer screenings by X
- Increase nonfat and sugar-free drink choices in X school district vending machines by X percent
- See X number of exercise-tracking apps downloaded
- Get legislation changed to require cigarette companies to include color photos of negative smoking effects on their packaging
- Get smoking eliminated from public places in certain geographic areas
- Require suspected concussions to be evaluated by medical professionals before high-school athletes are allowed to reenter games
- Have X percent of people check the "organ donation" box when renewing their drivers' licenses

Be careful, however, when setting goals in cases where there is no clear health behavior that is best for the group you are targeting. For example, men with elevated prostate-specific antigen test results could decide to have radiation therapy, another treatment, or do nothing. These choices will depend on their values. In this case, the goal would more likely be ensuring that these men understand the risks and benefits of their choices, rather than targeting a specific behavior (Fischhoff et al. 2011).

USE RESEARCH TO DESIGN CAMPAIGNS

Formative research is used to *form* the communication, helping designers choose content, format, and delivery strategies through participatory (user-centered) design (Fischhoff et al. 2011). Formative research emphasizes the target audience's understanding of the risk. Campaign planners have long practiced two kinds of research for message design: (1) research to determine audience predispositions and (2) research to pretest messages and materials for comprehensibility and response.

Some types of formative research include baseline surveys, usability testing, focus groups (Chapter 17), and the mental models approach (Chapters 2 and 9). A research-based approach takes the guesswork out of communication development, especially when the risk communicator is not part of the target audience and thus does not know how the audience thinks. For example, we know one professor whose research showed that wearing seat belts became more important to young men when they learned how their genitalia could be affected in vehicle accidents.

Baseline surveys can be helpful tools for formative research. These surveys are taken before a public health campaign to determine audience perceptions, knowledge, and even willingness to try certain interventions. For example, a mass media campaign in South Carolina was designed to promote public action to reduce abuse of children in families plagued by alcohol or drugs (Andrews et al. 1995). A public survey was conducted 1 year before the campaign to gauge opinions about child abuse and the respondents' likelihood of helping families where present. The results of this survey were used to design specific themes, messages, and materials for the campaign. One of the media forums was a televised talk show, which attracted higher viewership than the regular program, Entertainment Tonight, almost unheard of for a public affairs show. Most encouraging was a 62% increase in the number of people who called a phone service each month for information about how to help abused children.

> Baseline surveys can be helpful tools for formative research.

Usability testing is another important aspect of formative research and involves pretesting messages, visuals, and sometimes actions with the intended audience. A good example comes from the Stanford heart disease prevention project, a multicommunity campaign designed to improve cardiovascular health. One of the project's planned messages recommending all-season jogging was modified when it was realized that California winters can be too rainy and jogging paths can be too muddy for the less than highly motivated runners (Rogers and Storey 1987). The U.S. Environmental Protection Agency presumably did not pretest its advertisement to alert Americans to radon gas in the home, where children were shown turning into skeletons after they were exposed to radon (Moore 1997). The ad created such a protest that it was cancelled.

For ideas about how to research your audiences and their perceptions of the issues, see Chapter 8.

USE MULTIPLE METHODS TO REACH PEOPLE

In general, to raise awareness in an audience about a particular health risk or motivate them to take some action, audience members need to receive the information more than once. Different audiences also look for health information in different places, and various members may have varying learning styles. Given all that, it is important to use several channels, such as the news media, paid placements, independent coverage, and online venues, to ensure that your information is reaching those at risk.

News Media

More exposure means more opportunities to reach people. Base your selection of media channels on your audience analysis. Understand your community's media access points—where and how certain topics are covered in various media channels (Wallack et al. 1993). Consider going beyond traditional news programming. Television, for example, has magazine news shows, public affairs programming, and free speech or editorial announcements. Health topics may fit well with the lifestyle, financial, and business sections of newspapers as well as editorial content. Radio has talk shows, editorials, and public service announcements.

Remember that certain channels and formats are accessible only to some segments of the community. For example, in a campaign to lower cardiovascular disease in two California communities, public health experts found that a regular newspaper column on health was read by some members of the community and not others (Fortmann et al. 1995). In contrast, the child abuse prevention campaign in South Carolina, mentioned earlier, used an effective combination of media delivery channels and other formats. People heard about the problem of child abuse through television, billboards, posters, and print publications. After public service announcements were broadcast, people were invited to call a toll-free number for help or information.

A multimedia project directed at reducing teen pregnancy in the Ohio area used paid airtime on radio, supplemented by public service announcements in other media. People could call a hotline number for information and appointments. The campaign reached 80% of teens in Columbus over 7 weeks, resulting in 1000 calls per month at its peak (Taplin 1981).

The M.D. Cancer Center in Houston ran a campaign to reduce sun exposure behavior in several large cities in Texas from 1990 through 1992. In addition to radio and TV interviews, public service announcements in English and Spanish, and press conferences, the campaign also used 1-minute segments on six children's TV shows, live radio interviews with listener call-ins, and "Day at the Dome" baseball game publicity that included ticket giveaways on the radio before the game. The campaign reached more than 1 million people in three cities and resulted in a significant increase in people saying that they had taken actions to reduce the risk of skin cancer (Gelb et al. 1994).

Paid Placements and Independent Coverage

Media messages associated with public health issues can be either paid or independent. Paid placements, such as television and radio spots, enable you to control the content of the message, the audience it reaches, and its timing. The success of the effort is more easily measured because the audience is receiving a consistent, predetermined message. One disadvantage is that the message can be seen as biased or self-serving, for example, beer distribution companies warning about the dangers of underage drinking. Another disadvantage of mass media advertising is that it can require a large chunk of the campaign budget and may not necessarily target the population most at risk, especially if they do not use the traditional mass media channels.

Health campaign managers can also provide information to the media through press releases and public service announcements. Media representatives then decide whether and how they pass along the information to the public. This is a low- or no-cost option, but with reduced control by the health campaign managers. Campaign planners should know their media gatekeepers' interests and potential conflicts of interest, whether reporters may feel compelled to obtain an opposing point of view, and whether this works to the advantage or disadvantage of the campaign.

> Campaign planners should know their media gatekeepers' interests and potential conflicts of interest, whether reporters may feel compelled to obtain an opposing point of view, and whether this works to the advantage or disadvantage of the campaign.

Media outlets also can independently cover a health campaign or the issue the campaign is featuring. Such coverage can create a tremendous support for the issue at hand. *The Alabama Journal*, a regional newspaper, took an advocacy position in a series of stories published in 1987 on state infant mortality. The series focused on real peoples' lives, how Alabama officials had failed to address the issue, the economic and social costs to the state, and what other states had done to reduce the problem. The newspaper sent reprints to 5000 Alabaman opinion leaders, who used them as lobbying tools. Citizens and reporters kept the pressure on the governor and legislators, urging them to take action to reduce the problem. In the 2 years after the series was published, the state legislature instituted several policy changes to combat the problem, and infant mortality rates dropped. Researcher Kim Walsh-Childers (1994) concluded that the newspaper series, which won a Pulitzer Prize, was the critical factor in accelerating public support for policy changes and in creating pressure on legislators to make those changes.

A potential downside of independent coverage is possible inconsistency with the message of a paid campaign. One example is mass media coverage in the 1980s about the connection between aspirin and Reye's syndrome, a potentially fatal illness occurring in children with flu or other viruses (Soumerai et al. 1992). A commonly occurring theme in media coverage was the battle between the consumer groups and the aspirin industry, which was fighting the Food and Drug Administration's proposal for warning labels on aspirin.

Some practitioners use an approach they call media advocacy, wherein community organizations and ad hoc groups work with the media to promote health policies aimed at fundamental social change (Wallack et al. 1993, 1999). This approach can combine the best of paid placement and independent coverage, but it must serve the agendas of all those participating. See Chapter 16 for more information on working with the news media.

Online Interventions and Social Media

Research has found that online health interventions can influence behaviors such as reducing binge drinking, increasing exercise, and managing weight. The advantage of individual online intervention is the relatively low cost and broad reach, with the added motivation of personalization.

Such interventions typically engage individual users in "relationships" that resemble the support offered by dieticians, fitness trainers, and other professionals. The interventions usually inform users about the consequences of their behavior, help them set and achieve goals, teach them skills, and provide pressure to change. Feedback mechanisms are common, with many interventions using tailoring and personalization and offering services to track and report users' progress toward their goals. A review of 30 online interventions yielded the following guidelines (Cugelman et al. 2011):

> Online interventions typically engage individual users in "relationships" that resemble the support offered by dieticians, fitness trainers, and other professionals.

- **Blend interpersonal online systems with mass media outreach.** This combination may help individuals achieve personal goals that help them improve the quality of their lives and ultimately lead to healthier societies.

- **Use short interventions to cope with rapid attrition and loss of motivation over time.** Note that this recommendation may be less applicable to demanding change processes, such as tobacco cessation or weight loss.
- **Design goals strategically.** Design interventions around goals that appeal to the target audience while offering tailored support to help participants who may lack motivation or ability.
- **Use adherence systems.** For example, consider having the system e-mail or text users with a reminder message if they do not log into the system within a certain time period.

One caution when using electronic systems with individual users is to ensure consent and confidentiality. Particularly with health-related information, participants on both ends need documented ways to ensure that they have agreed to who has access to personal information and how it is protected.

The use of social media is gaining importance in public health campaigns. The U.S. Centers for Disease Control and Prevention has developed several integrated social media campaigns, targeting topics including H1N1 flu, the salmonella outbreak associated with peanut products, and annual seasonal flu vaccination. For an ongoing heart health campaign, Centers for Disease Control and Prevention (CDC) created a variety of embeddable tools that partners can use to share information, including audio podcasts, video, eCards, and text messages. Tools with portable content, such as widgets and online video, enable users to share messages and become health advocates. See Chapter 19, Social Media, in this book for more information on using social media to communicate risks.

Other Methods

For many years, risk communicators relied on the mass media as the primary method of spreading information about public health risks. Research on public health campaigns, however, shows that the mass media are not the only channels to which people pay attention and are not necessarily the most credible for changing attitudes and behaviors (Rogers and Storey 1987). For the greatest chance of success, the message must be reinforced through other communication channels such as opinion leaders and community groups.

The SmokeFree Resource Centre of the United Kingdom's National Health Service has had success with face-to-face events such as in shopping centers with phone follow-up; a Facebook page that brings together an online community of smokers, ex-smokers, and National Health Service advisers; and extensive resources for midwives and employers. An anti-alcohol abuse campaign targeting Michigan State University students used media interviews and public service announcements, flyers, e-mails to college students, letters to alcohol vendors, theater troops, and alcohol-free events (Witte et al. 2001). Interpersonal communication has been shown to play a crucial role in changing strongly held attitudes and motivating behavior change. The Stanford heart health campaign used community leaders and support groups to disseminate information and to persuade by example (Kim 1985).

EVALUATE SUCCESS

In evaluating public health campaigns, researchers typically attempt to measure one or more of the attributes shown in Table 23-1. Formative evaluation, discussed earlier, can help create a targeted risk communication campaign. Process evaluation can improve and

Table 23-1. Evaluation in public health campaigns*

Evaluation type	Definition/purpose	Example questions
Formative	Assesses the strengths and weaknesses of campaign materials and strategies before or during the campaign's implementation.	How does the campaign's target audience think about the issue?
		What messages work with what audiences?
		Who are the best messengers?
		What actions will the audience be capable of taking?
Process	Measures effort and the direct outputs of campaigns—what and how much was accomplished. Examines the campaign's implementation and how the activities involved are working.	How many materials have been put out?
		How many people have been reached?
Outcome	Measures effect and changes that result from the campaign.	Has there been any affective change (beliefs, attitudes, social norms)?
	Assesses outcomes in the target populations or communities that come about as a result of strategies and activities	Has there been any behavior change?
	Also measures policy changes.	Have any policies changed?
Impact	Measures community-level change or longer term results that are achieved because of the campaign's aggregate effects on individuals' behavior and the behavior's sustainability.	Has the behavior resulted in its intended outcomes (for example, lower cancer rates and less violence in schools)?
	Attempts to determine whether the campaign caused the effects.	Has there been any systems-level change?

*Adapted from the National Cancer Institute (1992).

document its delivery, and outcome/impact evaluation can quantify its effects. Campaign managers should evaluate the effort partway through, if possible, to make any necessary midcourse corrections.

An important caution is to measure the desired outcome, such as behavior or knowledge, as directly as possible. Self-reported attitudes or intentions to act are notoriously unreliable because they may be based on a desire to help the researcher, or to appear well intentioned.

The most rigorous outcome evaluations compare the outcomes for a group that was part of a risk communication effort to a group that was not part of it but is otherwise equivalent (Rosen et al. 2006). The strongest design is a randomized controlled trial, with participants randomly assigned to a group receiving the communication or to a control group, making it possible to attribute differences in outcomes to the communication. Of course, sometimes randomized controlled trials are not practical, such as when behaviors are influenced by many uncontrollable factors or budgets are extremely limited. In such cases, a quasi-experimental design

> Self-reported attitudes or intentions to act are notoriously unreliable because they may be based on a desire to help the researcher, or to appear well intentioned.

can still be useful, such as systematically exploring differences between people who were and were not exposed to the communication. A less rigorous, but typically less costly, outcome evaluation is pre- and postcommunication testing: Knowledge or behaviors are measured in the same population before and after the communication campaign.

Evaluation is difficult because many factors other than media coverage contribute to change at the personal and policy level. However, some typical evaluation methods for campaigns are media monitoring, website statistics, ad assessments, case studies, and surveys.

Lawrence Wallack, professor at the University of California-Berkeley and Director of the Berkeley Media Studies Group, advocates broad avenues of inquiry to assess campaign effects. These avenues include surveying and observing individuals in the target population; examining institutional records of individual behavior; interviewing those who interact with the individuals; and investigating institutional changes in the legal, business, industrial, or education systems. Official statistics, such as drunk driving arrests, sales data, and hospital emergency room data, also can be used as indicators of success (Wallack 1981).

Some of the advice in Chapter 20 may be helpful in evaluating public health campaigns.

CHECKLIST FOR PUBLIC HEALTH CAMPAIGNS

For public health campaigns, ensure that:

☐ Campaign goals and objectives are attainable.
☐ The campaign has been designed using formative research, including usability testing of information materials.
☐ Communication channels have been selected based on audience analysis.
☐ More than one communication channel is being used.
☐ Online interventions and social media are used appropriately.
☐ The effectiveness of the campaign has been assessed with one or more of these types of evaluation:
   ☐ Formative
   ☐ Process
   ☐ Outcome
   ☐ Impact
☐ The campaign is evaluated for midcourse corrections.
☐ Desired outcomes are measured as directly as possible, rather than relying on self-reported attitudes or intentions to act.

REFERENCES

Andrews, A. B., D. G. McLeese, and S. Curran. 1995. "The Impact of a Media Campaign on Public Action to Help Maltreated Children in Addictive Families." *Child Abuse and Neglect*, 19:921–932.

Cugelman, B., M. Thelwall, and P. Dawes. 2011. "Online Interventions for Social Marketing Health Behavior Change Campaigns: A Meta-Analysis of Psychological Architectures and Adherence

Factors." *Journal of Medical Internet Research*, 13(1):e17. http://www.ncbi.nlm.nih.gov/pmc/articles/PMC3221338/ (accessed January 22, 2013).

Fischhoff, B., N. T., Brewer, and J. S., Downs 2011. "Communicating Risks and Benefits: An Evidence-Based User's Guide." U.S. Food and Drug Administration, U.S. Department of health and Human Services, Silver Spring, Maryland. http://www.fda.gov/downloads/AboutFDA/ReportsManualsForms/Reports/UCM268069.pdf (accessed January 22, 2013).

Fortmann, S. P., J. A. Flora, M. A. Winkleby, C. Schooler, C. B. Taylor, and J. W. Farquhar. 1995. "Community Intervention Trials: Reflections on the Stanford Five-City Project Experience." *American Journal of Epidemiology*, 142:576–586.

Gelb, B. D., W. B. Boutwell, and S. Cummings. 1994. "Using Mass Media Communication for Health Promotion: Results from a Cancer Center Effort." *Hospital and Health Services Administration*, 39(3):283–293.

Kim, Y. 1985. *Opinion Leadership in a Preventive Health Campaign*. Unpublished doctoral dissertation, Stanford University, Stanford, California.

Moore, C. C. 1997. *Haunted Housing: How Toxic Scare Stories Are Spooking the Public Out of House and Home*. Cato Institute, Washington, DC.

National Cancer Institute. 1992. *Making Health Communication Programs Work: A Planner's Guide*. U.S. Department of Health and Human Services, Washington, DC.

Rogers, E. M. and J. D. Storey. 1987. "Communication Campaigns." In C. Berger and S. H. Chaffee, eds., *Handbook of Communication Science*. Sage Publications, Newbury Park, California, pp. 419–445.

Rosen, L., O. Manor, D. Engelhard, and D. Zucker. 2006. "In Defense of the Randomized Controlled Trial for Health Promotion Research." *American Journal of Public Health*, 96:1181–1186.

Soumerai, S. B., D. Ross-Degnan, and J. S. Kahn. 1992. "Effects of Professional and Media Warnings about the Association between Aspirin Use, Children, and Reye's Syndrome." *The Milbank Quarterly*, 70:155–183.

Taplin, S. 1981. "Family Planning Communication Campaigns." In R. E. Rice and W. J. Paisley, eds., *Public Communication Campaigns*. Sage Publications, Beverly Hills, California, pp. 127–142.

Wallack, L. 1981. "Mass Media Campaigns: The Odds against Finding Behavior Change." *Health Education Quarterly*, 8:209–260.

Wallack, L., L. Dorfman, D. Jernigan, *and M. Themba*. 1993. *Media Advocacy and Public Health: Power for Prevention*. Sage Publications, Newbury Park, California.

Wallack, L., K. Woodruff, L. Dorfman, and I. Diaz. 1999. *News for a Change: An Advocate's Guide to Working with the Media*. Sage Publications, Newbury Park, California.

Walsh-Childers, K. 1994. "A Death in the Family—A Case Study of Newspaper Influence on Health Policy Development." *Journalism Quarterly*, 71:8220–8829.

Witte, K., G. Meyer, and D. Martell. 2001. *Effective Health Risk Messages: A Step-By-Step Guide*. Sage Publications, Thousand Oaks, California.

ADDITIONAL RESOURCES

Bennett, P. and K. Calman. 1999. *Risk Communication and Public Health*. Oxford University Press, New York.

Coffman, J. 2002. "Public Communication Campaign Evaluation: An Environmental Scan of Challenges, Criticisms, Practice, and Opportunities." Harvard Family Research Project, Cambridge, Massachusetts. Prepared for the Communications Consortium Media Center, Washington, DC. http://www.mediaevaluationproject.org/HFRP.pdf (accessed January 22, 2013).

Desvousges, W. H. 1991. "Integrating Evaluation: A Seven-Step Process." In A. Fisher, M. Pavlova, and V. Covello, eds., *Evaluation and Effective Risk Communications Workshop Proceedings*. U.S. Environmental Protection Agency, Washington, DC, pp. 119–123. EPA/600/9-90/054.

Desvousges, W. H. and V. K. Smith. 1988. "Focus Groups and Risk Communication: The 'Science' of Listening to Data." *Risk Analysis*, 8(4):479–484.

Fischhoff, B. 1989. "Helping the Public Make Health Risk Decisions." In V. T. Covello, D. B. McCallum, and M. T. Pavlova, eds., *Effective Risk Communication: The Role and Responsibility of Government and Nongovernment Organizations*. Plenum Press, New York, pp. 111–116.

Kline, M., C. Chess, and P. Sandman. 1989. *Evaluating Risk Communication Programs: A Catalog of "Quick and Easy" Feedback Methods*. Rutgers University, Cook College, Environmental Communication Research Program, New Brunswick, New Jersey.

Santos, S. L. 1990. "Developing a Risk Communication Strategy." *Management and Operations*, November:45–49.

Smith, V. K., W. H. Desvousges, A. Fisher, and F. R. Johnson. 1987. *Communicating Radon Risk Effectively: A Mid-Course Evaluation*. U.S. Environmental Protection Agency, Office of Policy Analysis, Washington, DC.

RESOURCES

The field of risk communication continues to grow. A number of studies have been conducted, articles and books written, and seminars constructed that present information that can be useful to the risk communicator. Some of these resources are listed below. Resources are first grouped topically, by the type of communication (general, environmental, safety, and health) and then by purpose (care, consensus, and crisis communication). References used in the book and additional resources can be found at the end of each section and are only repeated here if they provide more general resource material.

GENERAL RISK COMMUNICATION RESOURCES

Chess, C., B. J. Hance, and P. M. Sandman. 1989. *Planning Dialogue with Communities: A Risk Communication Workbook*. Rutgers University, Cook College, Environmental Communication Research Program, New Brunswick, New Jersey.

Covello, V. T. and F. W. Allen. 1988. *Seven Cardinal Rules for Risk Communication*. OPA-87-020, U.S. Environmental Protection Agency, Washington, DC.

Covello, V. T., P. M. Sandman, and P. Slovic. 1988. *Risk Communication, Risk Statistics, and Risk Comparisons: A Manual for Plant Managers*. Chemical Manufacturers Association, Washington, DC.

Covello, V. T., D. B. McCallum, and M. T. Pavlova, eds. 1989. *Effective Risk Communication: The Role and Responsibility of Government and Nongovernment Organizations*. Plenum Press, New York.

Risk Communication: A Handbook for Communicating Environmental, Safety, and Health Risks,
Fifth Edition. Regina E. Lundgren and Andrea H. McMakin.
© 2013 The Institute of Electrical and Electronics Engineers, Inc., and Regina E. Lundgren and
Andrea H. McMakin. Published 2013 by John Wiley & Sons, Inc.

Davies, J. C., V. T. Covello, and F. W. Allen, eds. 1987. *Risk Communication: Proceedings of the National Conference on Risk Communication, held in Washington, DC, January 1986.* Conservation Foundation, Washington, DC.

Fischhoff, B., N. T. Brewer, and J. S. Downs, eds. 2012. *Communicating Risks and Benefits: An Evidence-Based User's Guide.* U.S. Food and Drug Administration. http://www.fda.gov/AboutFDA/ReportsManualsForms/Reports/ucm268078.htm (accessed January 23, 2013).

Hance, B. J., C. Chess, and P. M. Sandman. 1988. *Improving Dialogue with Communities: A Risk Communication Manual for Government.* New Jersey Department of Environmental Protection, Division of Science and Research, Trenton, New Jersey.

Hance, B. J., C. Chess, and P. M. Sandman. 1990. *Industry Risk Communication Manual.* CRC Press/Lewis Publishers, Boca Raton, Florida.

Kasperson, R. E. 1986. "Six Propositions on Public Participation and Their Relevance for Risk Communication." *Risk Analysis*, 6:275–281.

Lundgren, R. E. Consultant and Trainer in Risk Communication, Public Involvement, and Science and Strategic Communication. http://www.rlriskcom.com for more information (accessed January 23, 2013).

Morgan, M. G., B. Fischhoff, A. Bostrom, and C. J. Atman. 2002. *Risk Communication: A Mental Models Approach.* Cambridge University Press, Cambridge, United Kingdom.

National Research Council. 1989. *Improving Risk Communication.* National Academy Press, Washington, DC.

National Research Council. 1996. *Understanding Risk: Informing Decisions in a Democratic Society.* National Academy Press, Washington, DC.

Navy Environmental Health Center. No Date. *Risk Communication Primer: Tools and Techniques.* http://www.med.navy.mil/sites/nmcphc/Documents/policy-and-instruction/nmcphc-risk-communications-primer.pdf (accessed February 7, 2013).

Persensky, J., S. Browde, A. Szabo, L. Peterson, E. Specht, and E. Wright. 2004. *Effective Risk Communication: The Nuclear Regulatory Commission's Guidelines for External Risk Communication.* NUREG/BR-0308, U.S. Nuclear Regulatory Commission, Washington, DC.

Santos, S. L. 1990. "Developing a Risk Communication Strategy." Management and Operations, November:45–49.

Society for Risk Analysis. http://www.sra.org (accessed January 23, 2013).

Tucker, W. T. S. Ferson, A. M. Finkel, and D. Slavin, eds. 2008. *Strategies for Risk Communication: Evolution, Evidence, and Experience*, Vol. 1128. Annals of the New York Academy of Sciences, New York.

U.S. Environmental Protection Agency. 1987. *Risk Assessment, Management, and Communication: A Guide to Selected Sources.* EPA 1MSD/87-002, U.S. Environmental Protection Agency, Office of Information Resources Management and Office of Toxic Substances, Washington, DC.

U.S. Environmental Protection Agency. 2002. *Community Culture and the Environment: A Guide to Understanding a Sense of Place.* EPA 842-B-01-003, Office of Water, Washington, DC.

ENVIRONMENTAL RISK COMMUNICATION RESOURCES

Environmental Education and Training Partnership. http://www.eetap.org (accessed January 23, 2013).

Hance, B. J., C. Chess, and P. M. Sandman. 1988. *Improving Dialogue with Communities: A Risk Communication Manual for Government.* New Jersey Department of Environmental Protection, Division of Science and Research, Trenton, New Jersey.

Hance, B. J., C. Chess, and P. M. Sandman. 1990. *Industry Risk Communication Manual.* CRC Press/Lewis Publishers, Boca Raton, Florida.

Krimsky, S. and A. Plough. 1988. *Environmental Hazards: Communicating Risks as a Social Process*. Auburn House, Dover, Massachusetts.

North American Association for Environmental Education. http://www.naaee.net/ (accessed January 23, 2013).

Sachsman, D. B., M. R. Greenberg, and P. M. Sandman, eds. 1988. *Environmental Reporter's Handbook*. Rutgers University, Cook College, Environmental Communication Research Program, New Jersey Agricultural Experiment Station, New Brunswick, New Jersey.

Sandman, P. M. 1986. *Explaining Environmental Risk*. U.S. Environmental Protection Agency, Office of Toxic Substances, Washington, DC.

Sandman, P. M., D. B. Sachsman, and M. R. Greenberg. 1988. *The Environmental News Source: Providing Environmental Risk Information to the Media*. New Jersey Institute of Technology, Hazardous Substance Management Research Center, Risk Communication Project, Newark, New Jersey.

U.S. Environmental Protection Agency. 1992. *Community Relations in Superfund: A Handbook*. EPA/540/G-88/002, U.S. Environmental Protection Agency, Office of Emergency and Remedial Response, Washington, DC.

U.S. Environmental Protection Agency. 2002. *Community Culture and the Environment: A Guide to Understanding a Sense of Place*. EPA 842-B-01-003, Office of Water, Washington, DC.

SAFETY RISK COMMUNICATION RESOURCES

National Institute for Occupational Safety and Health. http://www.cdc.gov/niosh/ (accessed January 23, 2013).

National Safety Council. http://www.nsc.org (accessed January 23, 2013).

No author. 1998, revised. *Chemical Hazard Communication*. OSHA 3084, Occupational Safety and Health Administration, Washington, DC. http://www.osha.gov/Publications/osha3084.pdf (accessed January 23, 2013).

Occupational Safety and Health Administration. http://www.osha.gov (accessed January 23, 2013).

Society for Risk Analysis. http://www.sra.org (accessed January 23, 2013).

HEALTH RISK COMMUNICATION RESOURCES

Agency for Toxic Substances and Disease Registry. http://www.atsdr.cdc.gov (accessed January 23, 2013).

Agency for Toxic Substances and Disease Registry. No Date. *A Primer on Health Risk Communication Principles and Practices*. http://www.atsdr.cdc.gov/risk/riskprimer/index.html (accessed January 23, 2013).

Baram, M. S. and P. Kenyon. 1986. "Risk Communication and the Law for Chronic Health and Environmental Hazards." *Environmental Professional*, 8(2):165–179.

Bennett, P. and K. Calman. 1999. *Risk Communication and Public Health*. Oxford University Press, New York.

Centers for Disease Control and Prevention. http://www.cdc.gov (accessed January 23, 2013).

Cohen, A., M. J. Colligan, and P. Berger. 1985. "Psychology in Health Risk Messages for Workers." *Journal of Occupational Medicine*, 27(8):543–551.

Fischhoff, B. 1989. "Helping the Public Make Health Risk Decisions." In V. T. Covello, D. B. McCallum, and M. T. Pavlova, eds., *Effective Risk Communication: The Role and Responsibility of Government and Nongovernment Organizations*. Plenum Press, New York, pp. 111–116.

Hyer, R. N. and V. T. Covello. 2005. "Effective Media Communication during Public Health Emergencies: A WHO Handbook." WHO/DCS/2005.31. World Health Organization, Geneva, Switzerland. http://www.paho.org/cdmedia/riskcommguide/Effective%20Media%20 Communication%20Handbook.pdf (accessed January 23, 2013).

McCallum, D. B. 1995. "Risk Communication: A Tool for Behavior Change." *NIDA Research Monograph*, 155:65–89.

Office of Public Health Preparedness and Response, Centers for Disease Control and Prevention, Department of Health and Human Services. No Date. "Public Health Workbook to Define, Locate, and Reach Special, Vulnerable, and At-risk Populations in an Emergency." CS211575A. http://www.bt.cdc.gov/workbook/pdf/ph_workbookFINAL.pdf (accessed January 23, 2013).

Rowel, R., P. Sheikhattari, T. M. Barber, and M. Evans-Holland. 2010. *A Guide to Enhance Grassroots Risk Communication among Low-Income Populations.* Maryland Department of Health and Mental Hygiene, Baltimore, Maryland. http://www.diversitypreparedness.org/SiteData/ docs/GUIDE%20TO%20E/499ec1d6f838a558/GUIDE%20TO%20Enhance%20GRC%20 Updated%20Feb%2018%202010.pdf (accessed January 23, 2013).

U.S. Department of Health and Human Services, Public Health Service and National Institutes of Health. 1992. *Making Health Communication Programs Work: A Planner's Guide.* Office of Cancer Communications, National Cancer Institute, NIH Publication No. 92-1493, Washington, DC.

Witte, K., G. Meyer., and D. Martell. 2001. *Effective Health Risk Messages: A Step-By-Step Guide.* Sage Publications, Thousand Oaks, California.

World Health Organization/Pan American Health Organization. 2012. *Risk Communication: Building Capacity Under International Health Regulations.* http://cursos.campusvirtualsp.org/course/ view.php?id=139 (accessed January 23, 2013).

CARE COMMUNICATION RESOURCES

Agency for Toxic Substances and Disease Registry. http://www.atsdr.cdc.gov/ (accessed January 23, 2013).

Bennett, P. and K. Calman. 1999. *Risk Communication and Public Health.* Oxford University Press, New York.

Centers for Disease Control and Prevention. http://www.cdc.gov (accessed January 23, 2013).

Cohen, A., M. J. Colligan, and P. Berger. 1985. "Psychology in Health Risk Messages for Workers." *Journal of Occupational Medicine*, 27(8):543–551.

Levitson, L. C., C. E. Needleman, and M. A. Shapiro. 1997. *Confronting Public Health Risks: A Decision Maker's Guide.* Sage Publications, Thousand Oaks, California.

Sandman, P. M. 1993. *Responding to Community Outrage: Strategies for Effective Risk Communication.* American Industrial Hygiene Association, Richmond, Virginia.

CONSENSUS COMMUNICATION RESOURCES

Chess, C., B. J. Hance, and P. M. Sandman. 1989. *Planning Dialogue with Communities: A Risk Communication Workbook.* Rutgers University, Cook College, Environmental Communication Research Program, New Brunswick, New Jersey.

Consensus the Quarterly Newsletter. Massachusetts Institute of Technology—Harvard Public Disputes Program, Harvard Law School Program on Negotiation, Cambridge, Massachusetts.

Creighton, J. L. and J. W. R. Adams. 2002. *Cyber Meeting: How to Link People and Technology in Your Organization.* Xlibris Corporation, Philadelphia, Pennsylvania.

Flynn, J., P. Slovic, and H. Kunreuther. 2001. *Risk, Media and Stigma: Understanding Public Challenges to Modern Science and Technology*. Earthscan, London.

Hance, B. J., C. Chess, and P. M. Sandman. 1988. *Improving Dialogue with Communities: A Risk Communication Manual for Government*. New Jersey Department of Environmental Protection, Division of Science and Research, Trenton, New Jersey.

Kasperson, R. E. 1986. "Six Propositions on Public Participation and Their Relevance for Risk Communication." *Risk Analysis*, 6:275–281.

National Research Council. 1996. *Understanding Risk: Informing Decisions in a Democratic Society*. National Academy Press, Washington, DC. http://www.nap.edu/openbook.php?isbn= 030905396X (accessed January 23, 2013).

Ortwin, R. 1992. "Risk Communication: Toward a Rational Discourse with the Public." *Journal of Hazardous Materials*, 20:465–519.

Parkin, R., L. Ragain, R. Bruhl, H. Deutsch, and P. Wilborne-Davis. 2008. *Advancing Collaborations for Water-Related Health Risk Communication*. Jointly published by the American Water Works Association Research Foundation, American Water Works Association, and IWA Publishing, Denver, Colorado.

Walker, G. B. and S. E. Daniels. 1997. "Collaborative Public Participation in Environmental Conflict Management: An Introduction to Five Approaches." *Proceedings of the Fourth Biennial Conference on Communication and Environment*, State University of New York-Syracuse, New York.

Wilson, T. 1989. "Interactions between Community/Local Government and Federal Programs." In V. T. Covello, D. B. McCallum, and M. T. Pavlova, eds., *Effective Risk Communication: The Role and Responsibility of Government and Nongovernment Organizations*. Plenum Press, New York, pp. 77–81.

CRISIS COMMUNICATION RESOURCES

Barrantes, S. A., M. Rodriguez, and R. Perez. 2009. *Information Management and Communication in Emergencies and Disasters*. Pan American Health Organization, Regional Office of the World Health Organization, Washington, DC.

Caernarven-Smith, P. 1993. "Managing a Disaster." *Technical Communication*, 40(1):170–172.

Carney, B. 1993. "Communicating Risk." IABC Communication World, May:13–15.

Clawson, S. K. Date Unknown. "Crisis Communication Plan: A Blueprint for Crisis Communication." Northern Illinois University, DeKalb, Illinois. http://www3.niu.edu/newsplace/crisis.html (accessed January 23, 2013).

George, A. M. and C. B. Pratt, eds. 2012. *Case Studies in Crisis Communication: International Perspectives on Hits and Misses*. Routledge, New York.

Governor's Office of Emergency Services. 2001. *Risk Communication Guide for State and Local Agencies*. Office of Emergency Services, Sacramento, California. http://www.ca-sioc.org/ resources/RiskCommunicationsGuideforStateandLocal.pdf (accessed January 23, 2013).

Green, M., J. Zenilman, D. Cohen, I. Wiser, and R. Balicer. 2007. *Risk Assessment and Risk Communication Strategies in Bioterrorism Preparedness*, NATO Security through Science Series-A: Chemistry and Biology. Springer, Dordrecht, Netherlands.

Henry, R. 2000. *You'd Better Have a Hose if You Want to Put Out the Fire: The Complete Guide to Crisis and Risk Communication*. Gollywobbler Productions, Windsor, California.

International Association of Business Communicators (IABC). 1993. *Crisis Communication Handbook*. International Association of Business Communicators, Washington, DC.

Leiss, W. and D. Powell. 2005. *Mad Cows and Mother's Milk: The Perils of Poor Risk Communication*, 2nd ed. McGill-Queen's University Press, Montreal, Quebec, Canada.

Lerbinger, O. 1996. *The Crisis Manager: Facing Risk and Responsibility*. Lawrence Erlbaum Associates, Mahwah, New Jersey.

National Center for Food Protection and Defense. *Best Practices in Risk Communication*. http://www.ncfpd.umn.edu/Ncfpd/assets/File/pdf/NCFPDRiskCommBestPractices.pdf (accessed February 7, 2013).

Nuclear Plant Journal, EQES Inc. 799 Roosevelt Road, Building 6, Suite 208, Glen Ellyn, Illinois 60137-5925.

Missouri Department of Mental Health. 2006, August. *Disaster Communications Guidebook*. http://dmh.mo.gov/docs/diroffice/disaster/FINALGUIDEBOOKwCov.pdf (accessed January 23, 2013).

Texas Department of State Health Services. *Writing a Public Health Crisis and Emergency Risk Communication Plan*. http://www.dshs.state.tx.us/riskcomm/tools.shtm (accessed January 23, 2013).

U.S. Centers for Disease Control and Prevention. 2012. Crisis + Emergency Risk Communication Manual. http://emergency.cdc.gov/cerc/pdf/CERC_2012edition.pdf (accessed January 23, 2013).

U.S. Council for Energy Awareness, Annual Crisis Communication Workshop. 1776 I Street N.W., Suite 400, Washington, DC. 20006-3708, (202) 293-0770.

U.S. Department of Health and Human Services. 2002. *Communicating in a Crisis: Risk Communication Guidelines for Public Officials*. Substance Abuse and Mental Health Services Administration, Rockville, Maryland. http://www.hhs.gov/od/documents/RiskCommunication.pdf; http://store.samhsa.gov/product/Risk-Communication-Guidelines-for-Public-Officials/SMA02-3641 (accessed January 31, 2013).

U.S. Department of Homeland Security. *Risk Lexicon*. http://www.dhs.gov/dhs-risk-lexicon (accessed January 23, 2013).

U.S. Department of Justice. 2001. *OVC Handbook for Coping after Terrorism*. Office of Justice Programs, Office of Victims of Crime, Washington, DC.

GLOSSARY

A number of terms in this book may be used differently from standard usage in the field of risk communication. These terms are defined below, as are other terms related to risk communication.

alternative dispute resolution—Methods of settling disputes without litigation or administrative adjudication, often involving a neutral third party to solve a disagreement. Methods include facilitation, negotiation, and mediation.

audience—Those who may be affected by or perceive that they may be affected by a risk.

bioterrorism—The release of a disease-causing substance with the intent to inflict harm and increase fear for political or ideological reasons.

care communication—Communicating about a risk for which the risk assessment is completed and the results are accepted by the majority of the audience. This can include communication about industrial hazards and health risks.

community relations—Developing a working relationship with the public to determine the acceptable ways of cleaning up a Superfund site. This relationship is mandated by the Comprehensive Environmental Response, Compensation, and Liability Act.

consensus communication—Communicating risk to bring a number of groups or individuals to a consensus on how the risk should be managed. Often the extent and nature of the risk is not agreed on by the various groups when the communication effort begins.

crisis—A turning point that will decisively determine an outcome, for example, the rupture of a leaking underground storage tank.

Risk Communication: A Handbook for Communicating Environmental, Safety, and Health Risks,
Fifth Edition. Regina E. Lundgren and Andrea H. McMakin.
© 2013 The Institute of Electrical and Electronics Engineers, Inc., and Regina E. Lundgren and
Andrea H. McMakin. Published 2013 by John Wiley & Sons, Inc.

crisis communication—Communicating risk in the face of a crisis, such as an earthquake or a fire at a chemical plant.

crowdsourcing—Obtaining needed services, ideas, or content by soliciting contributions from a large, distributed group of people, especially from the online community, rather than from traditional employees or suppliers.

ecological risk—The hazards posed to specific components of the ecological system. This area of risk is getting renewed interest from the U.S. Environmental Protection Agency in regard to the Natural Resource Damage Assessments.

emergency—Sudden or unforeseen situation that requires immediate action, for example, a terrorist attack.

emergency communication—Communicating risk and appropriate responses in the face of an emergency such as a major disease outbreak.

facilitated deliberation—A facilitator leads groups of people in discussing common issues and recommending solutions, often for consideration by decision makers. Ranges from online discussion groups to thousands of citizens nationwide.

facilitation—A process that uses a facilitator to help groups accomplish their work. The facilitator uses skills and techniques that enable the group to clarify issues, generate ideas, prioritize goals or solutions, and solve problems.

hazard—Danger; peril; exposure to a situation that could cause loss or injury.

health risk communication—Communicating about how to prevent, mitigate, or manage hazards to human health (a kind of care communication).

health risks—Hazards to human health, usually from diseases or lifestyle factors.

interactive multimedia—Communication methods that give the user some control over the content, format, order of presentation, level of detail, language, delivery speed, sound, and/or other aspects. Sometimes involve conversations or questions and answers. Examples are multimedia CDs, computer-based kiosks, and live web-based seminars.

mediation—A process that uses a neutral mediator to help people resolve or better manage disputes by reaching agreements about what the parties will do differently in the future. Private caucuses between the parties and the mediator may be used to build support or trust, explore settlement options, or break down barriers to negotiation.

misperception—Something a person believes to be related to a risk probability or hazard outcome but that experts agree is irrelevant to the actual probability or outcome.

negotiation—A third person helps parties negotiate an agreement, sometimes recommending a particular settlement. The concept of "principled negotiation" rejects a "win–lose" mentality and is instead based on the premise that it is possible to meet one's own needs and those of others, and that conflict provides such opportunities.

public—People who may or may not be interested in the risk but who are not charged with communicating, assessing, or managing the risk.

public affairs—That division of an organization that is charged with the task of developing a positive relationship with the public.

public information—Information to communicate with the public as opposed to scientists or managers. Because the topic may not be a risk, public information and risk communication materials are not necessarily synonymous. However, most risk communication materials will be sent to the public.

public involvement—Involving the public and other interested groups and individuals such as activists groups, community leaders, regulators, and scientists in making some decision. Because the decision may not involve an environmental, safety, or health risk, public involvement is not synonymous with risk communication. The two can, however, overlap.

public participation—See public involvement.

public relations—The efforts involved in developing a positive relationship with the public, with the goal of getting the public to view your organization in a positive light.

risk—Probability of adverse outcome. Risk is inherent in any action, even in inaction.

risk assessment—Determining the risks posed by a certain hazard, usually to human health or the environment; can also include legal and financial risk.

risk/benefit analysis—Determining and weighing the relative risks and benefits of taking a certain action. It includes determining who receives the risks and benefits.

risk communication—The interactive process of exchange of information and opinions among individuals, groups, and institutions concerning a risk or potential risk to human health or the environment. Any risk communication effort must have an interactive component, if only in soliciting information about the audience in the beginning or evaluating success in the end.

risk decision—A decision about how to mitigate or prevent a risk.

risk management—Evaluating and deciding how to cope with a risk. Risk management may or may not include public participation.

risk message—Message that communicates information about the hazard, its probability, the potential outcomes, and actions that can be taken to manage the risk.

risk perceptions—The set of beliefs that a person holds regarding a risk, including beliefs about the definition, probability, and outcome of the risk.

social marketing—A type of care communication that adapts techniques from mass marketing and advertising to promote increases in knowledge, behavior change, and social change. In public health, social marketing is sometimes called health marketing.

social media—A group of interactive, online methods that integrate technology, words, pictures, videos, and audio with the concept of shared content, generated largely by users with easy-to-publish tools.

social networks—Websites that enables users to be part of a virtual community, in which they can post profiles, connect with friends, and make new acquaintances based on shared interests. Sometimes considered a subset of social media.

stakeholder—Person who holds a "stake," an interest in how a risk is assessed or managed.

stakeholder involvement—See stakeholder participation.

stakeholder participation—Involving those who hold an interest in the risk or in how the risk is assessed or managed.

technology-assisted communication—Technology (web pages, kiosks, mobile platforms, etc.) used as a conduit for risk communication information as opposed to simply a tool to create it.

terrorism—An act of violence intended to inflict harm and increase fear for political or ideological reasons.

INDEX

Page references followed by *f* denote figures. Page references followed by *t* denote tables.

Risk Communication: A Handbook for Communicating Environmental, Safety, and Health Risks,
Fifth Edition. Regina E. Lundgren and Andrea H. McMakin.
© 2013 The Institute of Electrical and Electronics Engineers, Inc., and Regina E. Lundgren and
Andrea H. McMakin. Published 2013 by John Wiley & Sons, Inc.